5th International Conference on Manufacturing Optimization, Industrial and Material Engineering (MOIME 2017)

IOP Conference Series: Materials Science and Engineering
Volume 215

Bali, Indonesia
1 - 2 April 2017

Editors:

Ford Lumban Gaol
Jeff Webb
Jun Ding

ISBN: 978-1-5108-4561-9
ISSN: 1755-1307

Printed from e-media with permission by:

Curran Associates, Inc.
57 Morehouse Lane
Red Hook, NY 12571

Printed by Curran Associates, Inc. (2017)

For permission requests, please contact the Institute of Physics at the address below.

Institute of Physics
Dirac House, Temple Back
Bristol BS1 6BE UK

Phone: 44 1 17 929 7481
Fax: 44 1 17 920 0979

techtracking@iop.org

Additional copies of this publication are available from:

Curran Associates, Inc.
57 Morehouse Lane
Red Hook, NY 12571 USA
Phone: 845-758-0400
Fax: 845-758-2633
Email: curran@proceedings.com
Web: www.proceedings.com

Table of contents

Volume 215

5th International Conference on Manufacturing, Optimization, Industrial and Material Engineering

1–2 April 2016, Bali, Indonesia

Accepted papers received: 23 June 2017
Published online: 27 July 2017

Preface

Papers

IOP Conf. Series: Materials Science and Engineering **215** (2017) 011001 doi:10.1088/1757-899X/215/1/011001

Preface

The 5th International Conference on Manufacturing, Optimization, Industrial and Material Engineering 2017 (MOIME 2017) already held in The Magani Hotel & Spa Legian, Bali - Indonesia from 01 - 02 April 2017

The MOIME 2017 conference is aimed to bring together researchers, engineers and scientists in the domain of interest from around the world. MOIME 2017 is placed on promoting interaction between the theoretical, experimental, and applied communities, so that a high level exchange is achieved in new and emerging areas within Material Engineering, Industrial Engineering and all areas that relate with Optimization.

We would like to express our sincere gratitude to all in the Technical Program Committee who have reviewed the papers and developed a very interesting Conference Program as well as the invited and plenary speakers.

This year, we received 121 papers and after rigorous review, 43 papers were accepted.

The participants come from 09 countries. There are 6 (six) parallel session and 3 Keynote Speakers.

It is an honour to present this volume *of IOP Conference Series: Materials Science and Engineering (MSE)* and we deeply thank the authors for their enthusiastic and high-grade contribution.
Finally, we would like to thank the conference chairmen, the members of the steering committee, the organizing committee, the organizing secretariat and the financial support from the conference sponsors that allowed the success of MOIME 2017.

The Editors of the MOIME 2017 Proceedings Editors

Dr, Ford Lumban Gaol,
Jeff Webb, Ph.D
Prof. Jun DING, Ph.D

Peer review statement

All papers published in this volume of *IOP Conference Series: Materials Science and Engineering* have been peer reviewed through processes administered by the proceedings Editors. Reviews were conducted by expert referees to the professional and scientific standards expected of a proceedings journal published by IOP Publishing.

IOP Conf. Series: Materials Science and Engineering **215** (2017) 012001 doi:10.1088/1757-899X/215/1/012001

A numerical application in four-point bent specimen

Lih-jier Young[1] and Bo-Han Yeh[2]

[1] Professor, Chung Hua University, Taiwan, R.O.C.

[2] Ph.D. Student, Chung Hua University, Taiwan, R.O.C.

E-mail: young@chu.edu.tw

Abstract. In this research a four point bent specimen with a crack will be employed to investigate the behavior of the crack by using the boundary element method numerically and the fracture mechanics theroetically due to the industrious important. Boundary element is employed because the reduction of the computational dimension. The accuracy of this approach is as high as 98.62% and shows the potential application of the boundary element in four-point bent specimen.

1. Introduction

Crack problems are always exist in industry. It will also cause many serious industrial injuries in this sophisticated technological era such as nuclear power plant, tunnel and aircraft. Therefore, how to improve these complicated problem is the major purpose of this research. Nowdays several laboratories are now dealing with the crack problem. H.L. Oliveria, Edson Denner Leonel [1] model multiple crack propagation in two-dimensional domains which simulate localisation and coalescence phenomena. M. Munsche et al [2] solve the transient elastodynamic crack analysis in two-dimensional (2D), homogeneous, anisotropic, and linear elastic solids by a hypersingular time-domain boundary element method. Lih-jier Young [3] reduce the weight of rescue robot with a cracked lateral plate.

Due to the reason of crack growth not only by normal and shear forces but bending moment four-point bent specimen is employed in this research. Figure 1 shows the dimension of this specimen.

2. Four-point bent specimen

2.1 Boundary conditions and numerical implementation

Four-point bent specimen was first discussed in [4] which dealing with the mixed mode fracture of the specimen. As described in [5] the first step in BEM solution is to divided the homogeneous media into two bodies B_γ ($\gamma = (1, 2)$ along the plane ot the crack referred to below as the interface of the center open crack problem as shown in Fig. 2. The interaction between two bodies is included through boundary conditions relating the displacements and stresses on either side of the interface. According to the different boundary conditions of the cracked four-point bent specimen as in Fig. 2 there are totally five regions in the specimen, i.e., the open crack region

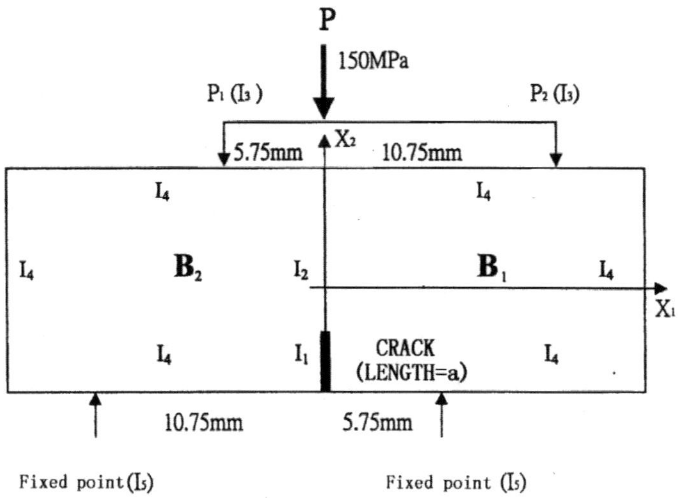

Fig. 1 Boundaries and dimensions of four-point specimen.

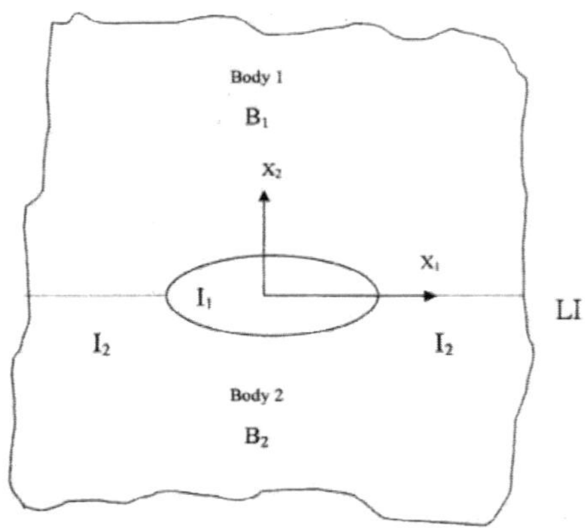

Fig. 2 Boundaries of center open crack problem.

(I_1), the ligament region (I_2), the loading region (I_3), the support region (I_4) and the free surface region (I_5). Let ${}_{\gamma}t_i$ and ${}_{\gamma}u_i$ ($\gamma = 1,\ 2$ and $i = 1,\ 2$) denote the ith boundary traction and displacement components, respectively, on the boundary B_{γ}. At points on this region the 2 displacement components and 2 stress components must be continuous. Therefore, the boundary conditions are $({}_2t_1)_{I_2} = -({}_1t_1)_{I_2}$ and $({}_2t_2)_{I_2} = -({}_1t_2)_{I_2}$, $({}_2u_1)_{I_2} = -({}_1u_1)_{I_2}$ and $({}_2u_2)_{I_2} = ({}_1u_2)_{I_2}$. This leaves $({}_1t_1)_{I_2}$, $({}_1t_2)_{I_2}$, $({}_1u_1)_{I_2}$ and $({}_1u_2)_{I_2}$, as the unknowns. At points on open crack region I_1 and free surface region I_4 stresses are free, .i.e, $({}_{\gamma}t_1)_{I_1,I_4} = ({}_{\gamma}t_2)_{I_1,I_4} = 0$. The unknowns are, of course, $({}_{\gamma}u_1)_{I_1,I_4}$ and $({}_{\gamma}u_2)_{I_1,I_4}$, where $\gamma = 1$ or 2 depending on whether that portion of I_4 is in B_1 or B_2. At points on loading region (I_3) a concentrate load $P = 150$ Mpa is applied to the bar at the top of the specimen which lead two point loads $P_1 = \dfrac{b_1}{b_1 + b_2} P$ and $P_2 = \dfrac{b_2}{b_1 + b_2} P$, where $b_1 = 10.75$ mm and $b_1 = 5.75$ mm, .i.e, $P_1 = 97.7273$ Mpa, $P_2 = 52.2727$ Mpa theoretically as shown in Fig. 2. The boundary conditions is therefore, $({}_1t_1)_{I_3} = ({}_2t_1)_{I_3} = 0, ({}_1t_2)_{I_3} = -52.2727$ Mpa, $({}_2t_2)_{I_3} = -97.7272$ Mpa. The unknowns are four displacement components, .i.e, $({}_1u_1)_{I_3}$, $({}_1u_2)_{I_3}$, $({}_2u_1)_{I_3}$ and $({}_2u_2)_{I_3}$. The boundary conditions of the support region (I_5), .i.e., two simplely support points in the bottom of the specimen is $({}_{\gamma}u_1)_{I_5} = ({}_{\gamma}u_2)_{I_5} = 0$. The unknowns are $({}_{\gamma}t_1)_{I_5}$ $({}_{\gamma}t_2)_{I_5}$, where $\gamma = 1$ or 2 depending on whether that portion of I_5 is in B_1 or B_2. Thus, at each pair of points on the ligament region (I_2) we have four conditions involving eight quantities. Four of those are eliminated algebraically using the boundary conditions, thus, leaving four unknowns at each point. Two coupled boundary integral integral equations, written as a function of position on the boundary of a body, enforce all of the field equations of elasticity for that body. The two equations foe each of the two artificially divided bodies are applied to each discretized point on the interface, tuus giving four equations and four unknowns at each pair of interface points. At each of the boundary points of either of artificially divided bodies consist of other than the common interface, i.e., the crack region (I_1), the loading region (I_3) , the free surface region (I_4) and the support region (I_5), there are four boundary quantities to be accounted for. The BEM consists of the discretization of the boundary surfaces and the numerical approximation of the boundary quantities in the set of equation obtained from the boundary integrals. Wev model the boundary, using straight-line element, centered about nodes at which the integrals of the 2D Green's function as in [5]. The final system of simultaneous linear algebraic equatioms for the unknown nodal displacements and stresses, can be obtained by using Gaussian elimination method.

2.2 Accuracy test

After pluging in the applied loading $P = 150$ Mpa in the BEM code with 200 elements for the non-crack specimen the numerical results show that $({}_1t_2)_{I_5} = 97.0163$ Mpa and $({}_2t_2)_{I_5} = 52.997$ Mpa, respectively.. Compare the therotical values as described above the accuracy of present method is 98.62%. Figure 3 shows the normal and shear stress distributions of the interface. Figure 4

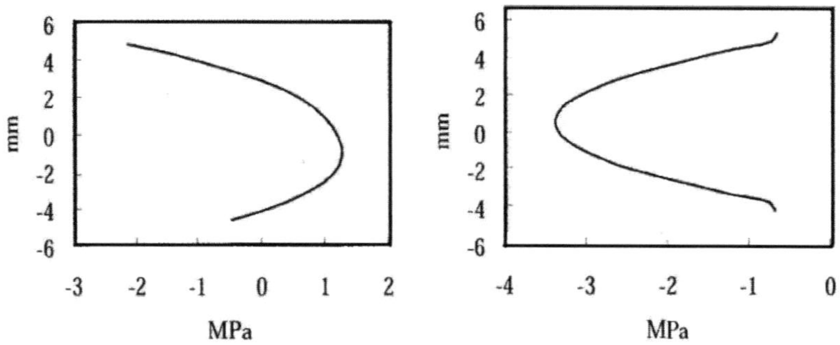

(a) (b)

Fig. 3 Normal (a) and shear (b) stress distributions of ligament portion of non-crack four-point bent specimen.

shows the normal and shear displacements of the interface. The displacement of the non interface points are shown in Fig. 5.

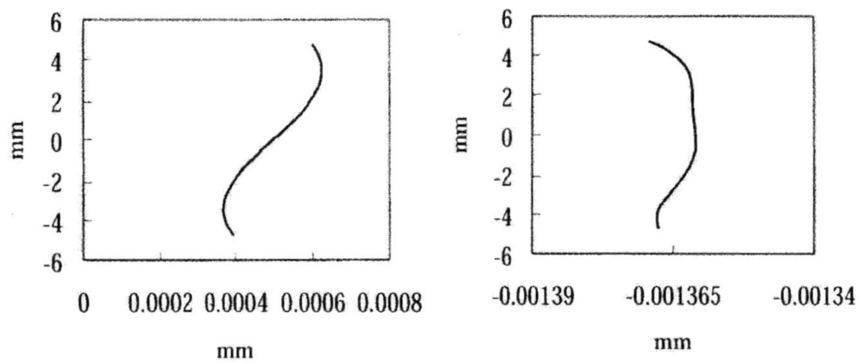

(a) (b)

Fig. 4 Transverse (a) and longitudinal (b) displacements of ligament portion of non-crack four-point bent specimen.

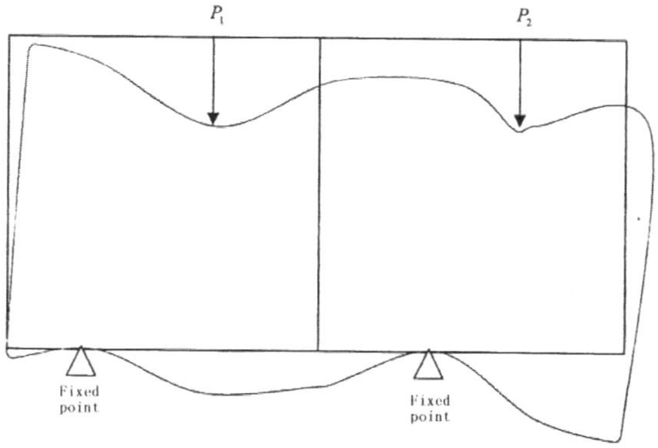

Fig. 5 Displacement of non interface points

3. Numerical results of the cracked specimen

3.1 Open crack four-point bent specimen

After substutung all boundary conditions and parameters into the BEM code the unknowns can be reached by Gauss Elimination. In the present model stress intensity factors are calculated by fitting the standard square root form for crack displacement at the elastic crack tip, in terms of stress intensity factor shown inverted for *K*, .i.e,

$$K = \frac{2u(x_t)G}{\kappa+1}\sqrt{\frac{2\pi}{x_t}} \tag{1}$$

where $\kappa+1=4(1-\nu)$ for plane strain

$\kappa+1=\frac{4}{(1+\nu)}$ for plain stress

2u is crack opening displacement (COD) for mode I or crack sliding displacement (CSD) for mode II which is calculated from the present BEM formulation,

G is the shear modulus,

x_t is the distancece between crack tip and the nearest nodal point on the crack. The closer x_t is to the crack tip the more accurate *K* is. The result shows the calculated value of K_I is 19.97 MPa$\sqrt{m}$. Compare with the analytical value 20.68 MPa$\sqrt{m}$ ($K_I = \frac{6M\sqrt{a}}{W^2}1.99$, when $\frac{a}{W}=0.3$, M=100Mpa) the accuracy is 96.57%.

Inspecting both displacements of the two surfaces of the crack we find that they are overlapped as shown in Fig. 6 which is impossible. Therefore, the boundary

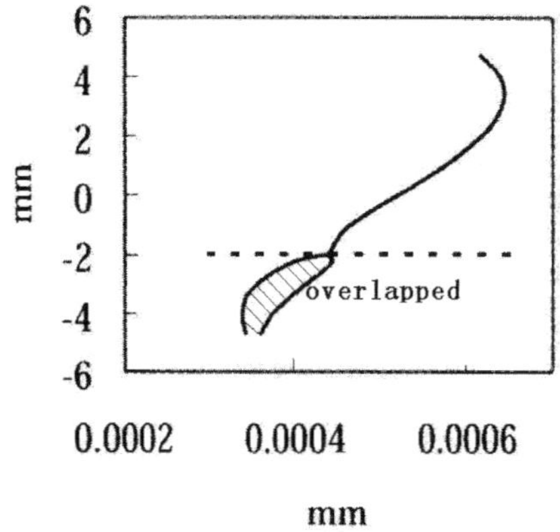

Fig. 6 Two surfaces overlapped of crack.

boundray condition of the open crack should chang into the partial contact boundary condition.

3.2 Partial contact four-point bent specimen

We divided the crack region I_1 into two subregions : the contack subregion I_{11} and not contact subregion I_{12}. The boundary condition of I_{11} is therefore, $({}_1t_1)_{I_{11}} = -({}_2t_1)_{I_{11}}, ({}_1u_1)_{I_{11}} = ({}_2u_1)_{I_{11}}$ with the unknowns $({}_1u_1)_{I_{11}}$, $({}_1u_2)_{I_{11}}$, $({}_1t_1)_{I_{11}}$, $({}_2u_2)_{I_{11}}$. However, the boundary condition of I_{12} wil be $({}_1t_1)_{I_{12}} = ({}_1t_2)_{I_{12}} = 0$, $({}_2t_1)_{I_{12}} = ({}_2t_2)_{I_{12}} = 0$ with the unknowns $({}_1u_1)_{I_{12}}$, $({}_1u_2)_{I_{12}}, ({}_2u_1)_{I_{12}}$, $({}_2u_2)_{I_{12}}$. The nodal number is shown in Fig. 7. From the result of Sec. 2.3 points 15~17 are contact and points 18~20 are not. The stress distributions are

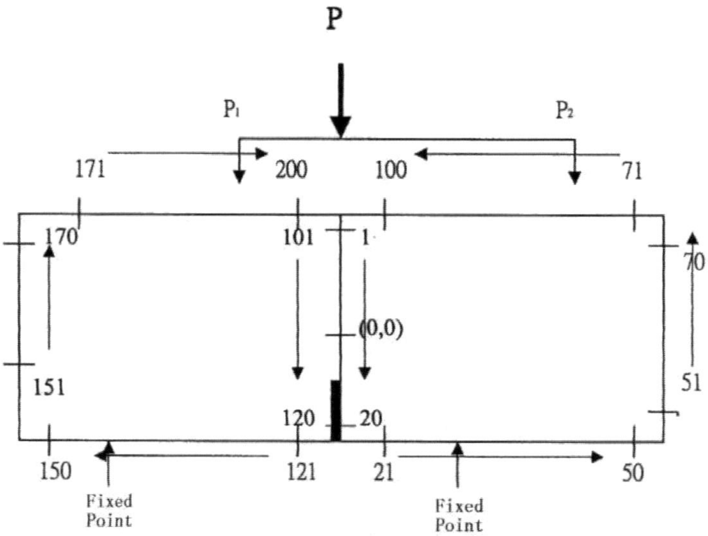

Fig. 7 Nodal number of the specimen.

shown in Fig. 8. The normal stress of the interface is similar to non-crack. There is no stress intensity occur at the crack tip. However, the shear force distribution is different. According to Fig. 8 (b) the shear stress of points 1~13 are the same with non-crack specimen. The shear stress near the crack tip is increase dramatically which is agreed with the theory of fracture mechanics, .i.e, the stress concentartion at the crack tip.

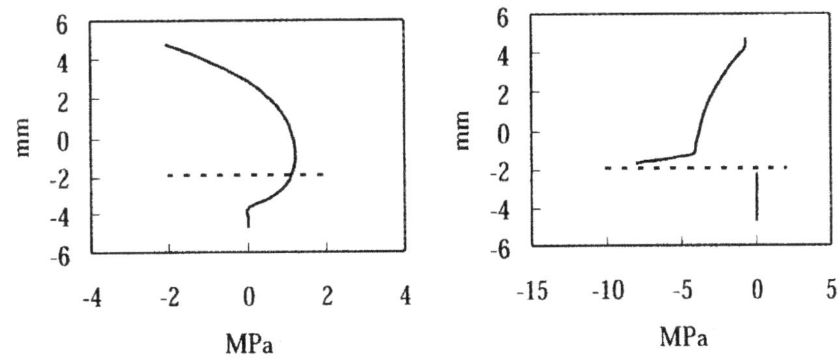

(a) (b)

Fig.8 Normal (a) and shear (b) stress distributions of interface of particial contact specimen.

The transverse displacement of the ligament region I_2 (1~13) shown in Fig. 9 (a) are quite the same as the non-crack specimen. However, the crack region I_1 is slightly different. The subregion I_{11} (15~17) keep the same as the non-crack specimen the subregion I_{12} (18~20) does not, however, as shown in Fig. 9(b). It's easy to understand that the crack open at point 18 which is not the crack tip. It means that the normal stress of this crack will not cause the stress intensity in normal direction at the crack tip and therefore the crack will not propogate.

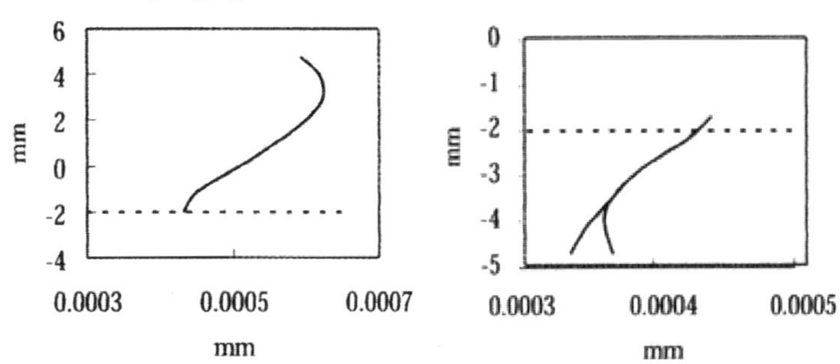

Fig. 9 Transverse displacement of ligament (a) and crack (b) portions of partial contact specimen.

The longitudinal displacements of the ligament region I_2 is small (0.00137 mm~0.00138 mm) shown in Fig. 10 (a). On the contrary the displacement of the crack region I_1 are much different. It can be seen from Fig. 10 (b) that the changing rate of the displacement is increasing near the crack tip. Therefore, the crack will be propogate in shear direction because of the shear stress concentration at the crack tip. K_{II} is the key factor which dominates the crack propogation of the four-point bent specimen.

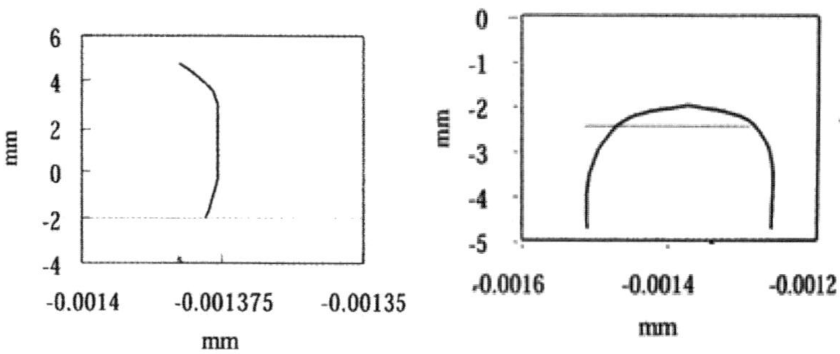

(a) (b)

Fig. 10 Longitudinal displacements of ligament (a) and crack portions (b) of partial contact specimen.

4. Conclusions

A 2-D BEM formulation predicting mode II intensity factor of the four-point bent specimen has been presented. Two differeft boundary conditions of the crack were offered to match the pratical case. Accuracy test has been performed by comparison of the BEM model with analytical solutions.The numerical results show the potential application of the BEM model to the effects of mode II stress intensity factor K_{II} which dominates the crack propogation of four-point bent specimen with a crack.

5. References

[1] H.L. Oliveria, Edson Denner Leonel (2013). Dual BEM Formulation Applied to Analysis of Multiple Crack Propagation. Key Engineering Materials, 5, 99-106.

[2] M. Munsche, Ch. Zhang, M. Kuna, S. Hirose, J. Sladek, V. Sladek (2008). A Hypersingular Time-Domain BEM for 2D Dynamic Crack Analysis in Anisotropic solids

[3] L.J. Young (2011). Numerical Application in Weight Reduction of Lateral Plate of Rescue Robot, Key Engineering Materials, 462-463, 762-767.

[4] M.Y. He, H.C. Cao and A.G. Evans (1990). Mixed mode Fracture: The Four-Point Shear Specimen, ACTA Metallurgica, 38, 5, 839~846.

[5] L.J. Young , Y.P. Tsai (1999). A Boundary Element Application for Mixed Mode Loading Idealized Sawtooth Fracture Surface. International Journal of Solids and Structures, 36, 3239-3252.

Slotting optimization of automated storage and retrieval system (AS/RS) for efficient delivery of parts in an assembly shop using genetic algorithm: A case Study

L Yue[1,2,3], Z Guan[1,2,3], C He[1,2], D Luo[1,2], U Saif[2,3],
[1]State Key Lab of Digital Manufacturing Equipment and Technology, Wuhan 430074, China
[2]School of Mechanical Science and Engineering, Huazhong University of Science and Technology, Wuhan 430074, China
[3]HUST-SANY Joint Lab of Advanced Manufacturing, Wuhan 430074, China

Abstract. In recent years, the competitive pressure on manufacturing companies shifted them from mass production to mass customization to produce large variety of products. It is a great challenge for companies nowadays to produce customized mixed flow mode of production to meet customized demand on time. Due to large variety of products, the storage system to deliver variety of products to production lines influences on the timely production of variety of products, as investigated from by simulation study of an inefficient storage system of a real Company, in the current research. Therefore, current research proposed a slotting optimization model with mixed model sequence to assemble in consideration of the final flow lines to optimize whole automated storage and retrieval system (AS/RS) and distribution system in the case company. Current research is aimed to minimize vertical height of centre of gravity of AS/RS and total time spent for taking the materials out from the AS/RS simultaneously. Genetic algorithm is adopted to solve the proposed problem and computational result shows significant improvement in stability and efficiency of AS/RS as compared to the existing method used in the case company.

Keywords: workshop distribution, slotting optimization, simulation method, automated storage and retrieval system, genetic algorithm

1. Introduction

For multi variety and small batch mixed flow assembly system, it is not only involved in the variety and quantity of parts, but also it is necessary for the parts to be delivered to the specified location at the time required. Due to this, the requirement of larger storage capacity and fast and accurate operations is reasonable. Therefore, AS/RS is widely used in the automotive, construction machinery and other industries. It becomes one of the signs in enterprise production automation and management information. But the application of AS/RS often need a huge investment in manpower, material resources and financial resources. It also requires lean management and some auxiliary control systems accordingly. So it is difficult to implement AS/RS in so many companies which are small or old. The initial warehouse operation management mode is to store goods by virtue of personal experience in general. The goods are often stored nearby, without considering the dynamic changes of material such as weight changes and customer demand changes which may often lead to uneven stress, slow operation of outcoming and incoming and low utilization. Therefore, slotting optimization came into being. Slotting optimization management refers to the reconfiguration of the storage assignment in the warehouse based on the current storage location assignment, so as to ensure the rational allocation of warehouse location. Reasonable slotting optimization plays an important role to reduce useless consumption, shorten the order fulfilment time, improve the efficiency.

In recent years, both domestic and foreign, the slotting optimization of AS/RS is one of the research hotspots. Thonemann and Brandeau [1] discussed the storage assignment method based on the turnover rate of goods location assignment and the principle of classification under stochastic environment. Hsieh and Tsai [2] studied storage assignment method for bill of material based on the classification, so as to realize the classification storage according to the region. Poulos et al. [3] proposed a new multi-objective genetic algorithm to solve the location allocation problem for the replenishment task demand of AS/RS. Muppani and Adil [4] presented a nonlinear integer programming model with an aim to reduce storage cost and the picking time simultaneously for the classification based reallocation of storage problem. Branch and bound method is used to solve this problem. Onut et al. [5] set up a multi-level warehouse shelf pattern in order to minimize the transportation costs, and the location is determined by the corresponding product turnover rate, moreover, particle swarm algorithm (PSO) is adopted to solve the problem. Zu et al. [6] design a kind of hybrid genetic algorithm to solve the problem which made turnover rate of goods and stability of goods shelves as the optimization objective. Li et al. [7] put forward the goods assignment optimization algorithm based on correlation for the slotting optimization problem with the effect of correlation on the goods picking efficiency in consideration; furthermore it is solved with an objective of minimizing the total picking time of goods. The above research results have laid the foundation for our further research, but these optimization models are not comprehensive including the influence of material demand fluctuation of mixed models. For this issue, research on the slotting optimization problem requires the integrated consideration of turnover rate of goods and stability of goods shelves and the different corresponding frequency of material demand of mixed models simultaneously. So this way, better scheme for the actual operation of AS/RS can be obtained and also the effective implementation of the assembly plan can be carried out.

2. Company problem description

A very famous pump truck manufacturer (hereafter Company X) in Changsha China has a final assembly workshop with the layout shown in Fig. 1. Three different kinds of pump trucks are produced in this shop floor which includes, pump truck, trailer pump and on-board pump truck. Beside the material distribution channel A and channel B, there are final goods assembly lines. The upper and left production line from west to east are final trailer pump assembly line and the on-board pump truck assembly line is adjacent. Under these two lines from top (north) to bottom (south), there are landscape belt and two parallel final pump truck assembly lines respectively. Beside the assembly line area, there are two visit channels. The two automated storage and retrieval systems (AS/RS) are both in the east (right) of the plant and are located in north (NAS/RS) and south (SAS/RS) respectively.

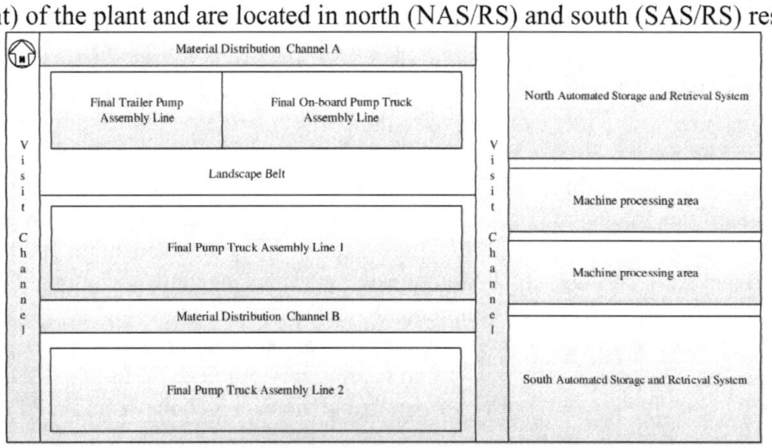

Figure 1. Final assembly workshop layout

The AS/RS mainly affords the material distribution service to the final assembly lines. The NAS/RS is used to store materials of pump trucks, while the materials of final trailer pumps and on-board pump trucks are deposited in SAS/RS. The materials of NAS/RS are transported from material entry of SAS/RS by the conveyor belt underground. The detail layout of AS/RS is shown in Fig.2.

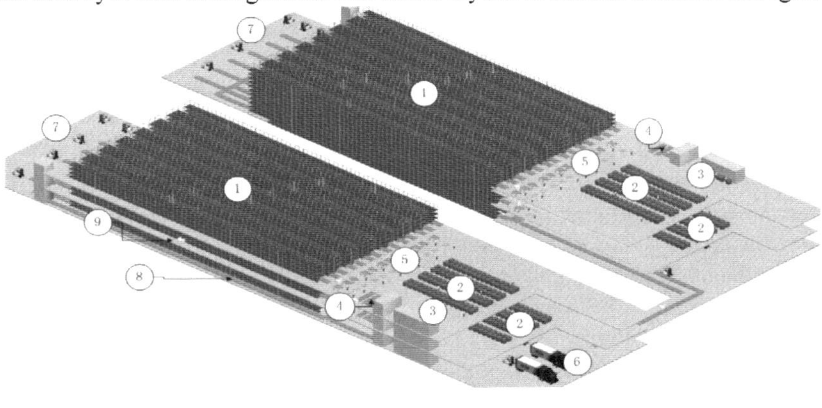

1. ASRS
2. Material dispatching area
3. Elevator depot sorting Area
4. Full container Exit
5. ASRS sorting Area
6. Entry of ASRS material in first floor
7. The place of container receipt of AGV
8. The conveyor belt of empty container recycling
9. Conveyor line of full container

Figure 2. The detail layout of AS/RS in X assembly workshop

(1) It can be seen from Fig.2 that the area ① is used to represent storage rack area. Both SAS/RS and NAS/RS have 14 storage racks and 7 roadways as clearance space for moving and transporting the parts from storage. Furthermore, there are totally 9016 goods allocations in SAS/RS while 6910 goods allocations in NAS/RS.

(2) Symbol ② indicates material sorting and collecting area of AS/RS which is used for materials packaging and waiting to be delivered to assembly lines.

(3) Symbol ③ denotes a cubic elevator depot sorting area. It is used to store materials which have weight equal or less than 1 kilogram. Unilateral has three independent elevator depots and each elevator depot has 130 layers of area to store materials.

(4) Symbol ④ indicates the full container exit area and the empty container recycling area. The full container goes to the transport region and waits to be transported to final lines. The empty container is lifted from first floor to third floor and then transported to empty container recycling area by conveyer.

(5) Symbol ⑤ is used to represent material sorting area and exit, the second and third floor of SAS/RS and SAS/RS both have material exits. Each floor has 12 exits for SAS/RS and for NAS/RS there are 6 exits.

(6) Symbol ⑥ denotes the material entry of AS/RS which is located in first floor of AS/RS. The material from suppliers and self-made components are unloaded, tested on quality and then placed in storage in this area.

(7) Symbol ⑦ indicates the automated guided vehicle (AGV) waiting area. The AGVs wait in this area and receive full container here. The full container is transported by conveyor and delivers it to the assembly line. On the other hand, the AGVs help to deliver the empty containers back to the conveyor.

(8) Symbol ⑧ and ⑨ indicate the conveyor of full container and empty container respectively. The conveyor of full container starts from the second floor to the third floor and used to deliver full container to AGV waiting area. The conveyor of empty containers located in the first floor is used to transport empty container from AGV waiting area to material sorting and collecting area of AS/RS.

3. Analysis of AS/RS operations based on simulation

3.1. Operation process of AS/RS for material delivering

AS/RS is the storage system used to store the material from supplier and self-made components in less space. The material required by the assembly lines is distributed in kitting methods in it. The stored material needs a sorting method to store material in different shelf. Fig. 3 indicates the second and third floor of the AS/RS which are in charge of four kinds of processes including AS/RS material sorting and sowing, elevator depot material sorting and sowing, empty container sowing and full container distribution etc. The second and third floor has an area for material sowing and packaging in order to put material into containers. All the material from AS/RS and other warehouse and elevator depot needs kitting distribution are packaged to deliver the material on time to final assembly lines. The storage and retrieval machine (SRM) in AS/RS gets the pallet of the material which is needed to supply from the shelf and place it on the sorting platform. The material is sorted and sowed by worker and the pallet on the sorting platform is transported back to the original location by SRM simultaneously. This process is shown by the red arrow in Fig.3. The sorting workers sort the material from elevator depot and use sorting trolley to deliver the material to sowing area for packaging as shown by blue arrow in Fig.3. The empty container from first floor is placed in empty container buffer. When the empty container is needed in material sowing area, the empty container is transported to the corresponding workstation area by forklift and the process is indicated in purple arrow in Fig.3. The full containers are distributed from sowing area to assembly lines on time. Part of the full containers is transported to the south conveyor of AS/RS and then transported to AGVs waiting area by conveyor, another part of the full containers is transported directly to AGVs waiting area by SRM and the process is indicated by green arrow in Fig.3.

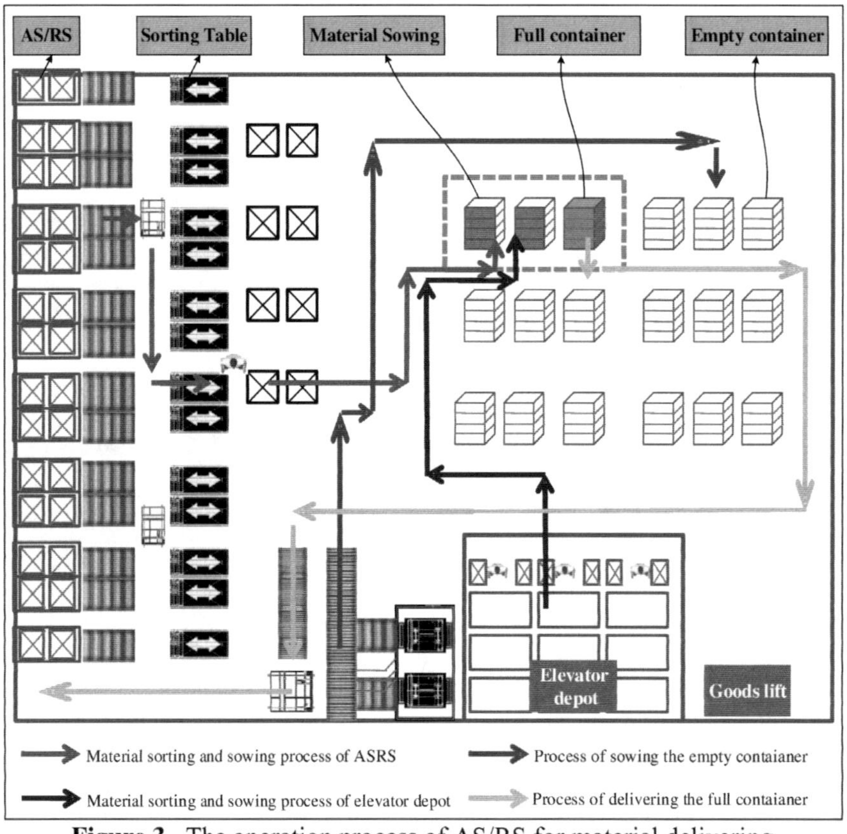

Figure 3. The operation process of AS/RS for material delivering

3.2. Simulation modeling and analysis of operation of AS/RS

In the case company, there is requirement of in time distribution of material and the number of containers needed to deliver the parts to assembly line in a distribution cycle (4 hours = 4 * production cycle time) is shown in table 1. The distribution cycle and required number of containers for assembly lines determine the pressure of material coming out of AS/RS.

Table 1. The number of containers needed to deliver in 4 hours

Assembly line	Types of container	Number of container	Number of container needed to be delivered
Pump truck line 1	25	27.1	78
Pump truck line 2	25	27.1	78
Trailer pump line	24	30	60
On-board pump truck line	17	31	31

The total number of distribution cycles when the material is coming out of AS/RS and moving into AS/RS including incoming of raw materials, outcoming of large-size materials, outcoming and incoming of full containers and outcoming and incoming of empty containers are indicated in table 2.

Table 2. Number of times of outcoming and incoming operations of AS/RS

AS/RS	Outcoming and incoming of materials	Incoming of raw materials	Outcoming of large-size materials	Outcoming and incoming of full containers	Outcoming and incoming of empty containers	Total number of times
SAS/RS	726	240	4	114	114	1276
NAS/RS	348	80	0	0	0	460

In the Case Company, the elevator can transport 25 containers per hour, i.e., the elevator can transport 100 containers in one distribution cycle at most. It can be seen from table 1 that it is needed to distribute 247 containers in every 4 hours, therefore elevator does not satisfy the demand of full container transportation. Therefore, it is required that part of full containers to be delivered by other ways.

It is needed about 4.65 hours to finish the sorting and collecting of AS/RS materials for 4 sets of material off-loading. Therefore, it cannot be finished that all the sorting and collecting of AS/RS materials are picked out from AS/RS in a distribution cycle (4 hours). The utilization ratio of RSMs for the case company is illustrated in Fig. 4. It can be seen from Fig.4, when using 4 sets of material off-loading from AS/RS, the utilization ratio of some RSMs are more than 90%, as a result, the RSM has higher probability to break down.

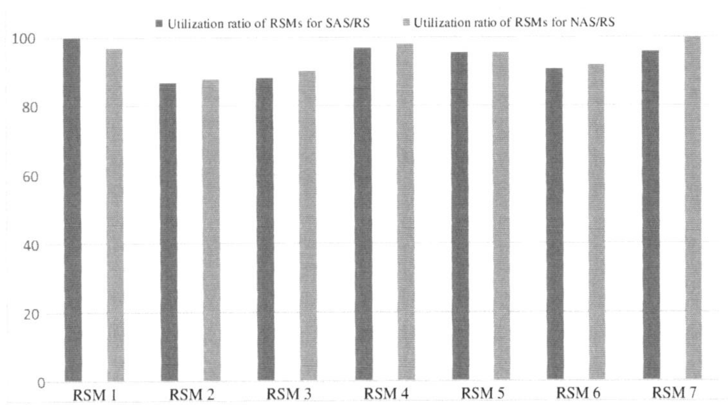

Figure 4. The utilization ratios of RSMs

According to the result of simulation model of AS/RS, it is hard to satisfy the material requirement of production line for the AS/RS due to inefficient and not optimized goods location of AS/RS. Moreover, the capacity of elevator cannot meet the demand of material coming put from AS/RS in current location method of AS/RS. As a result, a much better goods location method is needed to propose to improve the efficiency of AS/RS. Furthermore, in a distribution cycle, the off-loading time of material from AS/RS can be greatly reduced by optimizing the goods locations; therefore it is necessary to optimize the goods locations for AS/RS.

4. Genetic algorithm based slotting optimization

The current problem of AS/RS is a multi-objective programming problem and it is hard to solve the multi-objective problem using simulation. Genetic algorithm (GA) is a famous method to solve the multi-objective problems and is therefore used in current research.

4.1. Slotting optimization model for AS/RS

4.1.1. Encoding strategy

The material is only associated with the corresponding workstation in bill of material (BOM), but in real operation it is difficult to know the outcoming exit of material. However, in current study, the real data of company is connected with algorithm, and the material information is changed to genetic code information. The proposed encoding strategy for the current research is shown as below in Fig.5. The detail of different elements of the encoding strategy is explained in this section.

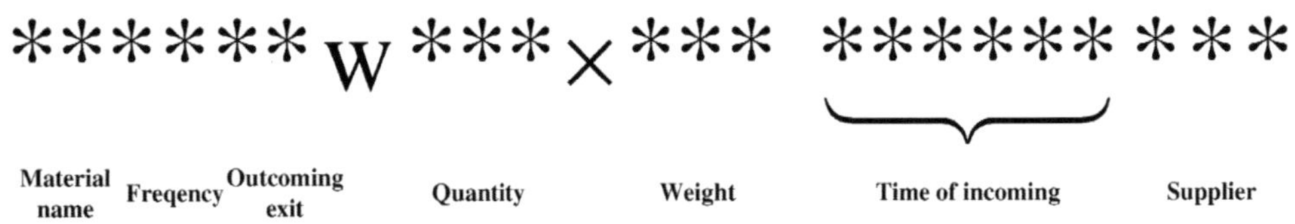

Figure 5. Encoding strategy of the current problem

It can be seen from Fig. 5 that there are different names given to the elements in the encoding strategy. The material name, frequency of the material, outcoming exit, quantity, weight, time of incoming of material and supplier information is included in the encoding strategy.

Material name: The current material name of company is indicated by English letters and numbers simultaneously is used here.

Frequency: The frequency of outcoming and incoming of material is used to judge the priority of outcoming and incoming of material in AS/RS. In order to show directly the code used here is the real frequency multiply by 100.

Outcoming exit: It is generally indicated by workstation information, but there is not only one outcoming exit and some material does not need sorting to deliver to production lines, therefore, it is not encoded here by workstation information. In current study, 00 is used to indicate the material which can both off-loaded from second and third floor, 01 is used to indicate the material which is moved to the assembly line, 02 is used to denote the material off-loading from second floor and 03 is used to represent material which is only off-loaded from the third floor of AS/RS.

W: it is used to distinguish digitals.

Quantity × weight: It is used to know the quantity and weight of the materials in container. The weight is for single material.

Time of incoming: Time of information is used to know the sequence of the arrival of materials to make sure which material arrives first to make sure the first in first out (FIFO) rule. Due to this, the material stays for less time.

Supplier: It is used to represent the set of material to make replenishment in time.

For example, the material spring steel wire is indicated by SY5250GJB8.5-47 in company. The weight is 0.15kg for each one spring steel wire. The material belongs to workstation 1, the frequency is 0.32, the incoming (uploading to AS/RS) time is May 7, 2016, and the No. of supplier is 125. According to the BOM, this material is small-size which need to package and the outcoming exit is located at second floor. Then the material is encoded as "THG3202W150×0.15 160507 125".

There are some special materials which are large-size and directly distributed to production lines. Due to different sequence of models to assemble, the frequency of these material changes in different periods. When encoding for these materials, it is required to be indicated by sequence of mixed models to assemble. The relation is shown in Eq.1.

$$P(\mathrm{i}) = \mathrm{p}(i) * m(i) \div M \tag{1}$$

Where $P(\mathrm{i})$ is the final frequency of material, $\mathrm{p}(i)$ is the frequency of material outcoming and incoming, $m(i)$ is the priority value of material i in a sequence. For common usage parts, in a sequence of mixed model to assemble, $m(i)$ is as same as M and the ratio of $P(\mathrm{i})$ and $\mathrm{p}(i)$ is 1.

For example, the material motor A with the weight 23.5kg, with the quantity 3, frequency of 0.06, required on workstation 2, the time of incoming is April 15, 2016 and supplier is 98. Suppose that the current production sequence is CACBA, the priority of the first C is 1, the second B is 0.8, the third C is 0.6, the fourth B is 0.4 and the last A is 0.2. According to this, the priority value of C can be obtained as 1 + 0.6 = 1.6, and for A the final priority value is 0.8 + 0.2 = 1. Therefore, the final frequency of motor A is 0.02 which is calculated by P(A)= （0.06×1）/（1.6+1+0.4）= 0.02. Then the material is encoded as "MDA0601W3×23.5 160415 98".

The encoding strategy described above is a general strategy. However, for the demand of real algorithm, the code is required to be simplified by only showing the outcoming exit, quantity × weight, final frequency of material and name. The time of incoming and supplier do not affect the result of algorithm.

4.1.2. The establishment of objective function for slotting optimization

According to the common principles of slotting optimization, the stress stability and efficiency of AS/RS are selected to make mathematical models. In current research, a row of shelves of AS/RS is studied. The AS/RS has *m* layer and *n* column, the layer at the bottom is the first layer, and the column nearest to outcoming exit is the first column. The vertical speed and horizontal speed are V_y and V_x respectively. The length of shelf is L and the height of it is H.

1) According to the stress stability principles, the position of the center of gravity in the vertical direction is required to be located in the lower part of the shelf to make it stable in its position when the material or part is removed from it. In order to facilitate the solution of the center of gravity of the entire shelf, it is assumed that the center of gravity of the each good is located in geometry center of slot. The objective function of stress stability is indicated in Eq.2.

$$\min f_1 = \frac{\sum_{i=1}^{m}\sum_{j=1}^{n} W_{ij} \times (i-0.5)H}{\sum_{i=1}^{m}\sum_{j=1}^{n} W_{ij}} \tag{2}$$

$$1 \le i \le m ; 1 \le j \le n$$

Where f_1 indicates the position of the center of gravity of shelf in the vertical direction. Using W_{ij} denotes the weight of materials in layer *i* and column *j*.

2) According to the principle of outcoming from and incoming to the nearest slot, the sum of all the material product of the frequency and time spent for taking the material out from the AS/RS should be minimized. Therefore, the sum is obtained from Eq.3 and Eq.4.

$$\min f_2 = \sum_{i=1}^{m}\sum_{j=1}^{n} t_{ij} \times p_{ij} \tag{3}$$

$$t_{ij} = \left[L \times (j-0.5)/V_x , H \times (i-1)/V_y \right] \tag{4}$$

Where f_2 indicates the sum of all the material product of the frequency and time spent for taking the material out from the AS/RS, and t_{ij} indicates the time spent for taking the material out from the AS/RS once.

In current study, the AS/RS has 10 layers and 13 columns. The first column and fourth column in layer 8 are the outcoming exit of full container, and the 13th column in layer 1 is the direct outcoming exit for certain materials. The time spent for taking the material out from the AS/RS once time is different when the outcoming exit changes. And the other three equations of t_{ij} are shown from Eq.5-Eq.7.

a. Equation for the outcoming exit 01

$$t_{ij} = \left[L \times (12.5 - j)/V_x , H \times (i-1)/V_y \right] \tag{5}$$

b. Equation for the outcoming exit 02

$$t_{ij} = \left[L \times (j - 0.5)/V_x , H \times (i-4)/V_y \right] \tag{6}$$

c. Equation for the outcoming exit 03

$$t_{ij} = \left[L \times (j - 0.5)/V_x , H \times (i-8)/V_y \right] \tag{7}$$

4.2. The steps of algorithm for slotting optimization

The slot allocation optimization is required to consider both the shelf stability and storage efficiency. This is a multi-objective optimization problem. The concept of Pareto optimal solution is combined with the improved genetic algorithm in the current research. The detailed steps are as follows:

Step 1: Encoding. According to the characteristics of goods allocation, the $p \times q$ matrix is used to encode the problem. If the *j*th column in row *i* is material *m*, it indicates that the material m is stored on *i*th colum and $(q+1-i)$ th layer.

Step 2: Calculating fitness function value. The fitness functions are the maximum value of objective functions minus the objective functions respectively.

Step 3: Selection operator. Before selecting the operator, there is need to consider coordination of the process of population group selection and classification operation. In the process of coordinate selection, the whole population is divided into two sub populations equally, and two sub populations of the same population are generated corresponding to the two objective functions.

Population classification: in order to get the optimal solution, it is needed to make the classification of the population according to the level of non -inferiority of the individual before the selection operation is carried out. The specific implementation method is as follows: all the non-dominated solutions in the population are divided into the same level, and assign 1 to the level value. Repeat the above process until all individuals in the population have been given the appropriate levels. In the selection process, the selection probability of the point of *i* level can be determined by the formula 9.

$$F_i = M(N_r - i + 1) \Big/ \sum_{i=1}^{N_r} (N_r - i + 1) P_{si} \tag{8}$$

Where, M is the population size, N_r is the number of levels of population group and P_{si} is the population size for level *I*.

Step 4: Crossover operation. In the genetic algorithm, two individuals are randomly selected for crossover operation. In the current multi-objective optimization problem, the mating restriction is introduced. Only when two individuals are not in the same group, the cross operation can be carried out. In this way, the better genes can be combined under different objective functions. Moreover, there is a larger probability to find an optimal compromise solution. According to the objective function value of the optimal allocation of goods directly depends on the location of goods on the shelves, the cross operation using a partially matched crossover method.

Step 5: Mutation operation. Mutation operator selects a part of the individual to implement mutation according to the probability. Select two locations randomly, and then change the goods in the two locations of AS/RS.

Step 6: Niche technology. In order to ensure that the optimization process does not converge to a certain part of the feasible region and make the populations evolve toward the direction of evenly distributed from the Pareto front, the niche technology is used here. Penalties are imposed to the individuals in the population which are aggregated into small pieces, so that their fitness value is reduced.

Step 7: Elite preservation strategy. In order to keep the best individual in the population evolution, parent generation and the sub generation population are combined to make one population. Later, sorting of the individual in the population is performed and penalty function is shared. According to the level of non-inferiority and level of individual fitness, the first 50% individuals in the population are made as a new population for the next cycle of the algorithm. The niche technology and optimization strategy can not only give solutions converging to Pareo front, but also make sure the solutions are distributed uniformly.

Step 8: Pareto solution set filter. The effect is to keep the non-dominated points in each generation and to remove the bad points of the solution set. The level 1 Pareto points in each generation are put into the Pareto filter. The size of the Pareto solutions set can be arbitrarily set. When the number of new points is more than the size of Pareto solution set, then make gain sorting is performed to remove the bad points. If the number of the remaining non -dominated points is still more than the size of the Pareto solution set, then the nearest point from the other point are deleted to improve the diversity.

4.3. Computational result

According to the above description, the shelf stability and storage efficiency are the two evaluation index. Genetic algorithm is coded in Visual C# and runs on an Intel Core i7, 3.4 GHz CPU, 4 GB RAM computer. The prepared material encoding table is directly imported into the algorithm for computational experiment, and the detailed parameters are as follows: L=1.3m, H=1.4m, V_x=2m/s, V_y =0.67m/s, m=10, n=13. The current real significant allocation for the AS/RS is shown in table 3.

Table 3. The significnat allocation before optimization

44	120	74	58	56	101	34	5	19	7	94	86	1
0	75	27	72	13	91	6	14	18	23	9	10	125
26	47	52	87	3	36	48	12	41	62	25	57	50
103	54	49	93	2	111	68	55	53	127	92	16	42
35	102	129	46	37	40	81	85	126	67	61	45	71
63	22	77	15	122	116	30	88	69	117	80	8	39
104	119	107	95	33	21	84	108	31	76	59	123	70
38	79	73	60	90	98	28	121	128	105	99	118	29
43	78	124	17	113	96	20	110	97	11	82	24	100
65	114	83	112	32	89	4	66	115	51	106	64	109

It can be seen from table 3 that the total frequency of material outcoming from and incoming into the AS/RS for the current situation is 155.2065, and the center of gravity is 7.0366. Through the proposed GA, after the 20 generation of crossover and mutation, more than one solution is obtained and one of the selected solution from the Pareto set is shown in table 4.

Table 4. The goods allocation after optimization

114	66	25	18	20	74	0	9	16	64	97	83	125
81	116	61	8	121	94	11	90	40	37	127	31	110
85	70	126	129	29	51	68	111	107	86	123	63	32
30	80	119	115	4	1	71	50	6	118	101	79	19
17	95	120	62	47	53	34	43	106	113	91	54	103
104	55	49	73	88	26	57	67	45	21	96	112	109
58	78	10	33	36	2	23	28	5	84	7	3	77

60	38	72	124	14	41	46	69	12	48	15	100	92
39	27	99	122	13	76	102	89	117	75	56	44	108
128	87	82	24	98	65	93	35	22	59	52	105	42

It can be seen from table 4 that the new storage allocation scheme gives AS/RS a frequency of 135.6705, the position of the center of gravity is 6.4057. Furthermore, from table 4 it can be seen that the frequency of new storage allocation scheme is reduced by about 12.9% as compared to previous company solution, significantly accelerates the storage efficiency and saves the precious time for the whole workshop distribution section, which improves distribution capacity to meet the demand of practical production. Although there is not a big change in the center of gravity, it improves the shelf stability and can prolong the service life of the AS/RS.

5. Conclusion

Current research introduces a final assemble workshop and the AS/RS system in it. According to the principles of AS/RS location assignment, the objective and constraint functions for the proposed problem are established. The problem of storage assignment is integrated with actual demand of mixed flow production style and a new mathematic model for slotting optimization is proposed. Moreover, a genetic algorithm with a new encoding strategy is proposed for the real company slotting optimization problem. Computational experiments are performed on real company data and computational result shows a big improvement in stress stability and efficiency of AS/RS as compared to the empirical solutions.

6. References

[1] Thonemann U W, Brandeau M L. Optimal storage assignment in an automated storage and retrieval systems with stochastic demands. Management Science, 1998, 44(1): 142-148.
[2] Hsieh S, Tsai K C. A BOM oriented class-based storage assignment in an automated storage and retrieval systems. International Journal of Advanced Manufacturing Technology, 2001, 17(9): 683-691.
[3] Poulos P N, Rigatos G G, Tzafestas S G, et al. A Pareto-optimal genetic algorithm for warehouse multi-objective Optimization. Engineering Applications of Artificial Intelligence, 2001, 14(6): 737-749.
[4] Muppani V R, Adil G K. A branch and bound algorithm for class based storage location assignment. European Journal of Operational Research, 2008, 189(2): 492-507.
[5] Onut S, Tuzkaya U R, Dogac B. A particle swarm optimization algorithm for the multiple-level warehouse layout design problem. Computers and Industrial Engineering, 2008, 54(4): 783-799.
[6] Zu Q, Cao M, Guo F, et al. Slotting optimization of warehouse based on hybrid genetic algorithm. In: Proceedings of 6th International Conference on Pervasive Computing and Applications, 2011, 19-21.
[7] Yin L, Jian S, Guo P. Slotting optimization based on SKUs correlations with serpentine routing policy. Journal of Zhejiang University (Engineering Science), 2012, 46(8): 1424-1430.

Acknowledgement

This work has been supported by MOST (Ministry of Science & Technology of China) the Funds for International Cooperation and Exchange of the National Natural Science Foundation of China (No.51561125002), the National Natural Science Foundation of China (No. 51275190, 51575211), and the Fundamental Research Funds for the Central Universities (HUST: 2014TS038).

IOP Conf. Series: Materials Science and Engineering **215** (2017) 012003 doi:10.1088/1757-899X/215/1/012003

Improved model of the retardance in citric acid coated ferrofluids using stepwise regression

J F Lin[1] and X R Qiu[2]

[1] Department of Industrial Design, Far East University, No.49, Zhonghua Rd., Xinshi Dist., Tainan City 74448, Taiwan (R.O.C.)
[2] Graduate School of Mechanical Engineering, Far East University, No.49, Zhonghua Rd., Xinshi Dist., Tainan City 74448, Taiwan (R.O.C.)

E-mail: jacklin@cc.feu.edu.tw

Abstract. Citric acid (CA) coated Fe_3O_4 ferrofluids (FFs) have been conducted for biomedical application. The magneto-optical retardance of CA coated FFs was measured by a Stokes polarimeter. Optimization and multiple regression of retardance in FFs were executed by Taguchi method and Microsoft Excel previously, and the *F* value of regression model was large enough. However, the model executed by Excel was not systematic. Instead we adopted the stepwise regression to model the retardance of CA coated FFs. From the results of stepwise regression by MATLAB, the developed model had highly predictable ability owing to *F* of 2.55897e+7 and correlation coefficient of one. The average absolute error of predicted retardances to measured retardances was just 0.0044%. Using the genetic algorithm (GA) in MATLAB, the optimized parametric combination was determined as [4.709 0.12 39.998 70.006] corresponding to the pH of suspension, molar ratio of CA to Fe_3O_4, CA volume, and coating temperature. The maximum retardance was found as 31.712°, close to that obtained by evolutionary solver in Excel and a relative error of -0.013%. Above all, the stepwise regression method was successfully used to model the retardance of CA coated FFs, and the maximum global retardance was determined by the use of GA.

1. Introduction

Citric acid (CA) coated Fe_3O_4 ferrofluids (FFs) have been conducted for biomedical application such as magnetic resonance imaging (MRI) and tumor hyperthermia. It has been known that the surface of the magnetic nanoparticles (MNPs) can be stabilized in an aqueous dispersion by the adsorption of CA. This process, as described by Sahoo et al. [1], occurs by the CA being coordinated via the carboxylate functionalities. Srivastava et al. [2] synthesized CA coated Fe_3O_4 MNPs of 6 nm particle size, which exhibits excellent magnetic properties. Effect of synthesis conditions on the properties of CA coated iron oxide nanoparticles was discussed [3].

In the present investigation, Fe_3O_4 MNPs were prepared by an improved co-precipitation of Fe^{3+}/Fe^{2+} salts and then the MNPs were stabilized against agglomeration by surfactant encapsulating of CA [4]. Afterwards, the CA coated Fe_3O_4 MNPs were used as the precursor of water-based FFs, and the retardance (magneto-optical property as the phase retardation of linearly birefringent medium, such as quartz and certain liquid crystals) of FFs were measured by a developed Stokes polarimeter [5]. It is known that Taguchi orthogonal design method may provide a powerful and efficient ways to find an optimal combination of factor levels that may achieve optimum. Hence, the Taguchi method with range analysis was employed to find the parametric combination for CA coated FF with high retardance readily [4].

Optimization and multiple regression of retardance in FFs were executed by Taguchi method and Microsoft Excel, and the *F* value of regression model was large enough [6]. However, the modelling of retardance executed by Microsoft Excel was not in a systematic way. Further, using Taguchi-based measured retardances as the training data, an artificial neural network (ANN) model was developed for the prediction of retardance in CA coated FF. Based on the well-trained ANN model, the predicted retardance at excellent program from Taguchi method showed less error of 2.17% compared with a multiple regression (MR) analysis of statistical significance [7].

The aim of this study was to use the stepwise regression for modelling retardance of CA coated FFs. The stepwise regression by MATLAB was executed successfully for modelling retardance of CA coated FFs. The regression model had highly predictable ability and had a high correlation coefficient of one. Moreover, two optimization techniques, including evolutionary algorithm (in Microsoft Excel) and genetic algorithm solver (in MATLAB® Optimization Toolbox), were used to find the global maximum retardance value. Then the CA coated FFs could be more useful in practical applications.

2. Method

In this study, the methods including Taguchi orthogonal design method, stepwise regression method, evolutionary algorithm and genetic algorithm were briefly introduced.

2.1. Taguchi method

Taguchi method is straightforward and easy to apply to many engineering situations, making it a powerful yet simple tool. Subsequently, the optimal synthetic condition of CA coated Fe_3O_4 FFs with high retardance was determined by the Taguchi orthogonal design method-$L_9(3^4)$, i.e. four parameters with three levels, respectively, and nine tests of FFs with 1000 g/L, correspondingly [4]. Influence parameters were (A) pH value of suspension after coating (4.5, 5, 5.5), (B) molar ratio of CA to Fe_3O_4 MNPs (0.03, 0.06, 0.12), (C) CA volume (10 ml, 20 ml, 40 ml), and (D) coating temperature (70 ℃, 80 ℃, 90 ℃). Procedure of the chemical co-precipitation method for producing CA coated FFs could be found [4]. The retardance (magneto-optical property) of FF was measured by a Stokes polarimeter with a feasible algorithm [5]. Results of the retardance under 64.5 mT were as shown in table 1 [4]. Moreover, the influence sequence between these parameters on the retardance and the excellent program (parametric combination) of maximum retardance were decided by a simple range analysis.

Table 1. Orthogonal design and test results of retardance.

No	A	B	C	D	Retardance (deg.)
A1	1	1	1	1	7.8228
A2	1	2	2	2	11.9872
A3	1	3	3	3	29.3618
A4	2	1	2	3	4.0525
A5	2	2	3	1	23.6294
A6	2	3	1	2	18.1662
A7	3	1	3	2	-0.4628
A8	3	2	1	3	4.5877
A9	3	3	2	1	20.2021

2.2. Stepwise regression

While dealing with large number of independent variables, it is of significant importance to determine the best combination of these variables to predict the dependent variable [8]. Stepwise regression method is a routine statistic technique used for variable selection and is a combination of forward and backward procedures. The method starts with an initial model which includes a subset of all the candidate variables. And then, a so-called stepping procedure which iteratively altering the model established at the previous step by adding or removing variables in accordance with the stepping criteria is used in order to add significant variables or remove insignificant variables.

The stepping procedure terminates when the model has been optimized, or when a specified maximum number of steps have been reached. The obtained model can be tested for validity through a variety of statistical methods. Usually, the evaluation of the model is made by using some criteria: correlation coefficient (R), adjusted square of correlation coefficient (adjusted R^2), standard error (SE), and F statistic [9]. The details on the use of stepwise model-building procedure could be found [8, 10].

In our previous study, optimization and multiple regression of retardance in FFs were executed by Taguchi method and Microsoft Excel, and the F value of regression model was large enough [6]. However, the modelling of retardance executed by Microsoft Excel was not in a systematic way, and the execution is time-consuming. Therefore, the stepwise regression method can be adopted to model the retardance of CA coated FFs efficiently.

2.3. Evolutionary algorithm in Microsoft Excel

We can use the evolutionary solver [11] in Microsoft Excel to solve optimization problems. A genetic or evolutionary algorithm applies the principles of evolution found in nature to the problem of finding an optimal solution. In a "genetic algorithm," the problem is encoded in a series of bit strings that are manipulated by the algorithm; in an "evolutionary algorithm," the decision variables and problem functions are used directly.

Most commercial solver products are based on evolutionary algorithms. The evolutionary solver combines genetic algorithm methods such as mutation, crossover, and natural selection with classic methods drawn from linear programming, nonlinear optimization, and pattern search, to find better solutions in less time. Further, on smooth nonlinear problems where the GRG (Generalized Reduced Gradient) solver can find only a locally optimal solution, the evolutionary solver can often find a better solution or even a globally optimal (or close to globally optimal) solution, which is described by Frontline System [11]. Therefore, the optimized parametric combination and the maximum retardance of CA coated FFs could be obtained by the evolutionary algorithm in Microsoft Excel, which is capable of solving smooth and non-smooth global optimization problems.

2.4. Genetic algorithm

The genetic algorithm (GA) technique is based on the natural process of evolution to solve optimization and search problems. There are three main operators in GA which are reproduction, crossover and mutation [12]. GAs were pioneered by Holland in the 1960s and 70s [13] and have emerged as a highly effective technique for solving a wide variety of optimization problems [14].

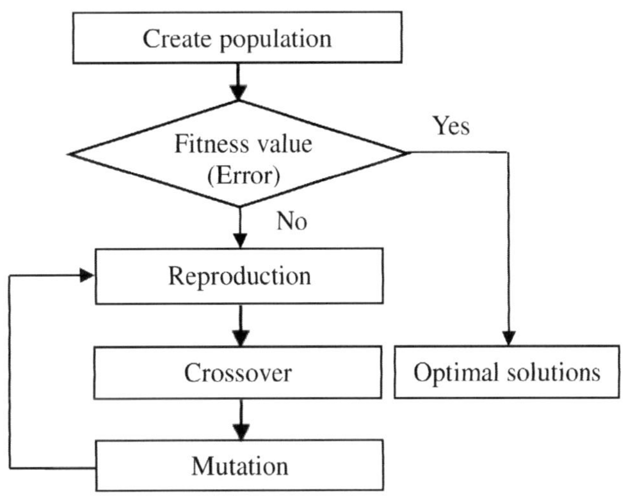

Figure 1. Flowchart of genetic algorithm.

Figure 1 presents a simple flowchart representation of a typical GA solution procedure [14]. As shown, the process commences by creating an initial population of random solutions to the problem of interest. Each potential solution is then evaluated using a fitness function, i.e. an objective function which quantifies the optimality of each solution such that it can be ranked against that of all the other potential solutions. If the termination criteria are not satisfied, multiple candidate solutions are selected for modification via reproduction, crossover and mutation operations in order to create a new population pool. The optimality of each member of the population pool is then re-evaluated using the fitness function.

3. Results and discussions

Results of stepwise regression (first order and second order) on the Taguchi-based measured retardances of CA coated FFs are provided below. And the optimized parametric combination and the maximum retardance of CA coated FFs are obtained by genetic algorithm, which are compared with those results obtained by the evolutionary solver in Microsoft Excel.

3.1. Results of stepwise regression

Based on the Taguchi-based measured retardances of CA coated FFs (as shown in table 1, the retardance value of the A7 sample is not considered due to the possible error), the results of first order stepwise regression executed by MATLAB are as shown in figure 2. We can see the meaningfulness degrees of input variables. This degree of meaningfulness is determined via being value of *p*-val (*P* value, *P* represents probability) below 0.05. Thus meaningfulness ranking of input variables is determined as variable 2 and variable 3, i.e. molar ratio of CA to Fe_3O_4 MNPs and CA volume, respectively. The variable 2 (molar ratio of CA to Fe_3O_4 MNPs) involved in this model represents the most significant influencing factor for the retardance of CA coated FFs. Based on the results obtained; equation of the first order stepwise regression model is given by equation (1):

$$Retardance = -5.59738 - 3.50353X_1 + 145.257X_2 + 0.455496X_3 - 0.227538X_4 \quad (1)$$

From the results of figure 2, the R^2 value is 0.911138. This indicates that the current equation is able to describe 91.1138 percent of the variation in the response variable; this doesn't guarantee the appropriateness of the first order model. In addition, *t* statistic (*t* Stat) in table 2 is equal to the ratio of respective standard error to the coefficient value and tests the significance of individual regression coefficients for a confidence interval as 95%. Under the normality assumption, for n number of observations, *t* statistic follows student's *t* distribution with (n-2) degrees of freedom (DF), and critical value of *t* can also be found out by using TINV function in Microsoft Excel which comes out to be 2.36. As can be seen that X_2 coefficient (4.4859> *t* critical value of 2.36) has the strongest effect on the regression model followed by the X_3 coefficient. *P* value in table 2 is the probability that for a particular confidence level as 95%, a random variable having a student's *t* distribution is greater than absolute value of observed *t* statistic and hence significant.

For testing whether the simple stepwise regression model gives any useful information about the response variable or not, as shown in figure 2. *F* statistic is the ratio of mean squares due to regression model and error, and is used to test the significance of regression model for a confidence level. The *F*-test is executed by comparing the observed *F* statistic with the appropriate critical value obtained from the standard *F* distribution table. Critical value of *F* for the current regression model can be found out from standard *F* distribution table by F(4, 4, 0.05) or by using FINV function in Microsoft Excel, which comes out to be 6.388. As can be seen that the observed *F* value of 25.6336 > 6.388, it implies the regression model is significant. Figure 2 shows the corresponding probability value *P*(*F*). It can be deduced that the regression model is statistically significant as *P* value (0.0023539, chance of 0.23539% that such model *F*-value could occur due to noise) is less than 0.05 (i.e. 95% confidence level). Compared to the results obtained by multiple linear regression in Microsoft Excel [6], the determination coefficient R^2 and the observed *F* value of the first order stepwise regression model are larger than those of 0.888 and 7.954, respectively [6]. It can be deduced that the first order stepwise regression model is statistically significant and the first order stepwise regression is more reliable than the multiple linear regression.

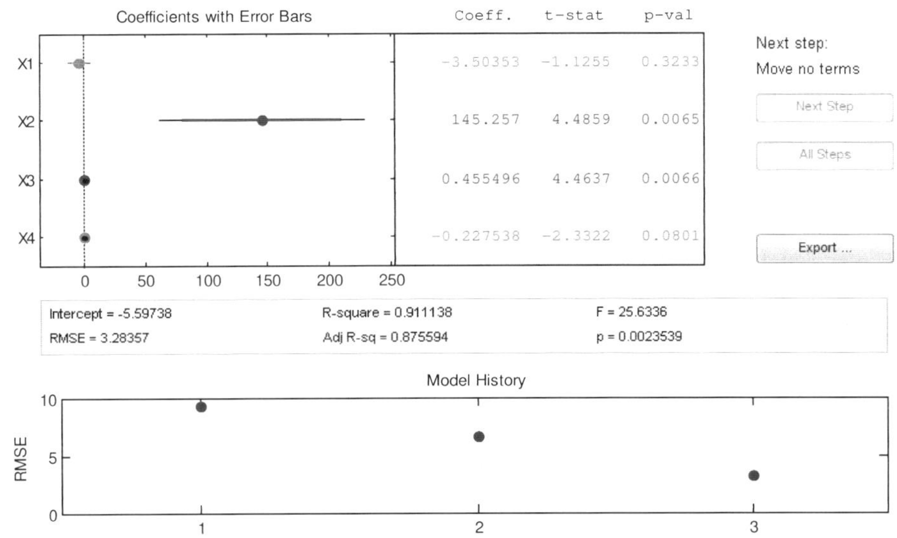

Figure 2. Results of stepwise regression (first order).

Table 2. Results of stepwise regression (first order).

	Coefficient	*t* Stat	*P* value
X_1 **(pH)**	-3.50353	-1.1255	0.3233
X_2 **(Molar Ratio)**	145.257	4.4859	0.0065
X_3 **(CA Volume)**	0.455496	4.4637	0.0066
X_4 **(Temperature)**	-0.227538	-2.3322	0.0801

Subsequently, we aim to increase the precision of regression model and precede the second order stepwise regression. The results of stepwise regression by MATLAB are shown in figure 3, the model has highly predictable ability owing to F of 2.55897e+7 and has a high correlation coefficient R of one, which is suggested that this regression model is statistically significant and gives useful information about the response variable. The average absolute error of predicted retardances is less just as 0.0044%. The equation for the second order stepwise regression model is given by equation (2):

$$Retardance=-59.2263+31.4321X_1-2.9987X_1^2+0.00801827X_3^2-0.000804091X_4^2-0.0458668X_1X_4+1.8636X_2X_4 \quad (2)$$

As shown in figure 3 and table 3, we can see the meaningfulness degrees of input variables. This degree of meaningfulness is determined via being value of p-val below 0.05. Thus meaningfulness ranking of input variables is determined as 13(X_2X_4), 7(X_3^2), 5(X_1^2), 1(X_1), 11(X_1X_4), and 8(X_4^2). As shown in figure 3, the RMSE (Root Mean Square Error) value is 0.00198777 and is much lower than that of 3.28357 in the first order stepwise regression as shown in figure 2. Further, the R^2 value is determined as one and is higher than 0.911138 in the proposed first order stepwise regression (in figure 2) and 0.9999998 in the multiple nonlinear regression model [6]. It is indicated that the ability of prediction in the second order stepwise regression model is excellent.

The significance t-test of each variable is listed in table 3. It is seen that the significant influence of each term on the retardance of CA coated FFs is given by: $X_1>X_3>X_4$. This indicates that the effect of coating temperature on the retardance is of the least significance. Although the effect of the coating temperature on the retardance is not significant, the cross correlations of coating temperature and other two parameters as pH value of suspension and molar ration of CA to Fe_3O_4 MNPs (X_2X_4 and X_1X_4)

are significant. The relative influence of cross correlations on the retardance of CA coated FFs (from examining the magnitude of the *P* value or from *t* Stat checking the significance level of the cross items) is given by: $X_2X_4 > X_1X_4$. Although the item of X_2 seems not significant in second order stepwise regression, the X_2 plays an important role in the influence of retardance judging from the X_2X_4. In addition, factors at higher order (i.e. X_3^2 and X_4^2) in equation (2) denote positive and negative quadratic relationships, respectively. Therefore, the effect of each factor on the experiment indices must be taken into account during the experiment.

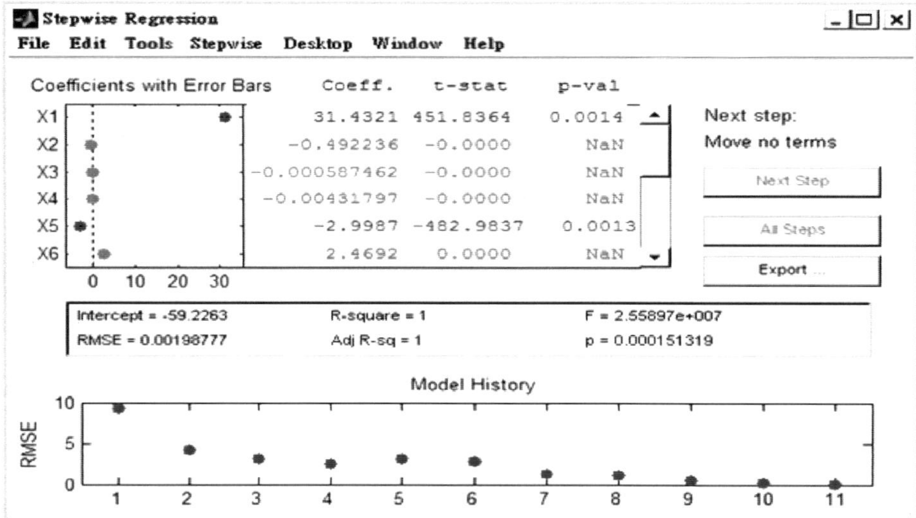

Figure 3. Results of stepwise regression (second order).

Table 3. Results of stepwise regression (second order).

	Coefficient	*t* Stat	*P* value
X_1	31.4321	451.8364	0.0014
$X_1^2 (X_5)$	-2.9987	-482.9837	0.0013
$X_3^2 (X_7)$	0.00801827	4644.0641	0.0001
$X_4^2 (X_8)$	-0.000804091	-86.9664	0.0073
$X_1X_4 (X_{11})$	-0.0458668	-156.2040	0.0041
$X_2X_4 (X_{13})$	1.8636	6389.4874	0.0001

3.2. Results of optimization using genetic algorithm

Based on equation (2), the evolutionary solver in Microsoft Excel is used to solve optimization problem firstly. The optimized parametric combination and the maximum retardance of CA coated FFs are obtained as [4.7037 0.12 39.99991 70.0021], corresponding to the pH of suspension, molar ratio of CA to Fe_3O_4, CA volume, coating temperature, and 31.716°, respectively. Also, the GA in MATLAB is used to solve the optimization problem. As shown in figure 4, the optimized parametric combination and maximum retardance of CA coated FFs are determined as [4.709 0.12 39.998 70.006] and 31.712°, close to those obtained by Microsoft Excel and a relative error of -0.013%. Above all, we successfully model the retardance in CA coated FFs by stepwise regression and the maximum global retardance is determined by the GA.

It is noted that the Fitness function as @FitFun75 is obtained by changing the sign of equation (2) and is used to find the minimum value. The results obtained by nine times of executions of GA in MATLAB are listed in table 4, and figure 4 is corresponding to the results of the second execution of GA.

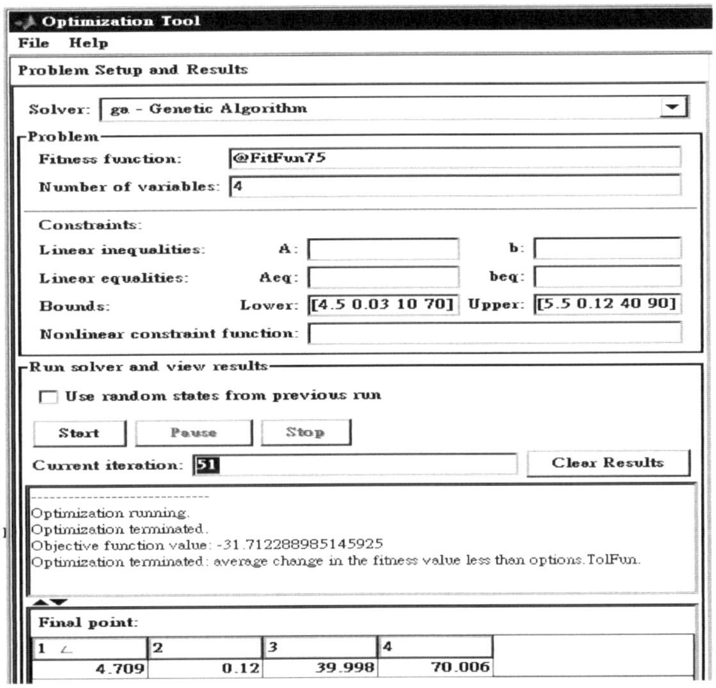

Figure 4. A graph output using the GA in MATLAB® optimization.

Table 4. Optimized parametric combination and maximum retardance.

No	pH	Molar Ratio	CA Volume	Coating Temp.	Retardance (deg)
1	4.693	0.12	39.999	70.005	31.714
2	4.709	0.12	39.998	70.006	31.712
3	4.706	0.12	39.994	70	31.684
4	4.722	0.12	39.999	70.024	31.710
5	4.673	0.12	39.997	70.213	31.680
6	4.697	0.12	39.998	70.001	31.714
7	4.693	0.119	39.998	72.979	31.273
8	4.71	0.12	39.998	70.108	31.703
9	4.72	0.12	39.999	70.487	31.664

In overall, Taguchi method is a discontinuous optimization approach, which can only obtain the local discontinuous optimal solution of pre-selected parameter level value, and cannot find out the global optimal solution [4]. The integration of Taguchi method with various approaches including numerical simulation, principal component analysis (PCA), artificial neural network (ANN), response surface methodology (RSM), and genetic algorithm (GA) could be used to enhance the efficiency of the optimization process. In addition, it is noted that GA based on a neural network model has been applied successfully to optimize complicated bioprocesses [4].

In future work, the predicted retardance value obtained by the trained ANN model [7] could be adopted as the fitness value, further, the GA operation such as select, crossover, and mutation are executed to find the global retardance of CA coated FFs. Moreover, the multiple-objective optimization including maximum retardance and low dichroism of CA coated FFs [4] or in other double-layer coated FFs is supposed to be performed in future. Therefore, the practical biomedical applications such as hyperthermia (magnetic inductive heating of cancer tumour) and magnetic resonance imaging (MRI) could be executed.

4. Conclusions

The stepwise regression method was successfully used to model the retardance of CA coated FFs. According to the results of second order stepwise regression, the *F* value was obtained as high as 2.55897e+7 and had a correlation coefficient of one; the developed regression model had highly predictable ability. Therefore, the retardance value, corresponding to each parametric combination, built by four parameters with three levels in Taguchi method, could be precisely determined by regression equation. Also, using the genetic algorithm in MATLAB, the optimized parametric combination was determined as [4.709 0.12 39.998 70.006]. The maximum retardance was found as 31.712°, close to that obtained by the evolutionary solver in Microsoft Excel and a relative error of -0.013%.

Acknowledgements

The financial support provided by the Ministry of Science and Technology in Taiwan under MOST 104-2221-E-269-006 and 105-2221-E-269-001 is gratefully acknowledged.

References

[1] Sahoo Y et al 2005 *J. Phy. Chem.* B **109** 3879
[2] Srivastava S et al 2011 *J. Colloid Interface Sci.* **359** 104
[3] Li L et al 2013 *Microel. Eng.* **110** 329
[4] Lin J F et al 2014 *J. Magn. Magn. Mater.* **372** 147
[5] Lin J F and Lee M Z 2012 *Opt. Commun.* **285** 1669
[6] Lin J F et al 2015 *Proc. SPIE* **9302** 930229
[7] Lin J F and Sheu J J 2016 *J. Magn. Magn. Mater.* **404** 201
[8] Cevik A et al 2010 *Adv. Eng. Softw.* **41** 611
[9] Jiao L and Li H 2010 *Chemometr. Intell. Lab. Syst.* **103** 90
[10] Hsieh K L and Lu Y S 2008 *Expert Syst. Appl.* **34** 717
[11] Frontline System 2016 http://www.solver.com/genetic-evolutionary-introduction
[12] Yusup N et al 2012 *Expert Syst. Appl.* **39** 9909
[13] Holland J H 1975 Adaptation in Natural and Artificial Systems (Cambridge, MA: MIT Press)
[14] Lin W L et al 2009 *IEEE J. L. T.* **27** 4136

IOP Conf. Series: Materials Science and Engineering **215** (2017) 012004 doi:10.1088/1757-899X/215/1/012004

Research on simulation based material delivery system for an automobile company with multi logistics center

D Luo[1,2], Z Guan[1,2], C Wang[1,2], L Yue [1,2], L Peng[1,2],
[1]State Key Lab of Digital Manufacturing Equipment and Technology, Wuhan 430074, China
[2]School of Mechanical Science and Engineering, Huazhong University of Science and Technology, Wuhan 430074, China

Abstract. Distribution of different parts to the assembly lines is significant for companies to improve production. Current research investigates the problem of distribution method optimization of a logistics system in a third party logistic company that provide professional services to an automobile manufacturing case company in China. Current research investigates the logistics leveling the material distribution and unloading platform of the automobile logistics enterprise and proposed logistics distribution strategy, material classification method, as well as logistics scheduling. Moreover, the simulation technology Simio is employed on assembly line logistics system which helps to find and validate an optimization distribution scheme through simulation experiments. Experimental results indicate that the proposed scheme can solve the logistic balance and levels the material problem and congestion of the unloading pattern in an efficient way as compared to the original method employed by the case company.

Key word: Simulation, material classification, logistics distribution strategy, Kitting, Kanban, Line Stocking

1. Introduction

In recent years, the production scale of China's automobile industry has increased the rapid development of logistics industry for automobile companies. In the process of automobile supply chain logistics, inbound factory logistics and production logistics of automobile parts are universally acknowledged as a key to begin function and continuous optimization of the logistics system [1]. In these industries, parts supply to the production system in appropriate level can significantly reduce the production time, reduce cost, improve the response to the market and improves flexibility in production. Furthermore, logistics balance in automobile parts assembling can help in the improvement of the supply chain in automobile industry. The problem of internal logistics in automobile manufacturing companies has complex structure and evaluation of these systems is critical to solve using exact mathematical modeling for its improvement. Therefore, simulation tools has been widely used as a modeling tool to measure the performance of manufacturing systems. In literature simulation tools has been implemented for evaluation of these manufacturing systems. For example, Wang [2] used Flexsim to establish a simulation model of collaborative logistic center (CLC) and optimized the model using theory of constraints (TOC). Zhou [3] used eM-Plant to simulate logistics distribution in assembly line and studied the relationship between completion time of production planning with buffer size of the material and the loading quantity and velocity of vehicle using design of experiments (DOE). Wan [4] studied the existing problem of logistics planning in assembly workshop and used AnyLogic to simulate each alternatives. Zhang [5] used a new object-oriented simulation software named Simio to model and analyzed their proposed model. Simio tool has been applied in automobile parts logistics in Yue [6] and Wang [7] work. They used GA combined with simulation software to optimize path planning in mixed model assembly line. Most of the literature indicates the significance of use of simulation methods for analysis and optimization of complex logistics system. However, the problems of distribution systems investigated in literature are related to different companies and their real situations.

Current research investigates logistics distribution system in an automobile plant in China. The current problem is focused to balance the part delivery to the final assembly line of the automobile assembly plant. The Case Company has problem in its current stocking rules and logistics distribution schedule which needs improvement. Therefore, current research proposed a logistics distribution optimization method based on a real case of automobile assembly plant, simulates the real situation of the logistics, design different experiments of simulation, and proposed an optimal solution of logistics distribution system for the company.

2. Company problem description

Current research investigates Third-Party logistics automobile company which provides services to designated car manufacturing company in China. The main business of the Case Company includes procurement, transportation, scheduling and distribution of parts, and so on. There are two primary warehouses in the case company which are Collaborative Logistics Center (CLC) and the Distribution Collaborative Center (DCC) respectively. The material suppliers transport the parts to the CLC where the supplied parts are stored in CLC. Later, these parts are transported to the DCC according to the products required in the manufacturing plan. The staffs of DCC distributes the required parts from DCC to line-side store from where they are used for assembly.

In the Case Company, there is 67storage area and 5shipping platforms in the CLC. Moreover, there are 5 receiving platforms associated with the shipping platforms in the DCC. The DCC is near the assembly line and three assembly lines mainly included in this automobile production plant including, Final assembly (AF), Assembly line of Engine (AE) and Welding line (WE). Layout and work flow chart of the case company is shown in Figure 1. There are two main problems in the studied case Company due to which the current distribution system is inefficient and unbalanced. The problems are unjustified stocking rules and unreasonable logistics distribution schedules.

(1) Unjustified stocking rules

The required material for automobile lines is large in quantity and diversity, as a result, different requirement is needed for different material when making distribution schedules. However, the current stocking rule used in the Case Company do not take this into consideration and a simple stoking rule is adopted for all materials which is not appropriate. The current method for logistics plans is not significantly operational , therefore, classification of materials is necessary to be dine during stocking.

(2) Unreasonable logistics distribution schedule

When planners calculates the operation time, the interval time between two processes is setting too long and all materials use the same scheduling method according to the same distribution strategy without considering the characteristics of the material and the fluctuation of the demand. Such a program leads to the generation of peak flow and causes congestion and vehicle's waiting line in the storage area.

In view of the present situation of the enterprise, it is important to design different distribution strategies for different corresponding materials and make diverse scheduling. According to the characteristics of the system, current research proposed a logistics distribution optimization method, and uses the simulation method to simulate the status of this logistics system, design the simulation experiments to get an optimal logistics distribution solution.

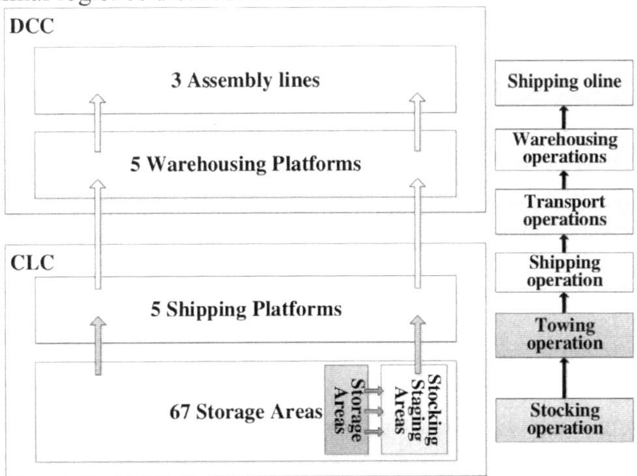

Figure 1. Layout and work flow chart

3 Logistics distribution method

For the current situation of Case Company, following logistics distribution optimization steps are employed in the current research.

- Classify the current parts
- Formulate the corresponding material distribution strategy according to the classification of parts
- Give the corresponding online plan for different distribution strategies

3.1 Method of material classification

The core idea of ABC classification is to distinguish between primary and secondary differences in parts of products based on different factors when making a decision and identifying a key factor that determine the role of things and most of the minor factors that have little effects on classifications. In order to maximize the benefits of the logistics distribution system, different material or goods are divided into three groups, i.e. ABC, according to the types of parts. Based on their classification, we can make different distribution strategies for each type of parts and their characteristics. However, traditional method do to consider parts usage rate and number of package volume and other factors during making the classification [8]. Therefore, to determine the classification of parts based on the features of parts and considering the factors that can affect the logistics system rather than relying on the occupation of cost, the volume of the parts and usage rate of the parts are significant and they are considered as a key factor in classification proposed in current research.

The volume is expressed in terms of package number. The larger package number of a batch of parts indicates volume is small and small package number of a batch of parts shows large volume. The package number can define the number of vehicles of one batch. The usage rate of parts is indicated by generality parameter. The calculation formulas of generality and package number are as follows:

$$G_i = N_i/\mathrm{T} \qquad (1)$$

Where G_i is generality for part i, N_i indicates the number of car type with the corresponding i part for one day, T indicates total number of car type.

$$\mathrm{P}_{ij} = \mathrm{P}_i/\mathrm{C}_{ij} \qquad (2)$$

Where P_{ij} is package number of i part for a j car, P_i indicates package number of part i, C_{ij} indicates number of part i which assemble in an automobile j.

In the proposed classification Class A is considered as the parts of low generality, and for Class B the parts with higher generality parameter bigger volume is considered. For Class C, and the parts with higher generality and smaller volume are considered the classification standard is shown in Table 1.

Table 1. ABC classification method of automobile parts

Type	Generality	Package number per car
A	0−0.8	Medium
B	0.4−1	Smaller than lot size
C	0.4−1	Bigger than lot size

3.2 Strategies for distribution of different category of materials

Due to the differences in the usage frequency of different material, the package number and features, different types of parts are required to be matched with different distribution strategies. This can improve the overall efficiency of distribution, reduce inventory and level the material flow.

3.2.1 Kitting

The usage frequency of A type of parts is not particularly high, but suitable for most containers which can use kitting distribution. Kitting distribution is used for packaging some of the demand of material into containers under packaging strategy and send it to the station according to the accurate quantity and time [5]. Kitting distribution is in synchronized with the requirement of the parts according to the assembly sequence. In this way, the quality of the assembly process can be improved, as well line inventory can be reduced, and help in the improvement of the vehicle stowage rate and reduce vehicle's transportation times.

Kitting distribution has two types. The first type of kitting distribution used fixed kitting for each station in which the all the parts that the station needs are distributed in current batch in place. The second type of kitting distribution is mobile kitting distribution in which all parts which a vehicle needs are pulled into a container that follows the production flow. In the automobile assembly lines of the case company, assembly line is suspension type that materials are hard to move following the bodywork, so it is better to use the fixed kitting distribution.

3.2.2 Kanban

B type of parts have a relatively high usage frequency in the case company, but with relatively large volume. These parts include engine, transmission, fuel tank and other large materials. Because of large volume, inadequacy distribution of B parts is one of the key factor to cause the line-side congestion. Kanban management is a means of JIT production mode which as a kind of production management technology to pursue the rationality, efficiency and flexibility of production process, has been used by many enterprises in the world. It is the basic features of Kanban that for specific parts, workers can dispatch the necessary products with necessary quantity in the required time to reduce inventory, costs and improve asset turnover. Therefore, we chose Kanban as the strategy for B type to effectively control the congestion of the line side and improve the overall distribution efficiency.

3.2.3 Line stocking

C type of parts have high frequency usage in the case company and small have volume. C type of parts includes Bolts, nuts and other standard parts. In order to increase the efficiency of distribution, decrease turnover work on the middleman, a method of transporting large quantities of materials to DCC at one times is adopted here, namely line stocking distribution. Line Stocking, batch transportation, is a distribution method in which a large quantity of material is delivered directly to the assembly line area without storing it in the intermediate warehouse. In general, the ordinary materials is temporarily stored in the DCC to release the storage area. However, C type of parts have some characters such as small volume, high use frequency and hardly occupying the assembly line space, so the line stocking distribution that operator directly deliver materials to the assembly line can substantially decrease workload of operators in the middleman and improve distribution efficiency.

3.3 Delivery time planning

The production of automobile assembly line runs on a uniform rate (i.e, cycle time of the line). The sequence of assembly of different product models for final assembly is given by master production planning. Production of assembly products at uniform cycle time is relatively easy to obtain the time at which material is required by different stations during assembly. The logistic process operation instructions are synchronized with the time of the delivery of material to different stations in the line. The generation process of the parts delivery time is depicted in Fig.2. Through the method shown in Fig. 2, distribution time of parts can be figure out and it can be used to make a logistics plan.

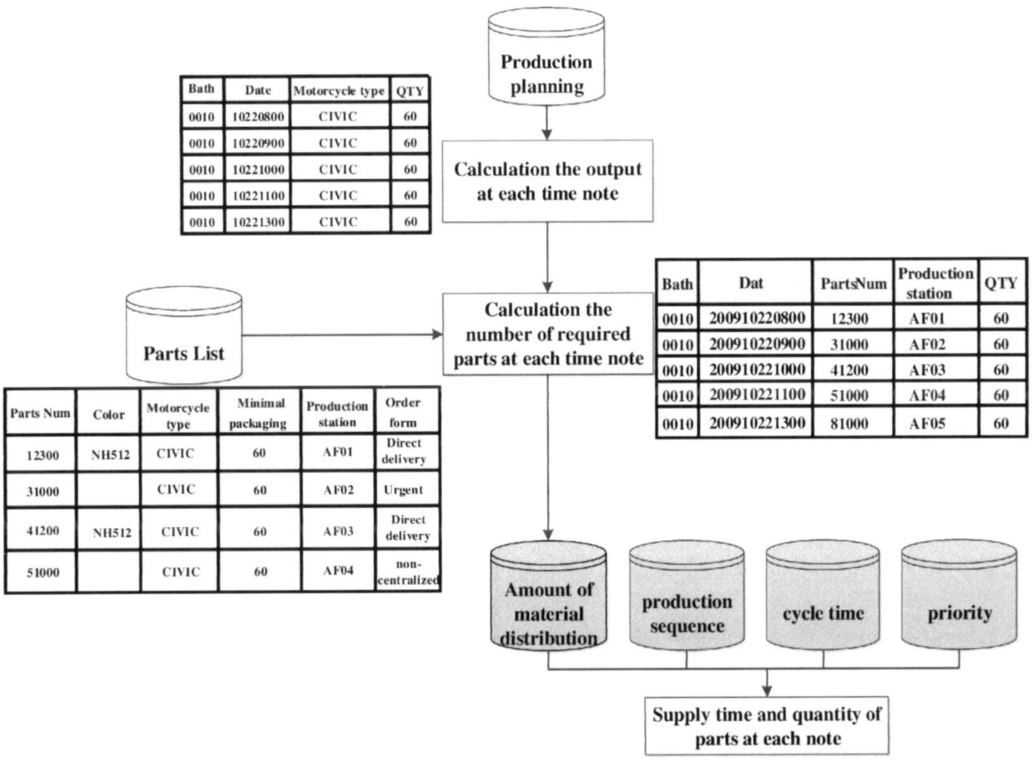

Figure 2. Generation process of the delivery time plan

According to actual situation and parts classification, there are different transport time schedule with various modes of material distribution strategies which are shown in Table 2.

Table 2. Time Settings for different distribution strategies

Distribution method	Kitting	Kanban	Line Stocking
Stocking time	120min	120min	120min
Towing time	60min	60min	60min
Transport time	60min	60min	60min
Unloading and storage time	75min	175min	25min
Time of moving to line-side	25min	25min	25min

Operation planning schedule of Kitting, Kanban and LS can be pushed back by the processing start time, such as online time, departure time, start stocking time, according to the operating time interval taken from Table. 2. The difference is that the distribution plan of LS parts directly sends all parts of the same kind required on the same day to assembly lines without stored in DCC in one times.

The plan figured out through this way is accurate to every second. In fact, the distribution plan only tells the operator a probably working time, so we need to set an appropriate interval time to generate distribution command. The rationality of current plan setting one hour as interval time requires improvement. Therefore, in the subsequent experimental design, we will test a variety of time intervals to choose the best setting.

4 Simulation experiments for evaluation of the distribution plan

4.1 Construction of the simulation model

Data import module is used to import various useful data into the system, which mainly include row materials table, the BOM table, production sequence table, vehicle flow table, table of parts requirement, table of stocking inventory, etc. All data information is managed through Excel, and there are conjunctions between some specific columns in order to convert Excel file to the data format which the simulation system can identify.

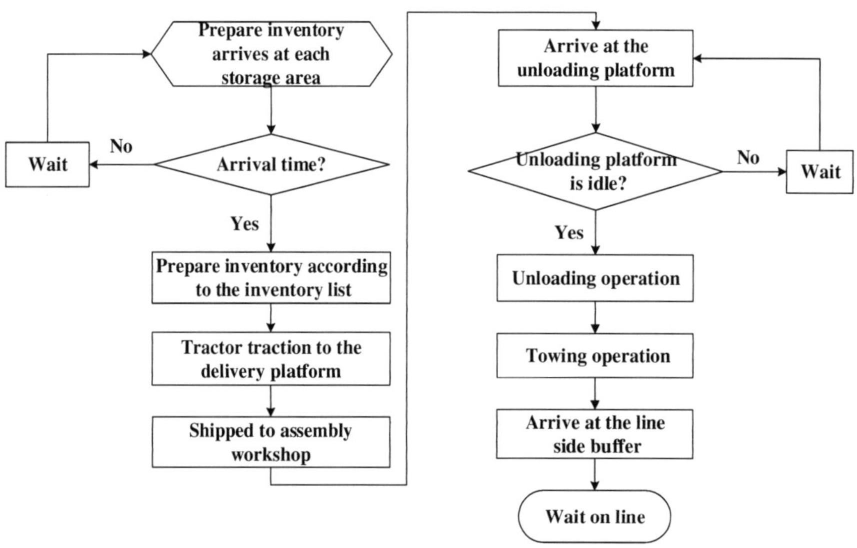

Figure 3. Logic diagram of dispatching parts

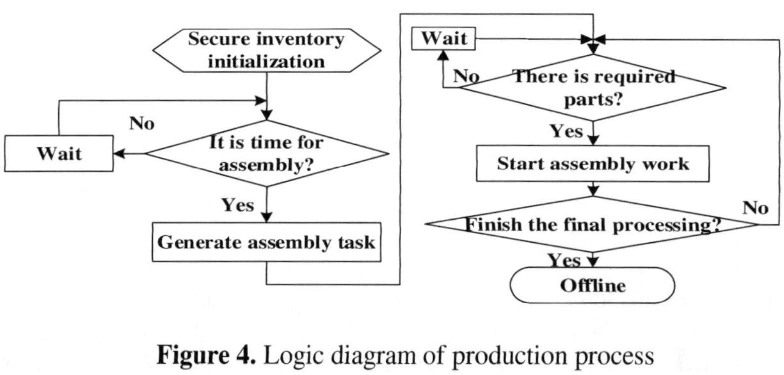

Figure 4. Logic diagram of production process

Based on the primary information and combined with Simio modeling features, simulation model is building to analyze automatic part dispatching. Moreover, system modeling concept is considered in the current research experiments. Fig.3 illustrated the logic diagram of dispatching parts from the CLC to the line-side for storage.

Production line modeling process describes the whole process from parts allocating on line to end of the line. The logic diagram of the model is depicted in detail in Fig. 4.

Three-dimensional simulation model for logistic distribution system according to the operations combined with actual logistics information is indicated in Fig. 5.

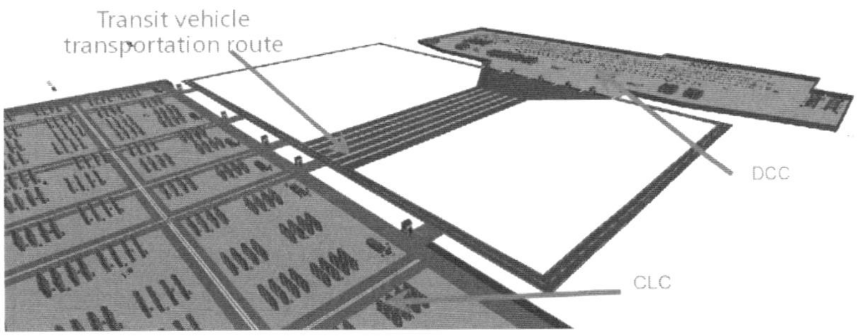

Figure 5. Three-dimensional simulation model of logistics system

4.2 Simulation experiments

4.2.1 Designing of alternative solutions

Different distribution strategies are made according to part groupings in different parameters conditions at different patterns of Time division and they are shown as alternatives in Table 3.

Table 3. Designing of alternative solutions

Time division (minute)	Simulation experiments	Kitting pattern	Kanban pattern		Line stocking pattern	
		Generality parameters	Generality parameters	Packing number	Generality parameters	Packing number
30	1	<70%	≥70%	≤60	≥70%	>60
	2	<60%	≥60%	≤60	≥60%	>60
	3	<50%	≥50%	≤60	≥50%	>60
	4	<40%	≥40%	≤60	≥40%	>60
45	1	<70%	≥70%	≤60	≥70%	>60
	2	<60%	≥60%	≤60	≥60%	>60
	3	<50%	≥50%	≤60	≥50%	>60
	4	<40%	≥40%	≤60	≥40%	>60
60	1	<70%	≥70%	≤60	≥70%	>60
	2	<60%	≥60%	≤60	≥60%	>60
	3	<50%	≥50%	≤60	≥50%	>60
	4	<40%	≥40%	≤60	≥40%	>60
90	1	<70%	≥70%	≤60	≥70%	>60
	2	<60%	≥60%	≤60	≥60%	>60
	3	<50%	≥50%	≤60	≥50%	>60
	4	<40%	≥40%	≤60	≥40%	>60

4.2.2 Simulation experiment conclusions

The inventory of DCC, the vehicle queue length, equilibrium of logistics and the number of vehicle deliveries are the main evaluation indicators of simulation optimization, as described in table 4.

Table 4. Evaluation indicators of simulation optimization

Evaluation indicators	Introduction
Warehouse inventory of DCC	Logistics volume of DCC(Collect freight yard, storage yard)
The vehicle queue length	The vehicle queue length of each cargo platform
(DCC, CLC) Equilibrium of logistics	Logistics volume is the total number of Flat car and Trolley per hour
	Standard deviation statistic of Logistics fluctuation
Number of deliveries	Number of vehicle deliveries and tractor deliveries

According to the designed experiments, the simulation results obtained are presented here. The comparison of warehouse inventory of DCC at different time intervals is indicated in figure 6. It can be seen from figure 6 that optimization schemes can significantly reduce inventory.

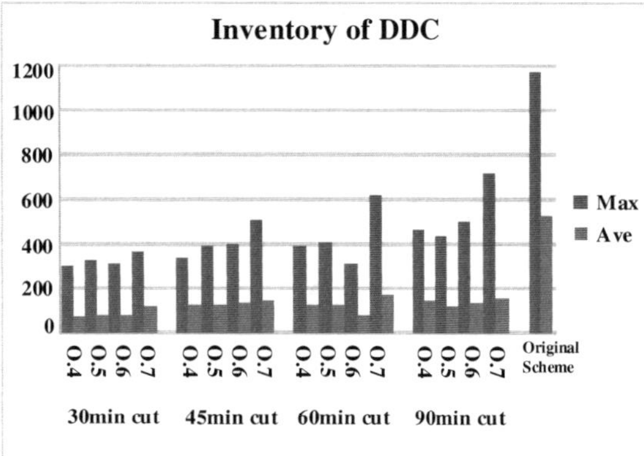

Figure 6. Comparison of inventory of DDC

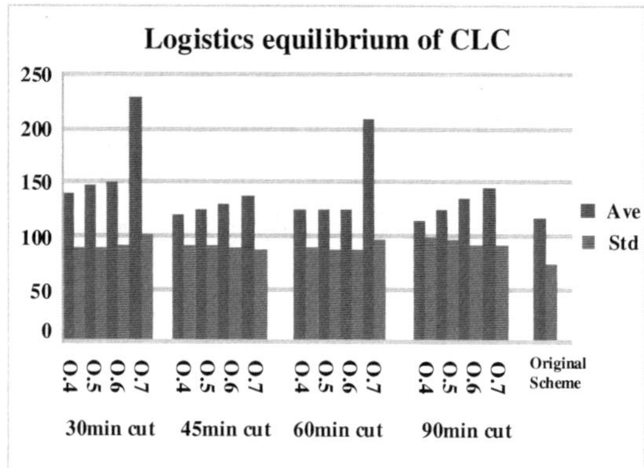

Figure 7. Comparison of CLC logistics equilibrium

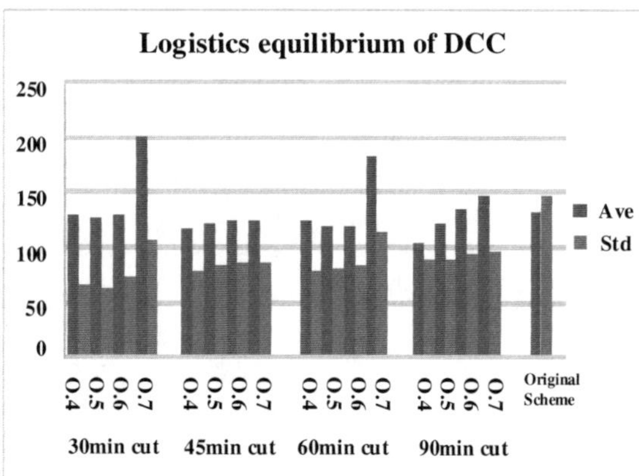

Figure 8. Comparison of DCC logistics equilibrium

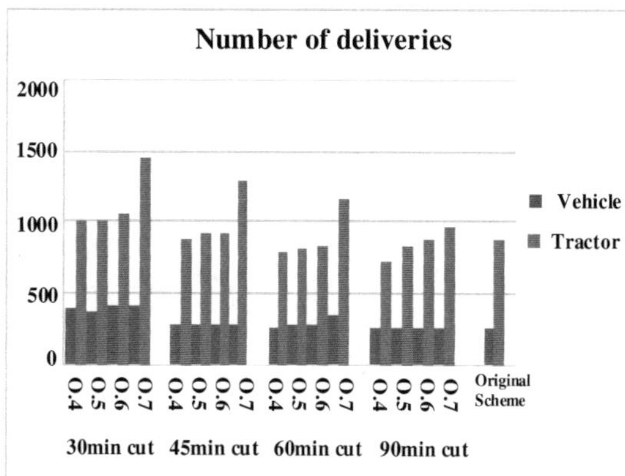

Figure 9. Comparison of deliveries number

The results of equilibrium of the logistics for CLC and DCC are indicated in figure 7 and 8. It can be seen from figure 7 and figure 8 that optimization schemes mainly reduce the logistics volume and improve the equilibrium of logistics with lower standard deviation. The results for number of deliveries is indicated in figure 9. And the results of queuing on unloading platform shows that the problem of warehouse inventory of DDC indicates that the waiting time after unloading is improved and there will be no need to wait for unloading after optimization. These results indicate that the congestion problem is significantly solved here. It can be seen from figure 9 that Kitting mode is the logistics balance but with higher workload and lower container 's loading, this problem will be discuss in following.

4.2.3 Optimal scheme

According to simulation results of different alternatives under different evaluation indicators in 3.2 section, the weight method and expert evaluation method are used to evaluate each plan, specific steps are below:

Step1: For each evaluation index, grade them into 16 scheme (score 1-16)

Step 2: Set weight for each evaluation indicator according to the actual need and expert evaluation

Step 3: Select the best program according to the comprehensive score

In this way, It can be find that the optimal scheme for inventory is to set 1 hours as the division unit, and generality parameter is set to 50%, and the material classification and distribution mode considered are as follows:

Kitting pattern： generality parameters＜50%

Kanban pattern： generality parameters≥50% ， packing stock ≤60

Line stocking pattern： generality parameters≥50%， packing stock＞60

A comparison between the optimal scheme and original scheme are made as following:

(1) Results of optimization of Warehouse inventory of DDC and equilibrium of logistics

The maximum and average values of warehouse inventory of DCC were significantly decreased， the maximum reduced from 1168 to 401, and average values reduced from 524 to 127. This results indicate that unloading platform congestion problem is solved through creating a logistics equilibrium model.

The comparison of the logistics volume data of each hour in a day for original and optimal scheme and are indicated in figure 10 and figure 11 for CLC and DCC respectively. It can be seen that fluctuations of CLC and DCC logistics volume flatten out. According to the diagram, we can know that logistics of CLC and DCC achieve equalization.

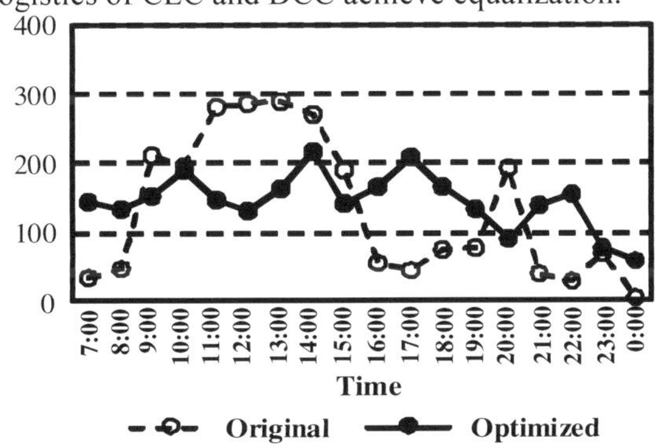

Figure 10. Comparison chart of CLC logistics volume

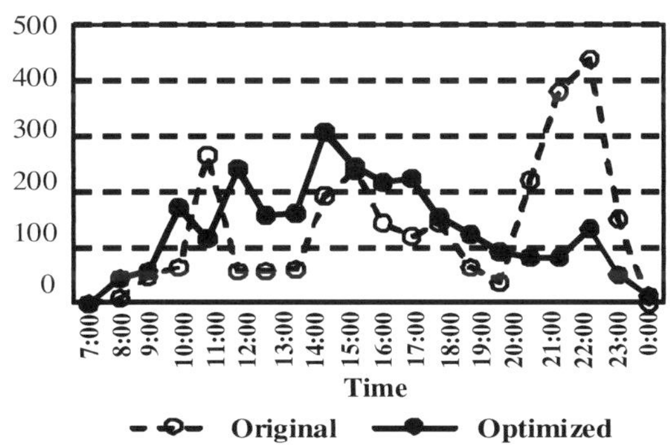

Figure 11. Comparison chart of DCC logistics volume

(2) Waiting time of unloading vehicle

In original scheme, 44% vehicles have waiting time is more than 30 minutes, but the results of optimization scheme basically do not exist and have waiting queue and the average waiting time is reduced by 20%.

(3) Optimization results of vehicle and tractor

The optimization results of the vehicle and tractor obtained from the proposed simulation model are indicated in table 5. The results shown in table 5 that the characteristic of Kitting mode is the logistics balance but with higher workload and lower container's loading. These results are because Kitting accounts for a large proportion, the optimization scheme will can increase the delivery number of vehicle but reduce the number of tractor and vehicle. So the optimization scheme can achieve the goal of reducing costs.

Table 5. Optimization results of vehicle and tractor

Scheme	Total number of vehicle delivery	Number of vehicle	Deliveries' number per vehicle and per day	Total number of tractor delivery	Number of tractor	Deliveries' number per tractor and per day
Original	265	40	6.625	878	48	18.291
Optimization	270	35	7.714	823	44	18.705

5 Conclusions

The paper studies the distribution method optimization of the logistics system in a Third-Part logistics company that provides professional service for the designated automobile manufacturing company in China. The proposed study classifies different parts according to the critical factors and determined appropriate logistics distribution mode and optimized the logistics distribution schedule. The simulation software Simio is applied to model and simulate the assembly line, which helps to find and validate an optimization distribution scheme through simulation experiments. Experimental results indicate that the proposed optimization scheme can solve the problems of logistics balance and congestion of the unloading platform in efficient way as compared to the original method used in the case company. In the future, there are two points needed to do more in-depth investigations. First, it's necessary to design containers for different types of parts. Another one is doing more experiments using intelligent algorithms to find optimal solution.

6 References

[1] M Xie. Research on Automotive Parts Inbound Logistics Distribution Problem under Inventory Constraints [D]. Chongqing University,2012.
[2] W Wei. Simulation and optimization on warehousing of distribution center integrating with theory of constraints [J]. Journal of Zhengzhou University of Light Industry：Natural Science, 2010, 25(6):71-75.
[3]J Zhou, L Fan. Optimization and Simulation of Assembly Line Material Distribution Based on eM.Plant [J]. Mechatronics, 2012,06:17-23.
[4] L Wan. Simulation Analysis and Optimization of the Logistic Dispatching in Assembly Production [D]. Huazhong University of Science and Technology,2012,01.
[5] K Zhang，R Wu，X Tao. Simulation Analysis of Queue System Based on Simio [J].Ordnance Industry Automation,2011,02:94-98.
[6] L Yue, Z Guan, C Wang, Z Zheng, X Shao. Logistics simulation analysis of assembly shop production line based on Simio[J]. Machinery manufacturing, 2013, 02: 83-88.
[7]C Wang, Z Guan, X Shao, U Saif. Simulation-based optimisation of logistics distribution system for an assembly line with path constraints[J].International Journal of Production Research, 2014:3538-3551.
[8] L Kong, J Lu, Y Zhan. Study on bottleneck shifting of production logistics based on queueing networks[J]. Journal of Zhejiang University of Technology, 2011, 06: 644-647.

Acknowledgements

This work has been supported by MOST (the Ministry of Science & Technology of China) under the grants No.2012AA040909, 2012BAH08F04, & 2013AA040206, and by the National Natural Science Foundation of China (Grants No.51035001, 50825503, & 71271156).

Research on remanufacturing scheduling problem based on critical chain management

Y Cui[1,2], Z Guan[1,2], C He[1,2], L Yue[1,2]

[1]State Key Lab of Digital Manufacturing Equipment and Technology, Huazhong University of Science and Technology, Wuhan 430074, China
[2]School of Mechanical Science and Engineering, Huazhong University of Science and Technology, Wuhan 430074, China

151704061@qq.com

Abstract. Remanufacturing is the recycling process of waste products as "as good as new products", compared with materials recycling, remanufacturing represents a higher form of recycling. The typical structure of remanufacturing system consists of three parts: disassembly workshop, remanufacturing workshop and assembly workshop. However, the management of production planning and control activities can differ greatly from management activities in traditional manufacturing. Scheduling in a remanufacturing environment is more complex and the scheduler must deal with more uncertainty than in a traditional manufacturing environment. In order to properly schedule in a remanufacturing environment the schedule must be able to cope with several complicating factors which increase variability. This paper introduced and discussed seven complicating characteristics that require significant changes in production planning and control activities, in order to provide a new method for remanufacturing production scheduling system.

Key word: Remanufacturing, DBR, critical chain, scheduling

1. Introduction

Remanufacturing is a life time cycle strategy that allows products which are no longer functional into the manufacturing process to be refurbished or disassembled into usable components or disposed. In other words, it is an industrial process that restores end-of-life goods to their original working condition. The largest remanufacturer in the world is The United States, and between 2006 and 2013, the value of U.S. remanufactured production grew by 15 percent to at least $43.0 billion, supporting 190,000 full-time jobs.

The remanufacturing-intensive sectors that account for the majority of remanufacturing activity include aerospace, consumer products, electrical apparatus, heavy-duty and off-road equipment, information technology products, vehicle parts, machinery, medical, furniture, restaurant equipment, and retreaded tires. Although In global market, country like China account for lots of remanufacturing activities, associated trade, and developing their own remanufacturing industries.

There is enormous complexity involved with developing remanufacturing operations. They are arguably more difficult than traditional manufacturing, such as forecasting the timing and quality of product returns and determining the optimal disassembly sequence(s) etc. Guide outlines the characteristics of significant and complex activities in production planning and control which are involved in remanufacturing: (1) the uncertain timing and quantity of returns, (2) the disassembly of returned products , (3) the uncertainty in materials recovered from returned items, (4) the need to balance returns with demands, (5) the stochastic routings for remanufacturing operations, (6) the complication of material matching restrictions, (7) the requirement for a reverse logistics network, and (8) highly variable processing times. Significant challenges are involved in remanufacturing scheduling, such as the selection of order release mechanisms, reverse logistics, lot sizes, and the priority of scheduling rules, restrictions in capacity, part commonality, the buffer inventories plan, multiple time periods scheduling, integration of forward and reverse manufacturing operations, etc.

Other researchers (e.g., Krupp, 1993; Brennan, Gupta,; Flapper et al., 2002) have noted other complicated issues and decisions involving remanufacturing scheduling, such as: the selection of lot sizes, order release mechanisms, priority scheduling rules, capacity restrictions, multiple products; the buffer size of inventories plan, scheduling over multiple time periods, integration of forward and reverse manufacturing operations, etc. and these are listed in Table 1.

Table 1 Remanufacturing scheduling complexities and issues

• Need for a reverse, rather than forward, logistics network and operations
• Stochastic demands
• Balancing returns with demand
• Single vs. multiple stage operations
• In line vs. off-line rework
• Resource availability and allocation
• Product structure considerations
• Sourcing decisions
• Uncertain timing and quantity of core returns
• Uncertainty in material recovery rate
• Uncertain routing for materials and parts
• Lot sizing
• One versus multiple products
• Order release mechanisms
• Capacity restrictions
• Priority scheduling rules
• Scheduling for single vs. multiple time periods
• Scheduling methodology employed

Guide (2000) describes a typical remanufacturing facility includes three main distinct operations: (1) disassembling,(2) remanufacturing/repair, and (3) reassembling.

IOP Conf. Series: Materials Science and Engineering **215** (2017) 012005 doi:10.1088/1757-899X/215/1/012005

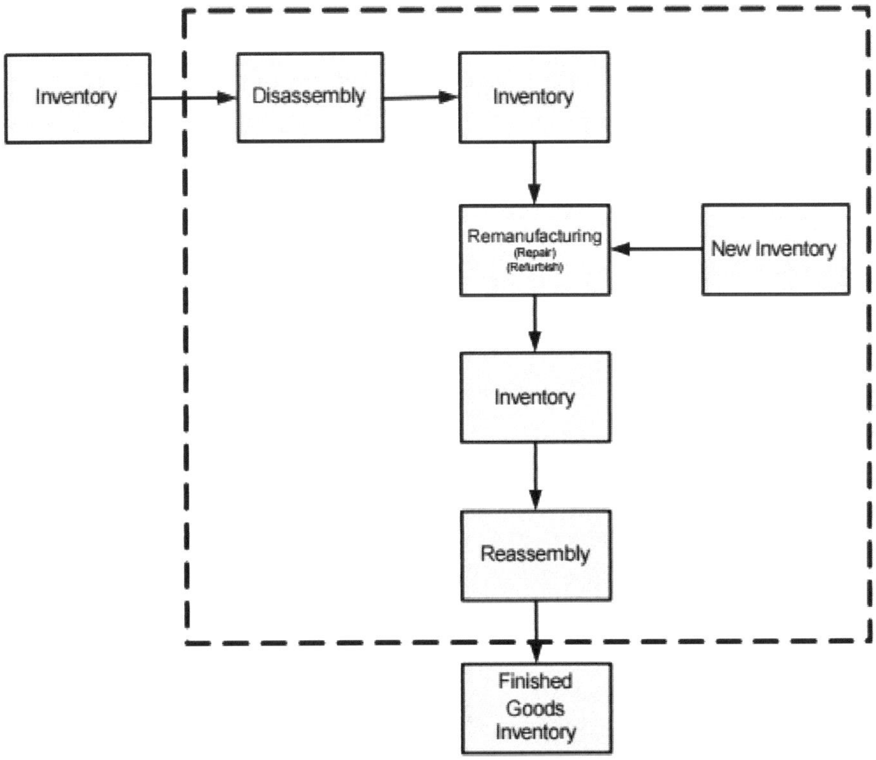

Figure 1. Remanufacturing Shop

In Figure 1 [1], the typical remanufacturing operations are following: First sorting the quality of reverse materials or products, enter in warehouse waiting for next progress. Disassembling separates the returned product into its components, or common materials. These are evaluated and determined to be acceptable for reuse, repairable, sold or discarded. Those reproducible parts and components are all enter in inventory for future recall or sent to the r next operations. After reconditioning to a usable state the modules or parts are inventoried awaiting use or sent directly to the reassembly processes, where they are reassembled into products for resale and readied for finished goods inventory or shipment. As highlights on the complicating characteristics, the scheduling and control of each of these operations is an extremely challenging task. Figure2 illustrates the three remanufacturing stages and their further analysis by production strategy. After the disassembling/remanufacturing/ reassembly of the product, there will be a preliminary check and classification. Parts and components will be sorted and identified in stock or wait for sale, which depends on the market needs. Classification is conducive to manage process and tracking, to enhance the level of information technology.

IOP Conf. Series: Materials Science and Engineering **215** (2017) 012005 doi:10.1088/1757-899X/215/1/012005

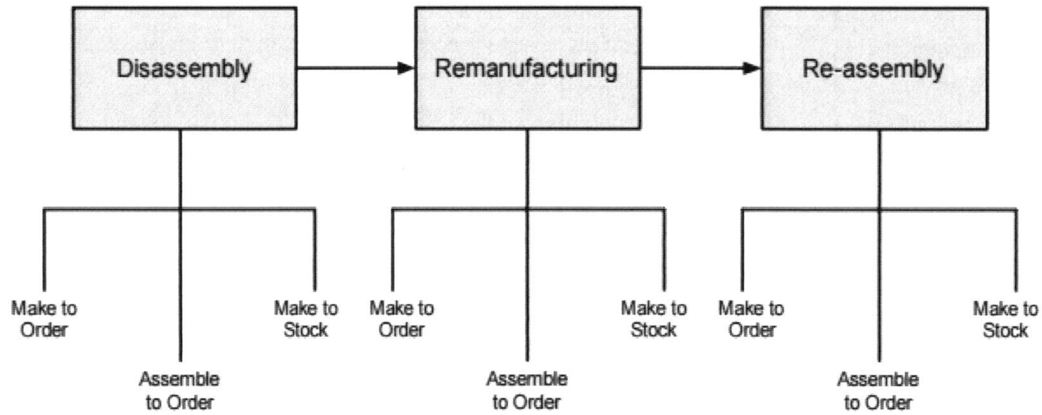

Figure 2. Remanufacturing stages and production strategy

Thus, this research is not only review the progress we have made in the scheduling and control of remanufacturing process, but also disassembly and reassembly operations, Actually, the integrated operations scheduling research also include single/multiple product, part commonality/no commonality, infinite/finite capacity, and deterministic/stochastic parameter organization, the purpose of remanufacturing research effort is to do more than just synthesis the progress made in the scheduling of disassembly, remanufacturing, and reassembly operations. To advance ability to schedule and control complexities mentioned in the literature; and focused in the whole close-loop chain. Our research is currently at review stage.

2. Remanufacturing literature

The purpose of this research is to synthesize the progress in the scheduling of disassembly, remanufacturing, and reassembly operations. Also to assess the ability to address the complexities mentioned in the literature; and to highlight additional research needs.

The recycling of waste products arrived at remanufacturing plant for a series of processes, such as cleaning, testing, surface repair, reprocessing, etc. Then remanufactured parts are assembled into products in the assembly shop. Three workshops will be involved in the production operation and scheduling problem, namely, disassembly scheduling, remanufacturing scheduling and reassembly scheduling. In Table 1, previous research on remanufacturing has addressed a variety of problems, and the relevant efforts be summarized to production planning and control. Furthermore a thorough discussion of this literature is available in Guide et al. 1999. While a number of topics have been reached in depth into the real remanufacturing scheduling by the existing body of knowledge, significant areas have not been addressed in previous works. The majority of these studies are limited to a specific functional activity, only a very limited number of case studies exist [2-3], and, such as scheduling. A number of characteristics that significantly complicate production planning and control activities may be inferred from the research literature. For instance, researchers proposed seven complicating characteristics in various forms. No one identifies more than three of the complicating characteristics, and most of the characteristics are only mentioned in passing. The seven characteristics are:

1. The uncertain timing and quantity of returns
2. The need to balance returns with demands
3. The disassembling degree of returned products
4. The uncertainty in materials recovered from returned items
5. The requirement for a reverse logistics network
6. The problems of stochastic routings for materials for remanufacturing operations and highly variable processing times.
7. The complication of material matching restrictions..

Remanufacturing operations are highly variable because the amount of time required to perform the necessary operations is a function of the part, or parts, causing the unit's failure and for that failed part, the cause and extent of the failure. Probabilistic routeings are the operations required to restore the part to functional operation capability and other standards. Probabilistic routeings are a subset of routeings required from a master set of possible operations. The likelihood of an operation being required is a function of the product service life and the work conditions the unit has been subjected to. In the model, the probability of an operation being required is assigned, based on the considerations above. For all probabilistic operations, the routeing represents a closed set of operations that occur in a fixed sequence, but each probability of each individual operation is a random variable. Because of those dynamic natures of scheduling policies, a basic structure of remanufacturing scheduling system was showed in figure 3.

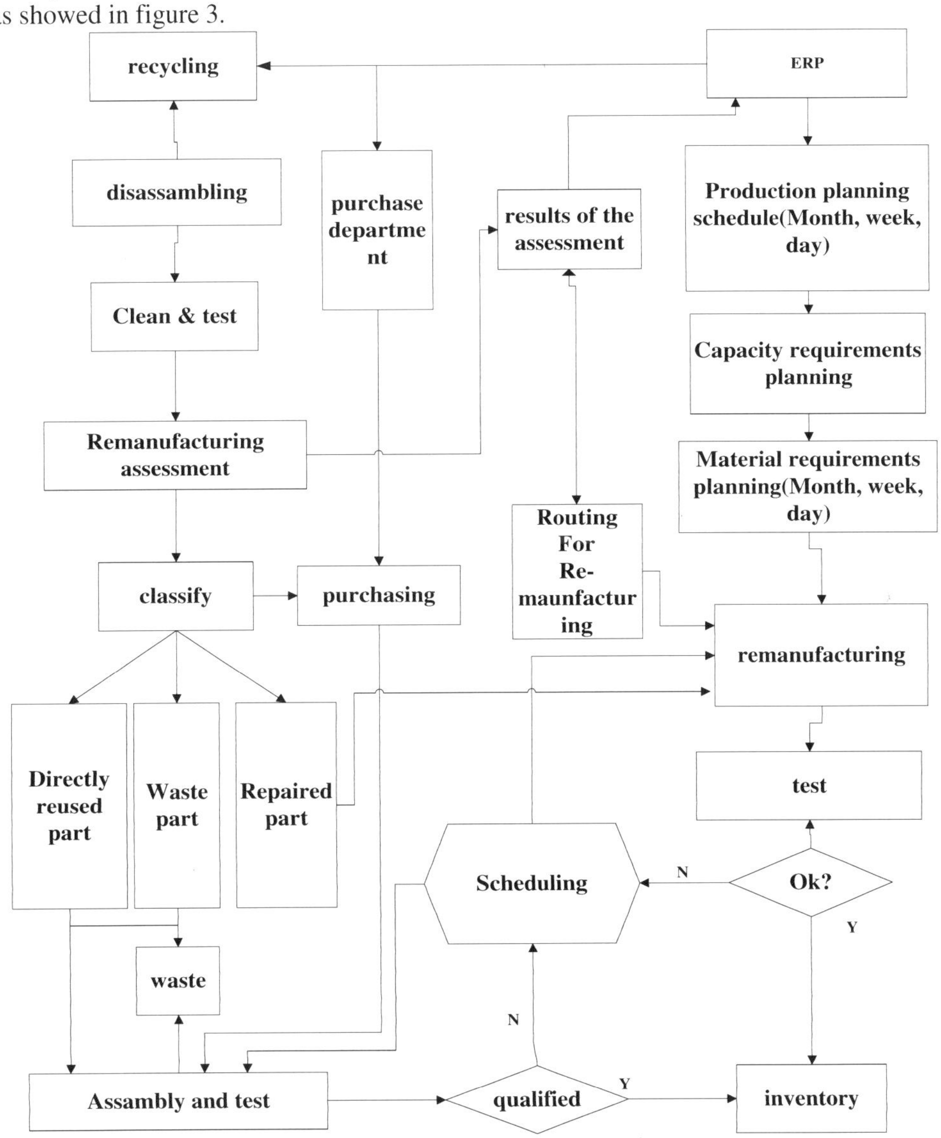

Figure 3. Remanufacturing scheduling system

Some scholars divided the remanufacturing model is into four different categories: OEM (Original Equipment Manufacturer) mode, independent manufacturer mode, OEM contract service model of remanufacturing and manufacturing OEM / independent manufacturer of mixed mode from the perspective of product life cycle [3], in general, the mainstream research field divided into two forms:

independent remanufacturing system (Independent Remanufacturing System, IRmS) and manufacturing / Remanufacturing hybrid system (Manufacturing/Remanufacturing Hybrid System, M/RmHS). For Independent manufacturers, they can do of any brand of product remanufacturing without OEM authorization, which can not only keep the original logo, but also create their own remanufacturing trademark. Independent remanufacturer has no direct relationship with the original product manufacturers, based on market demand forecast to develop their own production plan, control the production schedule. The characteristics of this model is to create more varieties, the general use of professional production line for mass production, scale and high efficiency, high resource utilization, low cost of production, its cost contains obvious superiority. OEM recycles or call-backs the old pieces through revers supply chain from dealers and repair stations, those products will be use in remanufacture then sale through the OME's after-sales service network. The characteristics of this model is it include the allocation of orders, balancing the production quantity, lot size, etc. between the new production and remanufacturing, to obtain the maximum profit.

3. Critical chain project management method

Compared with traditional manufacturing systems, there are a lot of uncertainties in the production scheduling of remanufacturing system [4]. In the manufacturing workshop, because the quality of recycling of waste products (such as the service life of wearing parts, etc.), the uncertainty caused by process parts re manufacturing process and process time, which makes the manufacturing workshop production scheduling is more complicated; and the processing route and time uncertainty caused by processing equipment the amount of task, it is easy to form the bottleneck of resources, and will continue to shift the bottleneck.

Under the condition of limited resources, how to arrange resources, solve the resource bottleneck problem, and fully consider the production process uncertainty on remanufacturing process scheduling, ensure the remanufacturing production smoothly, is a problem to be solved urgently.

The complexity of the manufacturing system is mainly shown in three aspects: the uncertainty of the process route, the uncertain processing time and the resource constrained problem. Taking into account the dismantling of each part after the cleaning, testing and other processes, to a certain extent, you can determine the process route.

At present, some research on the remanufacturing system has been optimized, such as Guide to carry out some research on Remanufacturing scheduling problem with static priority performance of remanufacturing system sharing workshop scheduling in the analysis. A hybrid genetic algorithm was proposed for the optimization of remanufacturing production planning, and the use of a priority random batch mechanism for its simulation [5]. Analysis of the effect of uncertainty to the quality of waste parts recycling remanufacturing production planning has great influence on the quality of that, remanufacturing enterprise profits by waste parts, and manufacturing enterprises are also affected by the number of used products [6].

Other applications, like an approximate optimal buffer setting method, are studied in the queuing problem of remanufacturing system. Some research testing to remanufacturing and repair as the research object, considering the uncertainties and products in remanufacturing system, to process, inventory constraints, using Lagrange relaxation method, get the demolition, repair and optimization of assembly operation time value. It has established a mathematical model of remanufacturing logistics, which is used to solve the uncertainty of product recovery time, quality and quantity. However, throughout these studies, we find that the process time and resource constraints are not considered in the process of re manufacturing scheduling.

Critical chain project management method (Critical chain method, CCM) is a new method developed in Guide on the basis of the theory of constraints (Theory of constraints, TOC), the key chain instead of the critical path, considering a sequence of work between precedence and resource constraints under the constraints of the project cycle, and through the project buffer, conveying buffer and resource buffer mechanism to eliminate the uncertainty the influencing factors of the project scheduling.

4. Problem description

Remanufacturing operations are composed of two types of routings: mandatory and probabilistic. As well as it highly variable, since the amount of time required performing the necessary operation is a function of the wear associated with a part. Remanufacturing Job Shop problem is similar to the classic one, remanufacturing production scheduling problem can be described as N parts to be processed, after clean and test process the remanufacturing routings have been identified, but the processing time is a random variable, obey a certain distribution; there are M machines at workshop. The purpose of scheduling is to arrange the optimal processing sequence and the start time of each work piece on each machine tool, in order to satisfy the process constraints of the machining process, the performance index of the remanufacturing system is optimal. The basic constraints in the processing process are as follows:

(1) All the routings must be processed on the specified machine and each process must be started before the process is completed.

(2) At a specific time a machine can only process a part.

(3) Different parts have the same priority.

The objective function is to minimize the mean flow time, namely to finish process in the shortest time.

5. Conclusion

Up to now, comparing reports in the field of remanufacturing scheduling is not easy for the variety of problems, objectives, and methods utilized. Defining a unanimously accepted set of problems with significant size and same complexities could make comparation more compelling.

The remanufacturing scheduling literature was classified into disassembly versus integrated scheduling, single versus multiple products. The complicating characteristics must be considered as a whole, rather than considering the effects of each characteristic separately. Remanufacturing firms must be able to manage complex tasks that are significantly different from tasks in a traditional manufacturing environment. After identify and describe those unique characteristics of scheduling and control for remanufacturing. Future work is suggested in development of approaches such as critical chain method which help reduce the variability in the whole remanufacturing chain, it may provide a dramatically huge challenge to reach trade-off multiple objectives, but result may in more useful solutions.

References

[1] Guide V.D.R Jr 2000 *J. Oper. Manag.* **18** 467-83.
[2] Ferrer G 1997 *Resour. Conserv. Recy.* **19** 221-55.
[3] Guide V.D.R Jr 2000 J. Oper. Manag.18(4): 467-483.
[4] Guide V.D.R Jr 2007 *Int. J. Prod. Res.* **34** 1081-91.
[5] Li J, González M and Zhu Y 2009 *Int. J. Prod. Econ.* **117** 286-301.
[6] Ferrer G and Whybark D C 2001 *Prod. Oper. Manag.* **10** 112-24.

IOP Conf. Series: Materials Science and Engineering **215** (2017) 012006 doi:10.1088/1757-899X/215/1/012006

PROCESS IMPROVEMENT OF REACTIVE DYE SYNTHESIS USING SIX SIGMA CONCEPT

Thanapat Suwanich[1] and Parames Chutima[1,2*]

[1]Department of Industrial Engineering, Faculty of Engineering
Chulalongkorn University, Bangkok, Thailand 10330

[2]Regional Centre for Manufacturing Systems Engineering, Faculty of Engineering
Chulalongkorn University, Bangkok, Thailand 10330

[1]Thanapat.Su@student.chula.ac.th, [2]cparames@chula.ac.th

Abstract. This research focuses on the problem occurred in the reactive dye synthesis process of a global manufacturer in Thailand which producing various chemicals for reactive dye products to supply global industries such as chemicals, textiles and garments. The product named "Reactive Blue Base" is selected in this study because it has highest demand and the current chemical yield shows a high variation, i.e. yield variation of 90.4% – 99.1% (S.D. = 2.405 and Cpk = -0.08) and average yield is 94.5% (lower than the 95% standard set by the company). The Six Sigma concept is applied aiming at increasing yield and reducing variation of this process. This approach is suitable since it provides a systematic guideline with five improvement phases (DMAIC) to effectively tackle the problem and find the appropriate parameter settings of the process. Under the new parameter settings, the process yield variation is reduced to range between 96.5% – 98.5% (S.D. = 0.525 and Cpk = 1.83) and the average yield is increased to 97.5% (higher than the 95% standard set by the company).

1. INTRODUCTION

Reactive dye is the synthetization via the chemical reaction of colour and reactive components. The differences between reactive and natural dyes in terms of chemical properties are the chemical resistance of water, reactivity used in dyeing application process, and variety of colours. The reactive dye is popular for dyeing on cellulose fibres because the shade and brightness are depended on the nature or structure of the reactive dye [1]. Recently, the reactive dye has developed the chemical properties that increase the reactivity to cellulose fibres and solubility in water. Most reactive dyes after completing the synthesis process are sold in a powder form. The synthesis process of the reactive dye has to be controlled for all parameters to match with the chemical property of the raw material. The control of the process parameter and condition depends on the reactive dye structure, synthesis operation and chemical property. The special property of the reactive dye is the key component. All textile industries need to concentrate on the process improvement to optimize the synthesis condition and parameters to be matched with the types of reactive dyes. The objective of the improvement process is to gain high chemical yield and low variation [2]. Currently, the variation of the chemical yield from the reactive dye synthesis process of the company is high and the yield is lower than the company's standard (95%). In this paper, the Six Sigma method is used to find the significant factors affecting the chemical yield and optimize the conditions suitable for controlling the parameters in each step of the reaction process. The Six Sigma DMAIC procedure comprises five key phases, i.e. Define, Measurement, Analyze, Improve and Control phases [3-5].

2. DEFINE PHASE

In this case study, the product named Reactive Blue Base is selected because it has highest demand and shown a high variation in the chemical yield, 90.4 – 99.1% (S.D. = 2.405, Cpk = -0.08). In addition, its average chemical yield is only 94.5% which is lower than 95% (company's standard). This product is synthesized via 3 process reactions, i.e. 1st Condensation, 2nd Condensation and Vinylation processes. 1st Condensation and 2nd Condensation involve the addition of the reactive substances and Vinylation is the process that converts the structure of in-reactive dye to reactive dye via the transformation process of chemical structure. In each process step, different conditions and controlled parameters are used depending on the type and chemical property of the raw material. The factors to control in each step of the process are temperature, additional time of chemicals, maintaining time and pH. The historical data of the chemical yield show S.D. = 2.40 and average = 94.5%. The chemical yield by production batch and descriptive statistics are shown in Fig.1 and Fig.2, respectively.

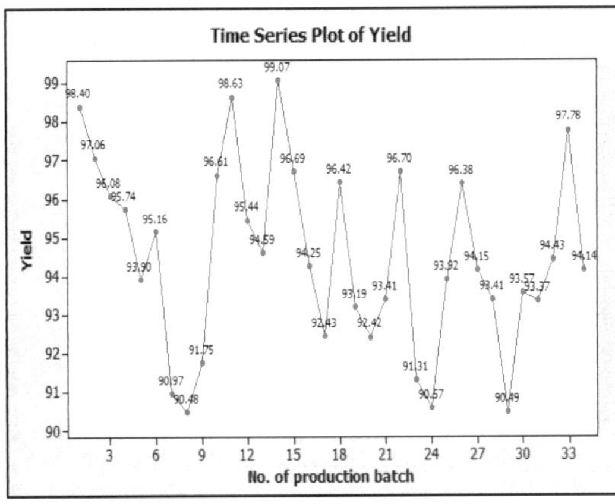

Figure 1. Chemical yield by production batch

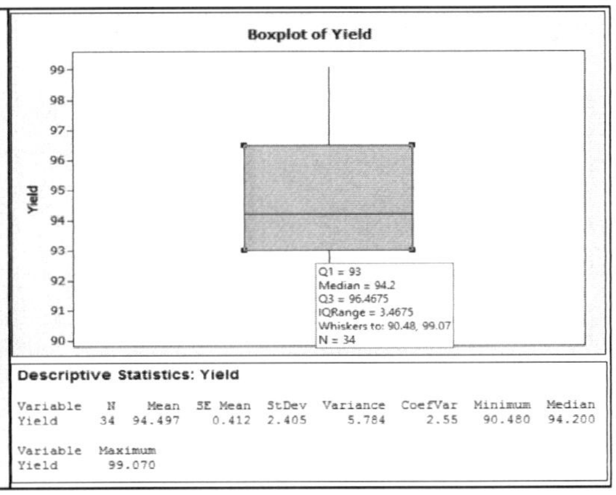

Figure 2. Descriptive statistics of yield

3. MEASUREMENT PHASE

The measurement phase includes three steps, i.e. variation analysis of raw material usage from different suppliers, measurement system analysis and problem identification analysis.

3.1. Variation analysis of raw material usage from different suppliers and lots

In the chemical reaction process, the quality of raw material is important to the acceptability of the final product. This research aims to reduce the variation of chemical yield in the reactive dye synthesis process. Therefore, variation analysis of raw material usage from different suppliers and lots are importance because the quality of raw material usage in this research is identified as a controllable factor (supplier selection). The quality of raw materials supplied by both suppliers (Supplier A and Supplier B) are analyzed in terms of chemical properties on lot-to-lot for each suppliers (total 30 lots each) by following the company specification (specification setup for the main component should be over 96%). The analysis results of the materials from Supplier A and Supplier B are shown in Fig.3 and Fig.4, respectively.

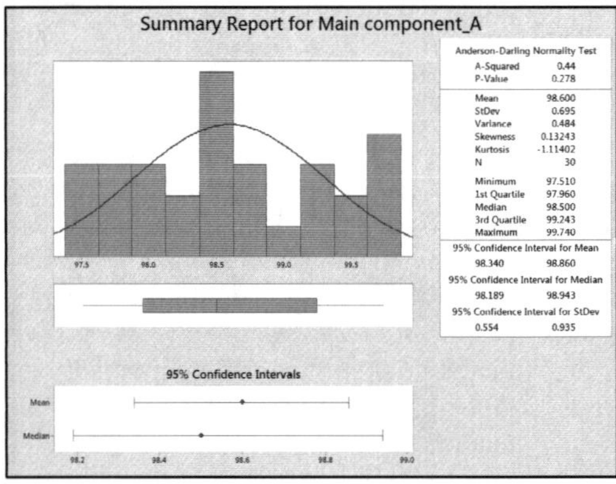

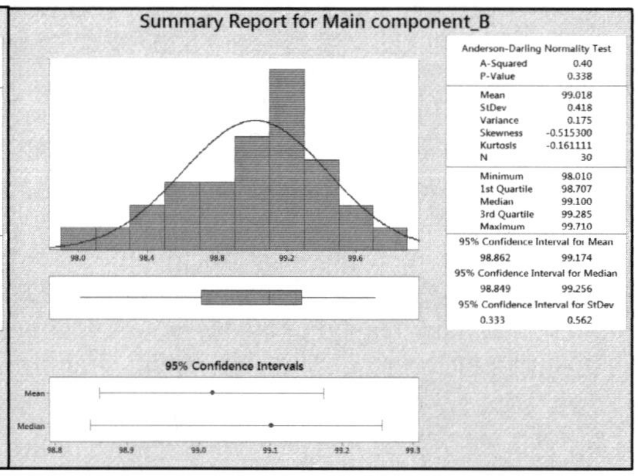

Figure 3. Raw material analysis from Supplier A **Figure 4.** Raw material analysis from Supplier B

The results indicated that the quality of lot-to-lot raw materials supplied by Supplier A is not significantly different (P-value = 0.278). This result is the same for Supplier B (P-value = 0.338). As a result, the raw materials from both suppliers can be used to synthesis the dye. In addition, this confirms that the quality of raw material usage is the controllable factor in this research.

3.2. Measurement System Analysis

The chemical instrument analysis called "high performance liquid chromatography (HPLC)" is used to analyze the percentage of the chemical composition during the reactive dye synthesis process. Another automatic measuring instrument, such as pH maintaining machine, is calculated before being used and tested to assess their precision and accuracy agents of the pH standard solution. For lab instruments, we can conclude that all measuring instruments are acceptable and ready to use for analyzing the accuracy and precision of the HPLC machine of chemical composition during the reactive dye synthesis. Three lab analysts are performed to test the chemical composition of the reactive dye synthesis process for 10 batches in total with two replicates [6]. The conclusion is shown in Table 1 which can be stated that the part-to-part variability is 99.41% and total gage R&R variability is 0.59%. Therefore, the measurement system is acceptable in terms of accuracy and variability.

Table 1. Result of measurement system variance components

Source	Variance Components	Contribution rate
Total Gage R&R	0.01936	0.59%
Repeatability	0.01924	0.58%
Reproducibility	0.00011	0.00%
Operators	0.00011	0.00%
Part-To-Part	3.27067	99.41%
Total Variation	3.29003	100.00%

3.3. Problem Identification Analysis

The team comprising several experts in the reactive dye synthesis process, especially for the Reactive Blue Base, brainstorms about the relevant factors that can affect the chemical yield. The cause-and-effect diagram with 5M1E methodology is applied in this study process. It is revealed that the number of tentative causes is 24 factors. These factors are used to analyze the relationship of causes and effects using cause-and-effect matrix by scoring these relationships with 0, 1, 3 and 9, respectively.

The factors with significant impacts on the chemical yield are selected and shown in Fig.5 (Pareto chart).

The results from the cause-and-effect matrix analysis indicates that 8 factors have high scores. These factors are further screened by using Failure Mode and Effects Analysis (FMEA) as shown in Fig.6. Four factors with high rating are selected for conducting the design of experiment (DOE) in the analyze phase, which is 66.17% of the total score.

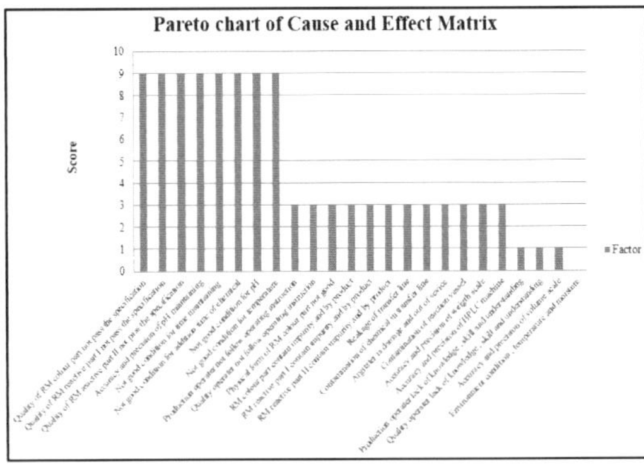

Figure 5. Pareto chart of Cause and Effect matrix

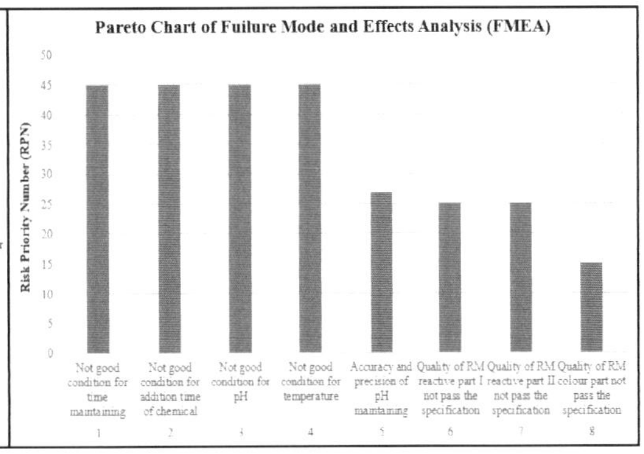

Figure 6. Pareto chart of FMEA

4. ANALYSIS PHASE

This phase involves the root cause analysis of the problem. From the Pareto chart analysis, 4 factors are selected to study in this analysis phase including time maintaining, addition time of chemical, pH and temperature. The Box Behnken design of experiments [7] with 4 factors and 3 levels are adopted. The experiments for reactive dye synthesis are separated into 2 steps, i.e. process study and optimize the factor. For 1st Condensation process, 3 factors to be experimented include pH, temperature and addition time of chemical (Table 2). In addition, 3 factors are tested on 2nd Condensation process including pH, temperature and time maintaining (Table 3). The response variable for each step of the processes is the Main component result is analysed by the HPLC machine and it must pass the specification in each process (Main component>90% for 1st Condensation process and Main component>75% for 2nd Condensation process).

Table 2. Factors level in Box Behnken design (1st Condensation process)

No.	Input factors	Units	Symbols	Levels of factors		
				-1	0	1
1	Temperature	Degree celsius	Temp	0	5	10
2	pH	-	pH	5.0	5.5	6.0
3	Addition time of chemical	Minute	Time dosing	5	10	15

Table 3. Factors level in Box Behnken design (2nd Condensation process)

No.	Input factors	Units	Symbols	Levels of factors		
				-1	0	1
1	Temperature	Degree celsius	Temp	10	20	30
2	pH	-	pH	7.0	8.0	9.0
3	Time maintaining	Minute	Time maintain	30	45	60

The experimental data are analysed by using the Minitab software and the results in each process step (1st Condensation and 2nd Condensation) are shown in Fig.7 and Fig.8, respectively. The statistical analysis of the results [8] shows that the main effect Temp, pH and Time dosing are significant factors for 1st Condensation process. For 2nd Condensation process, the statistical analysis shows that the main effects, i.e. Temp and pH, are significant factor, whereas the factor of Time maintain is not significant. In addition, two-way interactions of Temp*pH and Temp*Time dosing are also significant for 1st Condensation process.

Response Surface Regression: Result versus Temp, pH, Time dosing

The analysis was done using uncoded units.
Analysis of Variance for Result

Source	DF	Seq SS	Adj SS	Adj MS	F	P
Regression	9	1244.62	1244.62	138.291	77.93	0.000
Linear	3	787.78	120.88	40.293	22.71	0.002
Temp	1	556.11	12.01	12.006	6.77	0.048
pH	1	92.89	96.21	96.205	54.21	0.001
Time dosing	1	138.78	15.23	15.229	8.58	0.033
Square	3	396.54	396.54	132.178	74.49	0.000
Temp*Temp	1	258.99	295.57	295.570	166.56	0.000
pH*pH	1	98.38	107.55	107.551	60.61	0.001
Time dosing*Time dosing	1	39.17	39.17	39.170	22.07	0.005
Interaction	3	60.31	60.31	20.103	11.33	0.011
Temp*pH	1	38.13	38.13	38.131	21.49	0.006
Temp*Time dosing	1	21.76	21.76	21.762	12.26	0.017
pH*Time dosing	1	0.42	0.42	0.416	0.23	0.649
Residual Error	5	8.87	8.87	1.775		
Lack-of-Fit	3	6.00	6.00	2.000	1.39	0.444
Pure Error	2	2.87	2.87	1.436		
Total	14	1253.49				

S = 1.33212 PRESS = 102.462
R-Sq = 99.29% R-Sq(pred) = 91.83% R-Sq(adj) = 98.02%

Figure 7. ANOVA result of 1st Condensation

Response Surface Regression: Result versus Temp, pH, Time maintain

The analysis was done using uncoded units.
Analysis of Variance for Result

Source	DF	Seq SS	Adj SS	Adj MS	F	P
Regression	9	227.500	227.500	25.2778	14.97	0.004
Linear	3	120.820	63.495	21.1651	12.53	0.009
Temp	1	104.618	14.141	14.1407	8.37	0.034
pH	1	14.824	55.409	55.4091	32.82	0.002
Time maintain	1	1.378	1.138	1.1379	0.67	0.449
Square	3	104.855	104.855	34.9515	20.70	0.003
Temp*Temp	1	51.559	59.842	59.8425	35.44	0.002
pH*pH	1	52.205	53.060	53.0600	31.42	0.002
Time maintain*Time maintain	1	1.090	1.090	1.0900	0.65	0.458
Interaction	3	1.826	1.826	0.6086	0.36	0.785
Temp*pH	1	1.124	1.124	1.1236	0.67	0.452
Temp*Time maintain	1	0.632	0.632	0.6320	0.37	0.567
pH*Time maintain	1	0.070	0.070	0.0702	0.04	0.846
Residual Error	5	8.443	8.443	1.6885		
Lack-of-Fit	3	6.023	6.023	2.0075	1.66	0.398
Pure Error	2	2.420	2.420	1.2100		
Total	14	235.943				

S = 1.29943 PRESS = 101.806
R-Sq = 96.42% R-Sq(pred) = 56.85% R-Sq(adj) = 89.98%

Figure 8. ANOVA result of 2nd Condensation

5. IMPROVEMENT PHASE

In this phase, the significant factors in 1st Condensation and 2nd Condensation are further analyzed to find the optimal conditions by using the Response Optimizer in the Minitab software (Fig.9 and Fig.10). All parameters from the Response Optimizer are used as parameter settings for 20 batches of the pilot lab trial to observe the chemical yield in terms of variation and average yield. The chemical yields of the lab trials are shown in Fig.11 and the Descriptive Statistics is shown in Fig.12. The result demonstrates that the chemical yield from the synthesis process is in the range of 96.9% – 98.4% (S.D. = 0.436 and Cpk = 2.02) and the average of chemical yield is 97.6%.

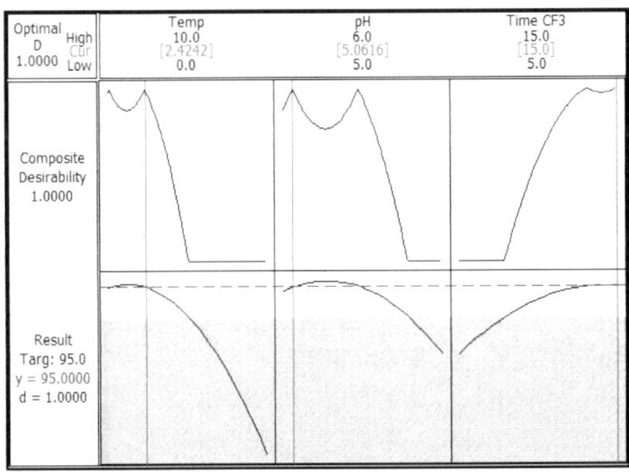

Figure 9. Parameters setting for 1st Condensation

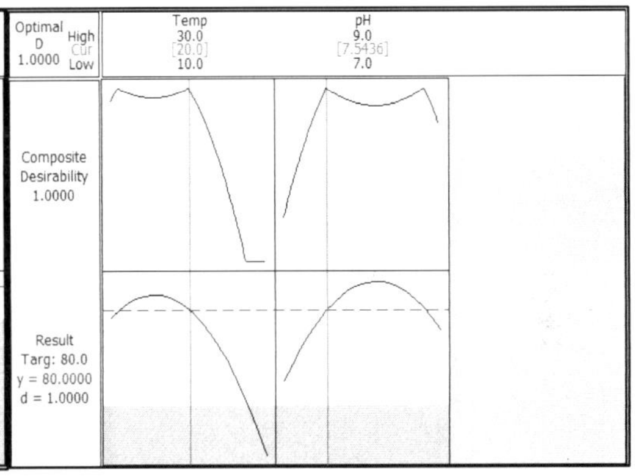

Figure 10. Parameters setting for 2nd Condensation

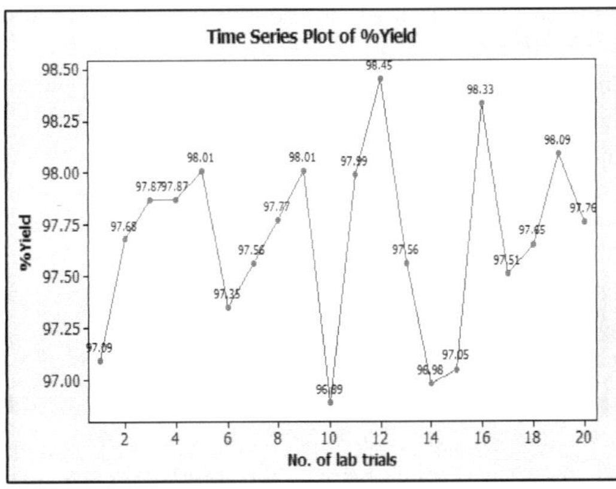

Figure 11. Chemical yield by lab trials

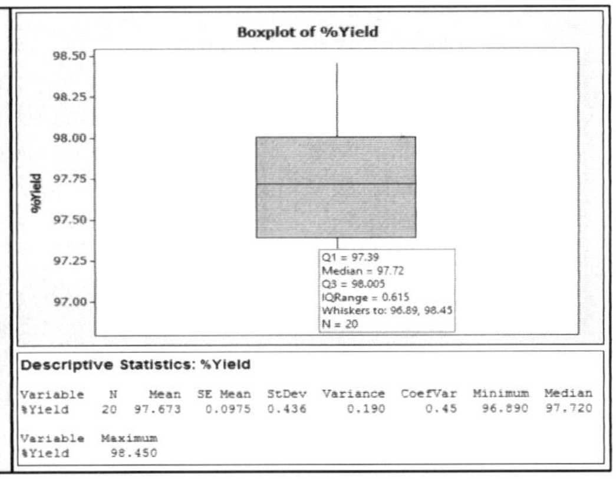

Figure 12. Descriptive Statistics of yield

6. CONTROL PHASE

All levels of the significant factors in each process step and parameter settings are implemented in the production unit. The data are collected from the synthesis process for 34 batches. It is found that the chemical yields from the synthesis process is in the range of 96.5% – 98.5% (S.D. = 0.525 and Cpk = 1.83) and the average chemical yield is 97.5%. Clearly, the chemical yield of the Reactive Blue Base is improved as expected. To maintain the quality of the chemical yield after improvement process, the operating instruction for operational control is documented and training of operators is provided. In addition, the concerned persons are trained to recognize and understand the importance of all parameters control for the reactive dye synthesis process to ensure that the performance improvements are maintained. The comparison of the chemical yield before and after the improvement process are shown in Fig.13.

IOP Conf. Series: Materials Science and Engineering **215** (2017) 012006 doi:10.1088/1757-899X/215/1/012006

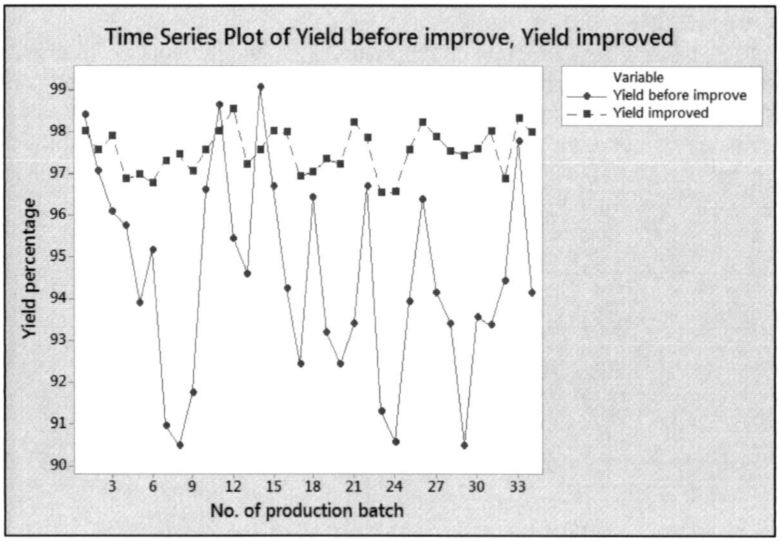

Figure 13. Comparison of the chemical yields

7. CONCLUSIONS

The Six Sigma concept is applied to improve the process of the reactive dye synthesis. The DMAIC methodology which incorporating the Box Behnken experimental design are employed to find the root causes that can affect the chemical yield of the Reactive Blue Base synthesis process and also indicates how to set the relevant process parameters. The results show clearly that after completing the process improvement each process step can reduce the variation of the chemical yield and increase the average yield to over 95% as expected. In addition, the results and knowledge gained from this case study can also be extended to another products of the company.

References

[1] D.P. Chattopadhyay and R. Chaudhary, 1997, "Chemistry and application of reactive dyes", Man-Made Textiles in India, 40, 495-504.
[2] R.M. Christie, 2001, "Colour Chemistry", Royal Society of Chemistry,UK.
[3] R.G. Schroeder, K. Linderman, C. Liedtke and A.S. Choo, 2008, "Six Sigma: Defination and underlying theory", Journal of operations Management, 26, 536-554.
[4] K. Linderman, R.G. Schroeder, S. Zaheer and A.S. Choo, 2003, "Six Sigma: a goal-theoretic persoective", Journal of operations Management, 21, 193-203
[5] M. Kim, Y.H. Lee, I.S. Han and C. Han, 2003, "Quality Improvement in the Chemical Process Industry using Six Sigma Techique", Process Systems Engineering 2003.
[6] A. AIAG, 2002, "Measurement Systems Analysis (MSA)", Reference Manual, Third Edition, The Automotive Industries Action Group.
[7] D.C. Montgomery, 2008, "Design and analysis of experiments", U.S. John Wiley & Sons, Inc.
[8] D.C. Montgomery and G.C. Runger, 2010, "Applied statistic and probability for engineers", U.S. John Wiley & Sons, Inc.

IOP Conf. Series: Materials Science and Engineering **215** (2017) 012007 doi:10.1088/1757-899X/215/1/012007

Electro-deposition painting process improvement of cab truck by Six Sigma concept

Kitiya Kawitu[1] and Parames Chutima[1,2*]

[1]Department of Industrial Engineering, Faculty of Engineering
Chulalongkorn University, Bangkok, Thailand 10330

[2]Regional Centre for Manufacturing Systems Engineering, Faculty of Engineering
Chulalongkorn University, Bangkok, Thailand 10330

[1]kitiya.kawitu@student.chula.ac.th, [2]Parames.C@chula.ac.th

Abstract. The case study company is a manufacturer of trucks and currently facing a high rework cost due to the thickness of the electro-deposited paint (EDP) of the truck cab is lower than standard. In addition, the process capability is very low. The Six Sigma concept consisting of 5 phases (DMAIC) is applied to determine new parameter settings for each significant controllable factor. After the improvement, EDP thickness of the truck cab increases from 17.88μ to 20μ (i.e. standard = 20 ± 3μ). Moreover, the process capability indexes (Cp and Cpk) are increased from 0.9 to 1.43, and from 0.27 to 1.43, respectively. This improvement could save the rework cost about 1.6M THB per year.

1. INTRODUCTION

The electro-deposited process (EDP) is the one of most important painting processes used in automotive industry to apply coatings to metal fabricated products. It has been widely used to coat bodies and parts of passenger cars, trucks and heavy equipment. The reason why they need EDP coating because it is protected corrosion and being one of the main primers that makes the surface smooth before going through the topcoat process. The data from January - June 2016 of the case study company showed that the highest rework cost about 1.9M THB is belong to EDP layer of truck cabs having lower thickness than the standard which impacts directly to the quality of products.

To solve the issue, this research applies the Six Sigma methodology because it is enables organizations to improve the quality, processes capability and finally it can be reduced the rework cost. The Six Sigma methodology is consists of 5 phases. In Phase 1, Define, important problem is selected and clearly defined. In Phase 2, Measure, the measurement system analysis is analyzed in terms of its accuracy and precision. In Phase 3, Analysis, all key process input variables are brainstormed and statistically tested whether they are significantly affected the problem. In Phase 4, Improve, the Design of Experiments technique is performed to find out the optimal settings of the key process factors. Finally, in Phase 5, Control, a confirmatory experiment is performed to confirm the result after improvement, and then the control plan and work instruction are updated.

2. DEFINE PHASE

Define phase is the step to articulate the problem, goal and project scope. Regarding the historical data of the EDP process collected from January 2.016 to June 2016 in Fig. 1 show that 89.91% or 1.9 MTHB of rework cost comes from the EDP lower thickness issue. The forecast data in Fig. 2 shows that 60% of main products is belong to the product Model SR. To meet customer requirement and company standard (EDP thickness of whole cab = 20 ± 3μ), this paper will consider on the EDP lower thickness issue by focusing on the whole cab area as shown in Fig. 3 and the main model is belong to the SR model.

IOP Conf. Series: Materials Science and Engineering **215** (2017) 012007 doi:10.1088/1757-899X/215/1/012007

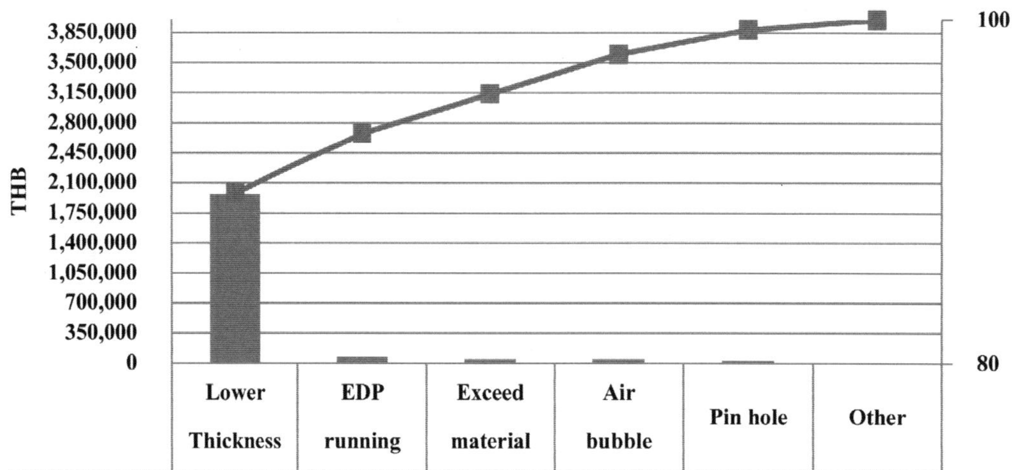

	Lower Thickness	EDP running	Exceed material	Air bubble	Pin hole	Other
Total cost	1,965,810	75,758	50,505	50,505	31,080	12,866
Percent'	89.91	93.37	95.68	97.99	99.41	100.00

Figure 1. Rework cost of EDP Painted Cab

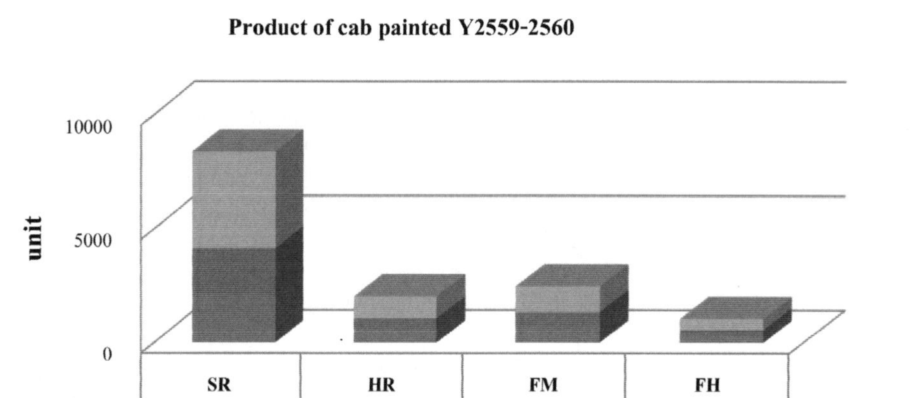

	SR	HR	FM	FH
Forecast Y2560	4230	960	1120	490
Q3-Q4 Y2559	1948	492	604	226
Q1-Q2 Y2559	2202	588	756	339

Figure 2. Production forecast data

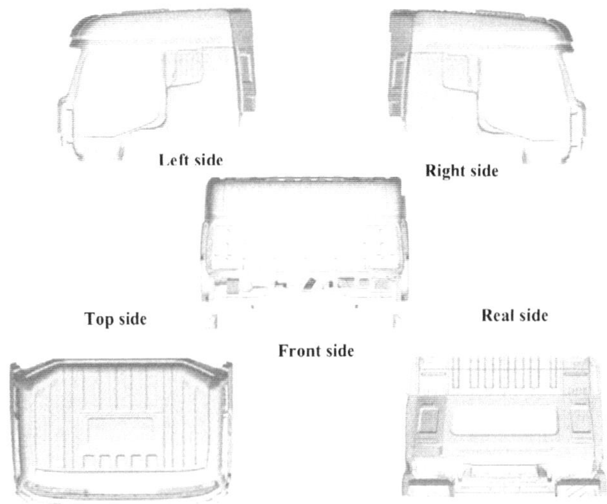

Figure 3. Identify of truck cab painted area

3. MEASUREMENT PHASE

Two main steps focused in this phase are measurement system analysis (MSA) and root cause selection for further elimination. A large number of potential root causes of the problem are identified in Fig. 4 by the fishbone diagram [1].

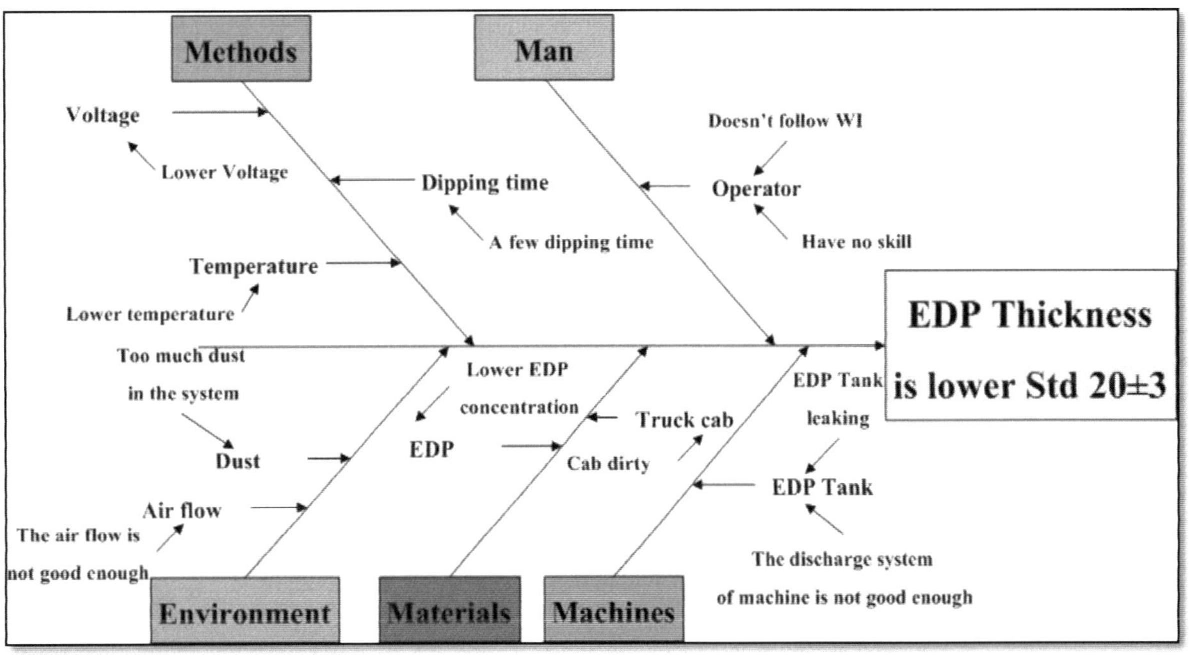

Figure 4. Fishbone diagram

IOP Conf. Series: Materials Science and Engineering **215** (2017) 012007 doi:10.1088/1757-899X/215/1/012007

3.1 Measurement system analysis (MSA)

MSA is an experimental and mathematical method of determining how much the variation within the measurement process contributes to overall process variability. This approach is used to analyse the repeatability and reproducibility of 2 operators who perform self-inspection on the finished good for 15 units. The result shows that %SV is 10.85 (Figs. 5-6). Since %SV is between 10 - 30 percent error, it could be concluded that the measurement system is acceptable.

Gage R&R

Source	VarComp	%Contribution (of VarComp)
Total Gage R&R	0.01335	1.18
Repeatability	0.01297	1.14
Reproducibility	0.00038	0.03
Operator	0.00038	0.03
Part-To-Part	1.12059	98.82
Total Variation	1.13394	100.00

Process tolerance = 6

Source	StdDev (SD)	Study Var (6 × SD)	%Study Var (%SV)	%Tolerance (SV/Toler)
Total Gage R&R	0.11553	0.69319	10.85	11.55
Repeatability	0.11389	0.68336	10.70	11.39
Reproducibility	0.01939	0.11632	1.82	1.94
Operator	0.01939	0.11632	1.82	1.94
Part-To-Part	1.05858	6.35148	99.41	105.86
Total Variation	1.06487	6.38920	100.00	106.49

Number of Distinct Categories = 12

Figure 5. Gage R&R result

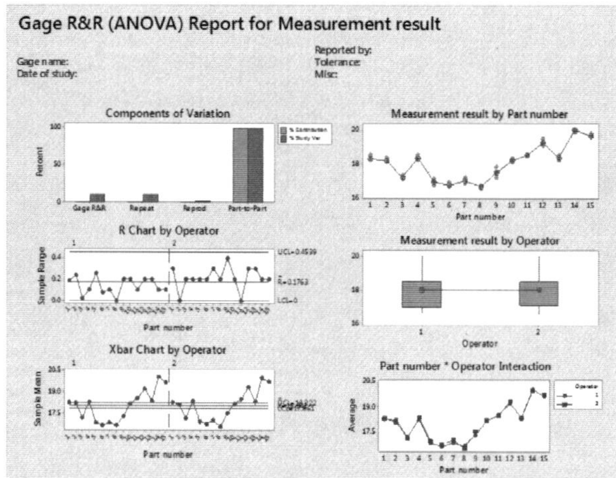

Figure 6. Gage R&R report

3.2 Problem Identification Analysis

A fishbone diagram or cause-and-effect matrix is a visualization tool for categorizing the potential causes of the problem in order to identify the root causes [2]. It's useful in brainstorming sessions to lead the discussion. After the working team has brainstormed all the possible causes for the problem, the facilitator helps the group to rate the potential causes according to their levels of importance. All potential root causes with their corresponding scores are ranked according to the Pareto concept (80:20) as show in Fig. 7. From 80% of the parameter that could be the root causes of the EDP issue. They are include seven factors as flowing lower voltage, a few dipping time, lower temperature EDP, tank leaking, lower EDP concentration, cab dirty and the discharge system of machine is not good enough as shown in Table 1. Three factors, i.e. lower voltage, a few dipping time and lower temperature EDP, are selected for further study in the analysis phase. Four remaining factors are not selected because tank leaking factor is can secure with the alarm system, lower EDP concentration is very high cost impact, and cab dirty and discharge system of machine not good enough are tested with hypothesis test and the result show that they are not significant.

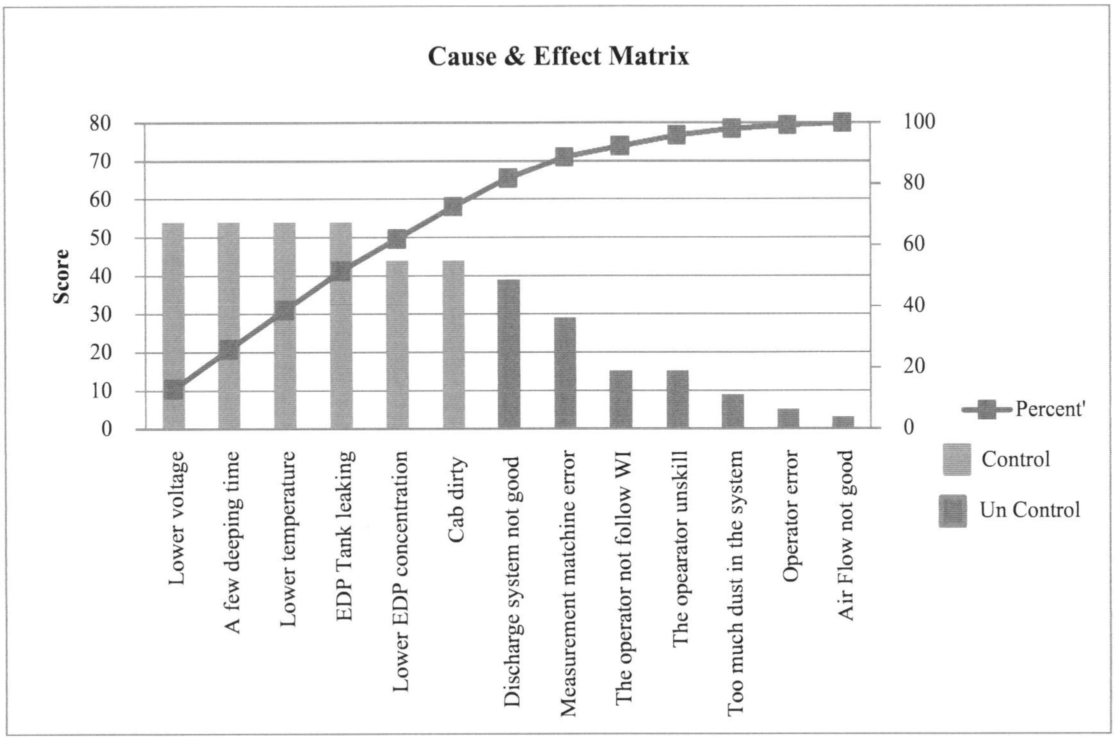

Figure 7. Pareto chart of Cause and Effect Matrix

Table 1: Potential Root Cause

Potential Root Cause		
Factor	**Implement**	**Remark**
Lower voltage	Yes	N/A
A few Deeping time	Yes	N/A
Lower temperature	Yes	N/A
EDP Tank leaking	No	Andon alarm
Lower EDP concentration	No	High Cost impact
Cab dirty	No	Hypothesis Test (no significant)
The discharge system of machine is not good enough	No	Hypothesis Test (no significant)

4. ANALYSIS PHASE

This phase applies the experimental design method by adopting 2^k fractional factorial design to prove which factors have significant effect to the EDP issue [3]. From the previous phase, the factors involved in the experiment are voltage, dipping time and temperature. In other words, 2^k fractional factorial design is performed with 3 factors and 2 levels each as shown in Table 2.

Table 2: factors level of 2^k fractional factorial design

No	Factors	Units	Levels of factors	
			Low (-1)	High (1)
1	Voltage	V	280	320
2	Dipping Time	Min	4	6
3	Temperature	°C	27	29

After the tests are conducted, the results from the experiments [4] are analysed using the Minitab software as shown in Fig. 8-9. All of 3 factors, i.e. voltage, dipping time and temperature, obtain P-value less than 0.05, meaning that all main effects are significant at 95% confident interval. The interaction between voltage and dipping time has P-value less than 0.05, meaning that interaction between voltage and dipping time is also significant at 95% confident interval, but interaction between voltage and temperature, dipping time and temperature are found not significant at 95% confident interval. As a result, all significant factors need further investigation to find their suitable settings in the next phase.

Factorial Fit: thickness versus vol, temp, dip time

```
Estimated Effects and Coefficients for thickness (coded units)

Term                 Effect     Coef  SE Coef      T      P
Constant                      17.713   0.2161  81.95  0.000
vol                   8.550    4.275   0.2161  19.78  0.000
temp                  1.525    0.763   0.2161   3.53  0.008
dip time              5.175    2.587   0.2161  11.97  0.000
vol*temp              0.500    0.250   0.2161   1.16  0.281
vol*dip time         -2.050   -1.025   0.2161  -4.74  0.001
temp*dip time        -0.475   -0.237   0.2161  -1.10  0.304
vol*temp*dip time    -0.850   -0.425   0.2161  -1.97  0.085

S = 0.864581    R-Sq = 98.63%    R-Sq(adj) = 97.43%
```

Figure 8. ANOVA result of 2^k fractional factorial design test

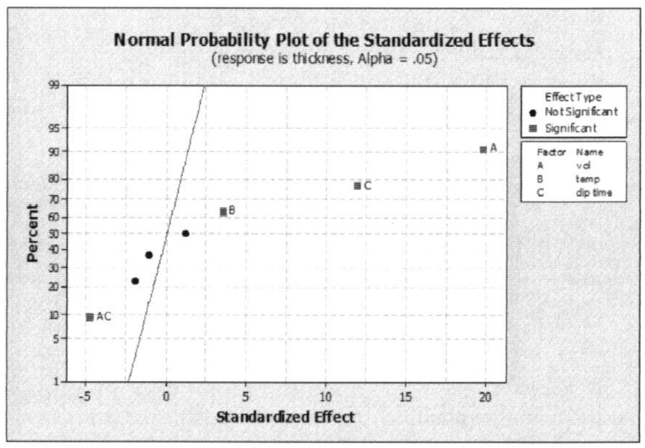

Figure 9. Normal Probability Plot of the standardized effect.

5. IMPROVEMENT PHASE

According to the analysis phase, 3 main factors including voltage, dipping time and temperature have statistically affected to the EDP issue [5]. The propose of this phase is to find new setting point for each factor by using Response Optimizer in the Minitab software (Central Composite Design Concept) as shown in Table 3.

Table 3: factors level of response optimizer (CCD)

No	Factors	Units	Levels of factor				
			-2	-1	0	1	2
1	Voltage	V	260	280	300	320	340
2	Dipping Time	min	3	4	5	6	7
3	Temperature	°C	26	27	28	29	30

IOP Conf. Series: Materials Science and Engineering **215** (2017) 012007 doi:10.1088/1757-899X/215/1/012007

The results of the experiments show in Fig. 10 that the new setting point of voltage is 300 V, dipping time is 5.6 min and temperature is 30°C. In addition, under the new setting points the EDP thickness is increased to 20µ which meet the EDP thickness standard of 20±3µ.

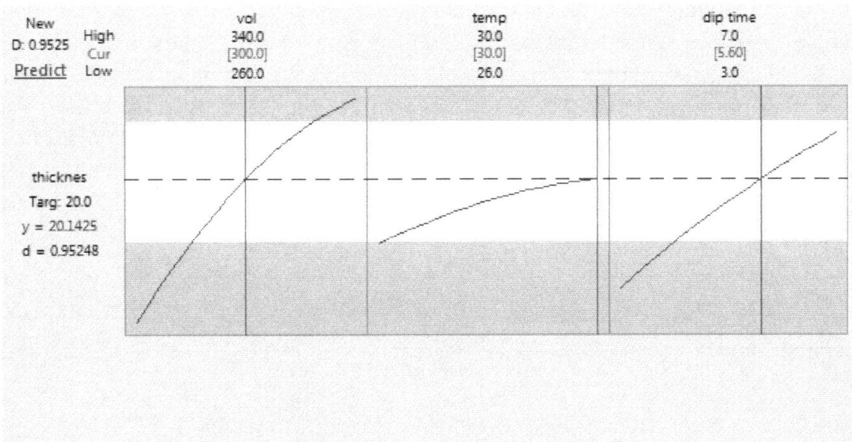

Figure 10: Response optimizer result of new setting point.

6. CONTROL PHASE

The new settings of the significant factors are implemented in the current process and it can prove that the quality of product is improved, i.e. the EDP thickness increases from 17.8µ to 20µ which meet the company standard. The process capability is improved from 0.90 to 1.43 and Cpk is increased from 0.27 to 1.43 in Fig. 11-12. To control the quality of the product and process to be sustainable after the improvement, the new work instruction for operational and control chart are written. Moreover, related operators are retrained to have a clear idea about how he should work.

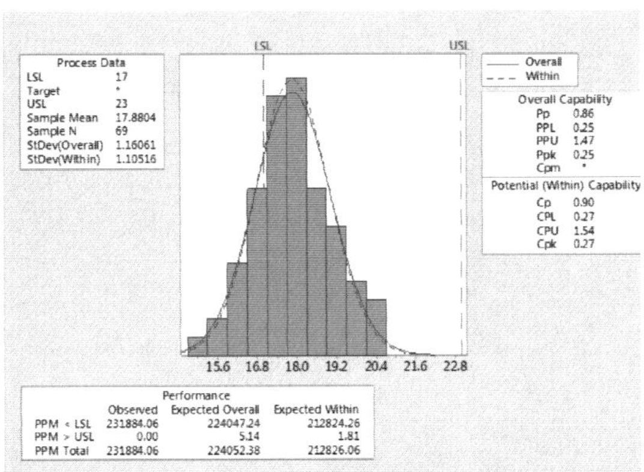

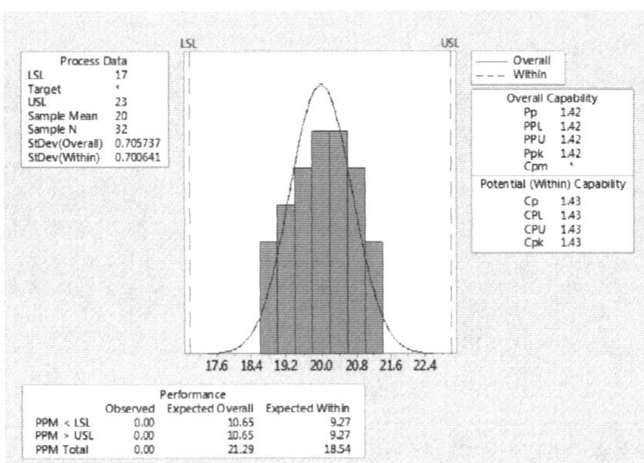

Figure 11: Process capability before improvement. **Figure 12**: Process capability after improvement.

7. CONCLUSIONS

This research presents the application of the Six Sigma method to improve the Electro Deposition Painting (EDP) process in an automotive business. The research follows the DMAIC steps to perform the process and product improvements. In define phase, the project goal is to have the average EDP thickness of 20μ with minimized rework cost. In the measurement phase, MSA is analyzed and found that %SV or P/TV is 10.85 which means the measurement system is acceptable. Three main factors including voltage, dipping time and temperature are selected to perform the C&E matrix. In the analyse phase, 2^k fractional factorial design is conducted and the results show that main effects of voltage, dipping time and temperature are significant. In the improvement phase, the Response Optimizer in the Minitab software (CCD Concept) is used to find the new setting points. It is found that the new setting point produces the EDP thickness at 20μ. In the control phase, the new work instruction for operational and control chart are created. After improvement, the EDP thickness is increased from 17.8μ to 20μ, Cp is increased from 0.90 to 1.43 and Cpk is increased from 0.27 to 1.43. More importantly, the company can save money about 1.6M THB per year from the rework cost.

IOP Conf. Series: Materials Science and Engineering **215** (2017) 012007 doi:10.1088/1757-899X/215/1/012007

References

[1] AIAG, 2002, "Measurement Systems Analysis (MSA)", Reference Manual, Third Edition, The Automotive Industries Action Group.

[2] Reddy C.M, Gaston R.S, and Weikart C.M, 1998, Influence of surface pretreatment and electrocoating parameters on the adhesion of cathodic electro-coat to the Al alloy surfaces, in Department of Chemical Engineering, University of Missouri, 21, 203-216

[3] Vatistas, N., 1997, A new electro-coating technique to avoid cratering and increase voltage rupture, in Department of Chemical Engineering, University of Pista 28, 311-325

[4] Montgomery, D.C. 2008, "Design and analysis of experiments", U.S. John Wiley & Sons, Inc.

[5] Montgomery, D.C. and G.C. Runger, 2010, "Applied statistic and probability for engineers", U.S. John Wiley & Sons, Inc.

IOP Conf. Series: Materials Science and Engineering **215** (2017) 012008 doi:10.1088/1757-899X/215/1/012008

PRODUCTION TIME LOSS REDUCTION IN SAUCE PRODUCTION LINE BY LEAN SIX SIGMA APPROACH

Thitima Ritprasertsri[1] and Parames Chutima[1,2*]

[1]Department of Industrial Engineering, Faculty of Engineering
Chulalongkorn University, Bangkok, Thailand 10330

[2]Regional Centre for Manufacturing Systems Engineering, Faculty of Engineering
Chulalongkorn University, Bangkok, Thailand 10330

[2]E-Mail: cparames@chula.ac.th

Abstract. In all industries, time losses, which are incurred in processing are very important. As a result, losses are incurred in productivity and cost. This research aimed to reduce lost time that occurs in sauce production line by using the lean six sigma approach. The main objective was to reduce the time for heating sauce which causes a lot of time lost in the production line which affects productivity. The methodology was comprised of the five-phase improvement model of Six Sigma. This approach begins with defining phase, measuring phase, analysing phase, improving phase and controlling phase. Cause-and-effect matrix and failure mode and effect analysis (FMEA) were adopted to screen the factors which affect production time loss. The results showed that the percentage of lost time from heating sauce reduced by 47.76%. This increased productivity to meet the plan.

1. INTRODUCTION

Commercial food world has been modified from its original focus on the export of agricultural products which are the basic food for humans and animals primarily through privatization as a processed or dried product by applying unusually high technologies. There is no value added to the product lot. Nowadays more processed foods are exported. Industrial sauce manufacturing is one of the most popular processing industries because sauce helps to improve the taste and flavor of food to taste delicious. Manufacturing sauce has been growing steadily resulting in high competition. Therefore, manufacturers must produce products of high quality to get more market share. Strategy relies on the production of high quality. To lower production costs, losses which occur in the production process need to be reduced. Productivity needs to be increased and products supplied timely which are sufficient for customers' needs.

Lean Manufacturing (LM) is a systematic approach to eliminate waste and change processes. This is done by identifying and reducing waste with continuous improvement [1]. There are seven wastes identified by Shigeo Shingo, namely over production, defect, unnecessary inventory, inappropriate processing, excessive transportation, waiting/idle time and unnecessary motion [2]. LM aims to develop solutions by using the minimum amount of resources (human, material and capital) and to ensure the products reach customers on time [3]. Six Sigma, provides quality philosophy and is a statistical tool to monitor process performance. It aims to reduce the variability in the process and to eliminate errors [4]. Also it gives more value to customers and stakeholders with focus on improving product quality and company productivity. There are five stages called DMAIC method (define, measure, analyse, improve, control) [5]. Lean and Six Sigma each have their own strengths in different applications and the two methodologies can be combined or applied in different situations to achieve the best results.

In this case study, the important problem in the production process and impact on the factory at this time is productivity which is insufficient to meet with the demand due to time lost in boiling sauce. As a result, most shipments are unable to meet the delivery dates. This research aimed to reduce lost time which occurs in the process of heating sauce by using LM. Six Sigma method to identify significant factors which affected losses in productivity. It is expected that by implementing LM Six Sigma the company could reduce production cost, increase utilization of resources, reduce loss time, and reduce variability in the production system.

2. DEFINE PHASE

This research focuses on lost time in the process of manufacturing sweet chili sauce (name SCS-TRI21) which is the main product of the factory and a huge demand from customers. The production process of the SCS-TRI21 consists of mixing, pre-heating, heating, cooling down and pasteurisation. From the data report, it was found that the volume of production per hour during the months of February to August 2016 less than planned as shown in Figure 1, resulting in the volume of sauce production during the months of February to August 2016 was below the target in every month as shown in Figure 2. As a result, some products could not be shipped on time.

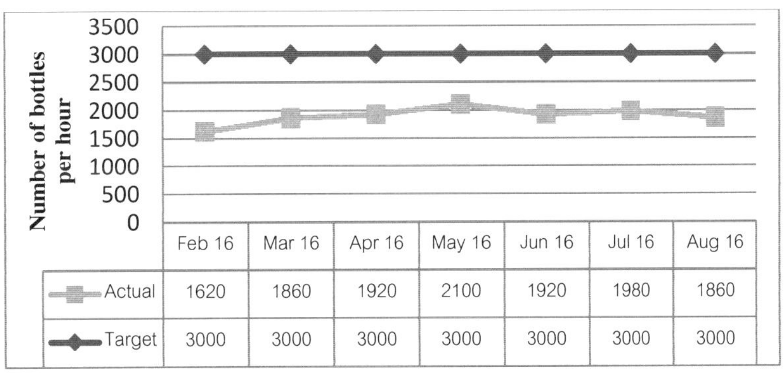

Figure 1. Actual production volumes per hour compared to the target during February to August 2016

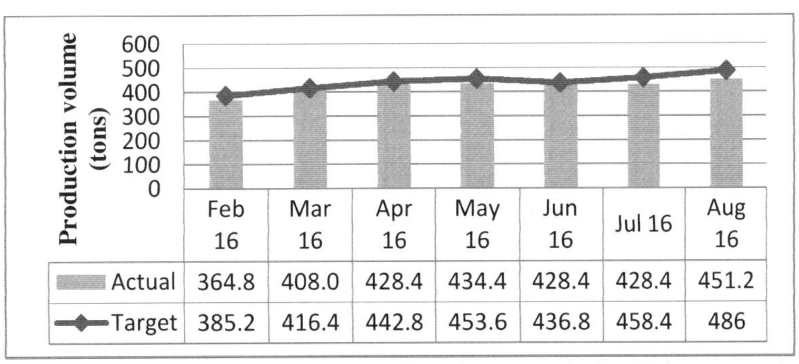

Figure 2. Actual production of sauce compared to the target during February to August 2016

3. MEASUREMENT PHASE

This phase was concerned with two main steps, which were waste measurement and problem identification analysis, which can be described as follows [6].

3.1. Waste Measurement

Concept 7 was applied in which waste can occur due to over production, defect, unnecessary inventory, inappropriate processing, excessive transportation, waiting/idle time and unnecessary motion [2]. The result of the waste in the production process of SCS-TRI21 is shown in Table 1. Obviously, the lost time, which is the most influence in the production, is waiting/idle time (42.27%) and inappropriate processing (20.32%).

Table 1. Summary of Waste Measurement in Production Process

No.	Types of waste	Frequency (%)
1	Over production	1.01
2	Defect	15.2
3	Unnecessary Inventory	5.58
4	Inappropriate Processing	20.32
5	Excessive Transportation	11.93
6	Waiting /Idle Time	42.27
7	Unnecessary Motion	3.69

3.2. Problem identification analysis

Brainstorming sessions were conducted with a team of experts from different departments involved in the production of the sauce. A system was implemented to identify factors which could affect the time lost in process. The cause and effect analysis concerned Man, Machine, Material, Method and Environment (4M1E) which were applied in this process [7]. It was revealed that the number of tentative causes was 12 factors as shown in Table 2. These factors were used for further analyzing the relationship between cause and related effect using the cause and effect matrix. By scoring these relationships 0, 1, 3 and 9 respectively, and analyzing data from the Pareto chart analysis the factors which significant impacted to the production time lost were identified as shown in Figure 3. Room temperature was found to have no effect.

Table 2. Cause and Effect Diagram with 4M1E
(Source : Brainstorming by the experts in the production of sauce)

No.	Area cause	Cause of problem	Total
1	Man	Employees lack attention to process control	3
2	Machine	Size of mixing tank	3
3		Long lifetime of mixing tank	1
4		Lack of preventive maintenance plan for mixing tank	1
5		Speed of agitator of mixing tank	9
6	Material	Type of raw material	3
7		Steam for heating	9
8	Method	Heating time	9
9		Heating temperature	9
10		Syrup temperature	9
11	Environment	Equipment not enough	3
12		Room temperature	0

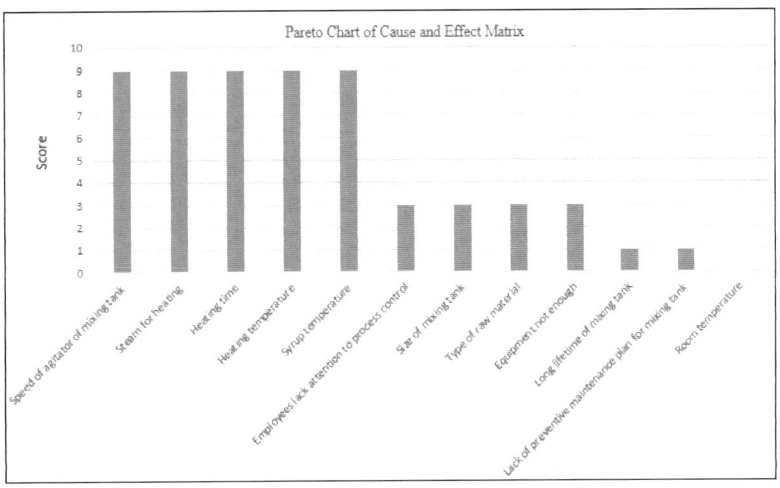

Figure 3. Pareto chart of Cause and Effect matrix

The results from cause and effect matrix analysis showed that 5 factors (speed of agitator of mixing tank, steam for heating, heating time, heating temperature and syrup temperature) have the highest scores. These factors were further screened by using Failure Mode and Effects Analysis (FMEA) as shown in Table 3.

Table 3. The Risk Priority Number of Each Root Cause

No.	Potential Failure	Risk Priority Number (RPN)
1	Speed of agitator of mixing tank	648
2	Steam for heating	175
3	Heating time	125
4	Heating temperature	900
5	Syrup temperature	900

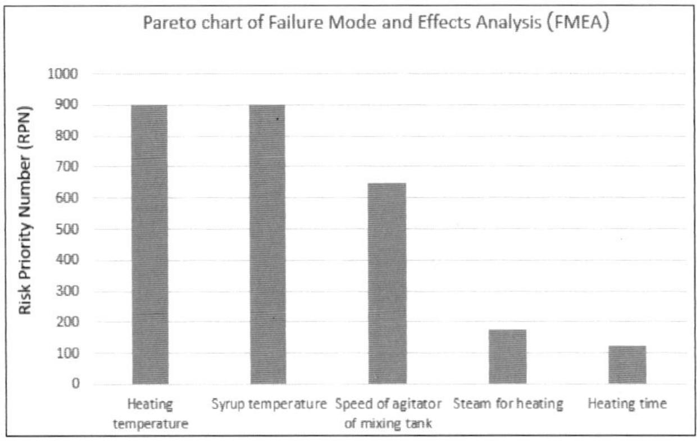

Figure 4. Pareto chart of Failure Mode and Effect Analysis

Figure 4 showed three factors with high rating in each category which were selected for the design of experiment (DOE) in the analysis phase. These three factors contribute 89.1% of the total score. Three factors which could affect the time lost in process are heating temperature (the temperature was in the boiling sauce), syrup temperature (the temperature was dissolved syrup before bringing to boil the sauce) and speed of agitator of mixing tank (agitator helps to mixed ingredient very well and heat transfer).

4. ANALYSIS PHASE

At this phase, the analysis was carried out to determine the causes of the problem. The 2^k factorial design is particularly useful in the early stages of experimental work when many factors are likely to be investigated. It provides the smallest number of runs with which k factors can be studied in a complete factorial design. Consequently, these designs are widely used in factor screening experiments [8]. Three factors were selected to study in this analysis phase including heating temperature, syrup temperature and speed of agitator of mixing tank. The level of each factor has been selected by the minimum and maximum adjustable values of the instruments used in the process as shown in Table 4. The eight trials were conducted within a completely randomized experiment. The response variable in this study was the time involved in the process.

Table 4. Factors and Level of Factors in 2^3 Factorial Design

No.	Factors	Units	Symbols	Levels of factor	
				Low (-1)	High (1)
1	Speed of agitator of mixing tank	RPM	A	30	45
2	Syrup temperature	Degrees Celsius	B	70	80
3	Heating temperature	Degrees Celsius	C	85	90

Figure 5 has shown the experimental data which were analyzed by using the Minitab software (version 17) and the results showed that the main effects were caused by syrup temperature (B) and heating temperature (C) and in three-factor interactions these were significant factors at 95% confidence interval, whereas factor A (speed of agitator of mixing tank) and interactions (AB, AC and BC) were not significant. One approach to the analysis of a single replicated factorial is to assume that certain high-order interactions are negligible [8]. Therefore, it could be concluded that the main effects B and C were significant at 95% confidence interval. As a result, all significant factors need further investigation to find their appropriate level settings.

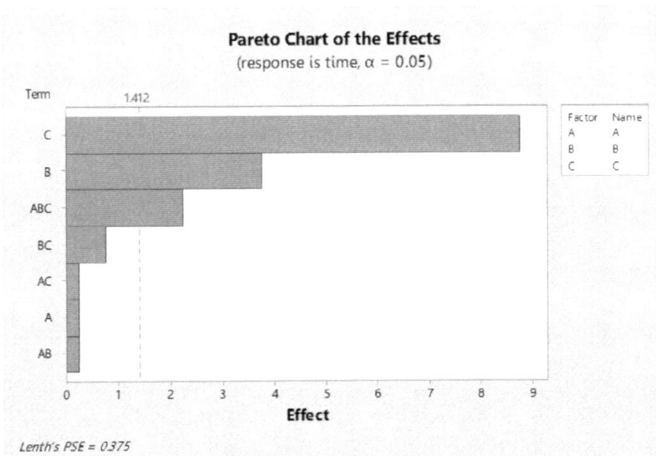

Figure 5. Pareto chart of effects

5. IMPROVEMENT PHASE

In this phase, two levels of the significant factors derived from the 2^k fractional factorial experiments were tested to find their appropriate levels which could reduce the lost time in process. Central Composite Design (CCD) was used in this experiment [8], to test the interaction of these factors and whether their impact was significant or not in lost time in process. The level of each factor was randomly selected, three levels of syrup temperature (B) and three levels of heating temperature (C) as shown in Table 5.

Table 5. Factors and Levels of Factors in Central Composite Design

No.	Factors	Units	Symbols	Levels of factor		
				Low (-1)	Medium (0)	High (1)
1	Syrup temperature	Degrees Celsius	B	70	75	80
2	Heating temperature	Degrees Celsius	C	85	87.5	90

The experimental data were analysed by using the Minitab software, which produced the statistics [9] shown in Figure 6 which revealed that the main effect of syrup temperature and heating temperature have the P-value less than 0.05. Therefore, it can be concluded that the main effect of syrup temperature and heating temperature affected the lost time in process at 95% confident interval.

Response Surface Regression: time versus Syrup temperature, Heating temperature

Analysis of Variance

Source	DF	Adj SS	Adj MS	F-Value	P-Value
Model	5	6010.77	1202.15	7.73	0.009
Linear	2	4725.28	2362.64	15.19	0.003
Syrup temperature	1	1131.09	1131.09	7.27	0.031
Heating temperature	1	3594.19	3594.19	23.10	0.002
Square	2	1279.24	639.62	4.11	0.066
Syrup temperature*Syrup temperature	1	1166.63	1166.63	7.50	0.029
Heating temperature*Heating temperature	1	36.80	36.80	0.24	0.642
2-Way Interaction	1	6.25	6.25	0.04	0.847
Syrup temperature*Heating temperature	1	6.25	6.25	0.04	0.847
Error	7	1088.92	155.56		
Lack-of-Fit	3	731.72	243.91	2.73	0.178
Pure Error	4	357.20	89.30		
Total	12	7099.69			

Model Summary

S	R-sq	R-sq(adj)	R-sq(pred)
12.4724	84.66%	73.71%	18.85%

Figure 6. Results of Analysis of Variance.

The significant factors in the previous step were analysed to find the optimal condition by using Response Optimiser of the Minitab software. Figure 7 shows the optimized value of the syrup temperature and the heating temperature are 77.6^{0}C and 91.0 ^{0}C, respectively, to consume the minimum lost time in the process.

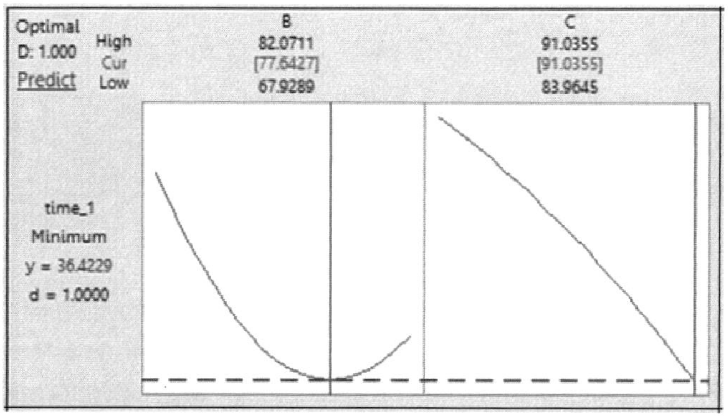

Figure 7. Response Optimiser of syrup temperature and heating temperature

6. CONTROL PHASE

In this phase, all settings of the significant factors derived from the previous phase were implemented in the process and the data relating to the time taken in the process were collected for 30 days. It was found that the volume of production per hour in the month of September 2016 was above target (3,300 bottles per hour) as shown in Figure 8 and the volume of sauce production in the month of September 2016 was above target (521 tons), comparing between the production volume before and after the improvement process as shown in Figure 9.

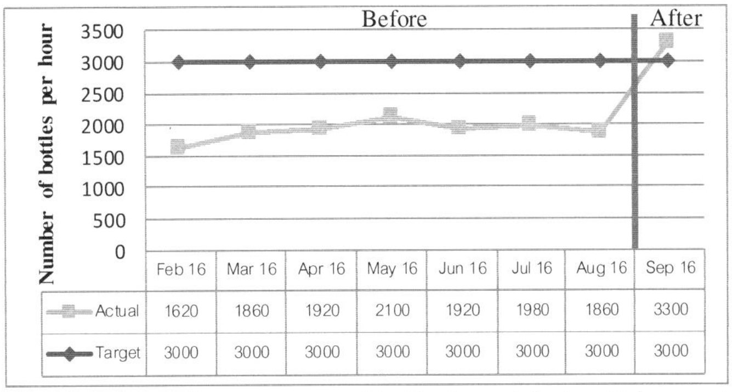

Figure 8. Production volumes per hour compared to the target during February to September 2016

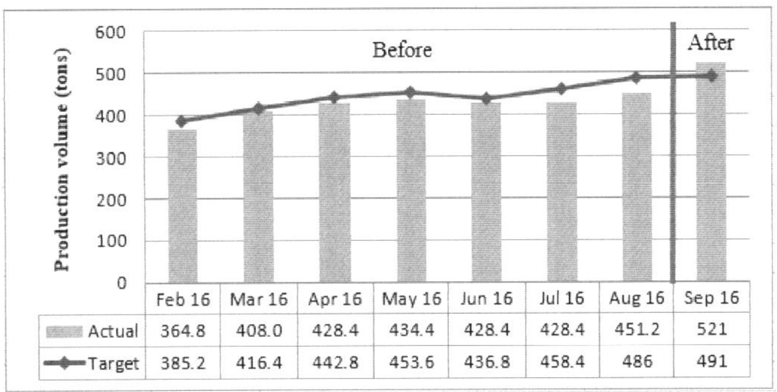

Figure 9. Production of sauce compared to the target during February to September 2016

The process improvements, which could reduce the time taken in each step, showed that syrup preparation was reduced by 3 minutes, heating was reduced by 8 minutes, cooling down was reduced by 51 minutes and pasteurization was reduced by 2 minutes. To sum up, the total time in production process was reduced by 47.76% as shown in Table 6.

Table 6. Production time compared between before and after improvement

No.	Process	Time (mins)	
		before improvement	after improvement
1	Syrup preparation	25	22
2	Heating	41	33
3	Cooling down	51	0
4	Pasteurisation	17	15
	Total	134	70

To maintain the production time of the process after improvement, the work instructions for operational control in the process were written and shared among all the production personnel that were involved in these activities. This helped the operators to manage and monitor the process systematically.

7. CONCLUSIONS

The Lean Six Sigma method was applied to reduce lost time which occurred in the sauce production line. Five steps of the Six Sigma were utilised in order to find the proper parameter settings of the significant factors and to indicate how to set the relevant process control parameters. It was found that the syrup temperature and heating temperature had significant effects on the problem of interest. After applying all new settings to the process, the result showed that the time in production process was reduced by 47.76% which could cut the cooling down process which caused the lost time. Consequently, productivity was increased to meet the plan. Thus LM Six Sigma can be applied to reduce the time in another part of the production process.

References

[1] Alacca, H., Ceylan, C., 2011 Value Chain anaysis using Value Stream Mapping, White Good Industry Application Department of Industrial Engineering, R. R.

[2] Taylor, D., Hines, P., 2000 Going Lean: A Guide To Implementation, Lean Enterprise Research Centre.

[3] Sule Girenes, S., 2013 Lean Six Sigma methodology and application, Science+Business Media B.V. 2011, **47**, 2107–2127.

[4] Brett, C., Queen, P., 2005 Streamlining enterprise records managements with Lean Six Sigma. Inf. Manag. J., **39**, 58–62.

[5] Gaspersz, V., 2007 Lean Six Sigma for Manufacturing and Service Industries, Gramedia Pustaka Utama, Jakarta.

[6] De Mast, J. and J. Lokkerbol., 2012 An analysis of the Six Sigma DMAIC method from the perspective of problem solving *International Journal of Production Economics*, **139**(**2**), 604-614.

[7] Yihua, M., Tuo, X., 2011 Research of 4M1E's effect on engineering quality based on structural equation model, 213-220.

[8] Montgomery. D.C. 2008 Design and analysis of experiments, U.S. John Wiley & Sons, Inc.

[9] Montgomery, D.C. and Runger, G.C. 2010. Applied statistics and probability for engineers, U.S., John Wiley & Sons, Inc.

Car Assembly Line Efficiency Improvement by Lean Principle

Suwalee Sawassalung[1] and Parames Chutima[1,2]
[1]Industrial Engineering Department, Chulalongkorn University, Bangkok 10330, Thailand
[2]Regional Centre for Manufacturing Systems Engineering, Chulalongkorn University, Bangkok 10330, Thailand

E-mail: [1]suwalee.saw@hotmail.com [2]cparames@chula.ac.th

Abstract: This research aimed to increase the efficiency of actual working time to compare to design standard time ratio (DSTR) as per analysing process of Lean System of the assembly line in a car manufacturer in Thailand. Currently, the case study factory and its group of factories, which have many branches all over the world, have competed with each other on quality, delivered time and production cost. The production cost which can reduce without affecting quality and acceptable by clients is the manpower cost. The index of competition is DSTR. The factory now has DSTR of 6.13 and DSTR of the assembly department is 4.24 which is very high comparing to other departments. The low DSTR indicates that the factory has good quality. The ways to solve the problem are to apply the following tools, i.e. Lean principle, Value Stream Mapping (VSM), Waste Analysis and ECRS. After implementing the above tools, the results showed that DSTR decreased from 4.24 to 4.06 or 4.25%.

1. Introduction

The case study factory is one of the top-three Japanese car assembly plants in Thailand that produces products as passenger car, sedan, SUV, etc. The production process consists of five steps (Fig.1). Due to the sluggish of world economy, the purchasing power of consumers tends to decrease substantially causing reduced production volume in all plants. As a result, the factories in the group start to benchmark and compete with each other about quality, delivered time and production cost so that they could be able to accept more demand since they are more efficient than the others. However, the policy of the group mentions that reduction in the production cost has to be done without affecting quality. Therefore, the group decides to focus on improving the cost of manpower. The index that the group used to indicate the manpower cost of factories is the Design Standard Time Ratio (DSTR). It is an index to show the ratio of value-added in all operations and can be used to compare productivity of the company at the global level. Refer to the ranking of DSTR, it was found that the factory has DSTR of 6.13 (Fig. 2). When considering the DSTR index, several production aspects could not be compared. Therefore, the factory re-computes the index based on the cars with the same sizes. The DSTR index can be calculated as follows.

$$DSTR = \frac{\text{Actual Working Time} \times \text{Man Power}}{\text{Design Standard Time} \times \text{Production Volumn}} \quad (1)$$

Figure 1. Production Process

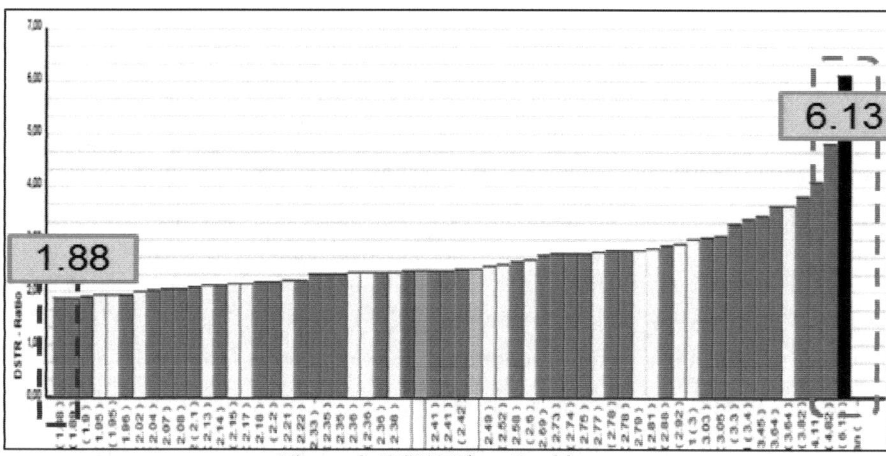

Figure 2. DSTR Plant Ranking

The other index that directly affects DSTR is the efficiency of the production line, namely the percentage of job balance (%JB). This index is depended on the percentage of work balancing efficiency that is used as a main factor and it can be calculated as follows.

$$\%Job\ balance = \frac{\text{Total Actual Time} \times 100}{\text{Tact Time} \times \text{Manpower}} \quad (2)$$

The relation between the equations (1) and (2) shows that the main factor is manpower. If we have more manpower, it will increase DSTR and decrease %JB.

From five steps of the production process, the processes that is used to calculate DSTR consisting of the body shop, paint shop and assembly shop. Historical data shows that the process which has highest DSTR is the assembly shop (DSTR = 4.24 and %JB = 73%) which is lowest compared to the other processes. The factor that affects DSTR of the assembly shop is that the assembly shop mainly uses manpower while the other processes mostly used machines to produce parts as shown in Table 1.

Table 1. Percentage of job balance for each shop

Dept	Body Shop	Trim&Chassis	Paint Shop						
			PBS	Seal	Sursand	Surfacer	Top Coat	Touch up	Under Coat
Actual time	722.95	360.5	13.6	41.51	18.88	19.21	23.86	18.92	18.45
Tact Time	2.48	2.28	2.45	2.5	2.6	2.72	1.66	2.59	2.5
Man Power	325	216	6	18	8	8	16	8	8
Job Balancing Ratio	89.7	73.2	92.51	92.24	90.76	88.28	89.83	91.31	92.25
			91.03						
DSTR	1.08	4.24	0.81						

The Lean Principle, that is create the manufacturing operation focused on continuous improvement [1], was considered by using Value Stream Mappings that is all actions (bolt value added and non-value added) currently required to bring a product through the main flows essential to every product. Therefore value stream map support people to see waste in process [2-3], it is found that the current value stream mapping of the production process has DSTR of 4.24 (Fig. 3)

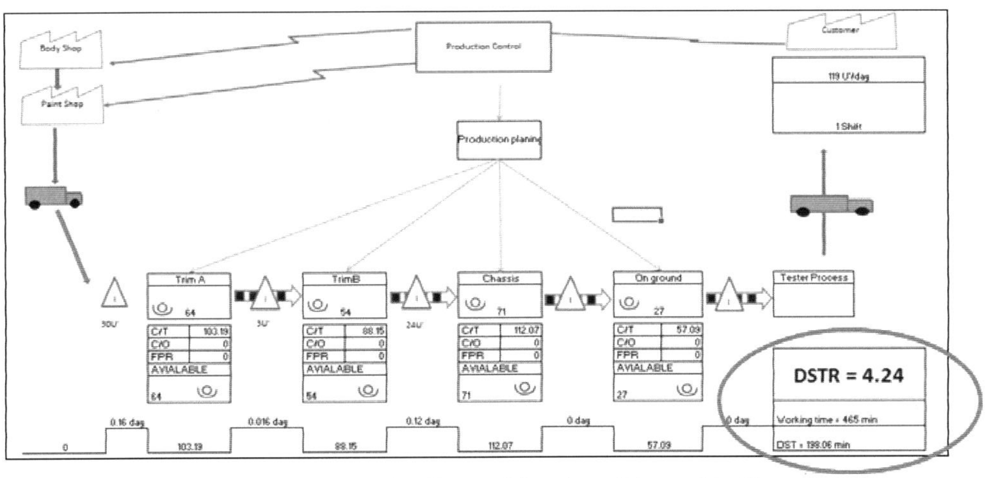

Figure 3. Current Value Stream Mapping of Assembly Shop

From the above problem, the factory decides to increase the actual working time of DSTR by using the following tools, i.e. Lean principle, Value Stream Mapping (VSM), Eliminate, Combine, Rearrange, Simplify (ECRS) and Waste Analysis (Seven types of wastes include Overproduction, Defects, Unnecessary Inventory, Inappropriate Processing, Excessive Transportation, Waiting and Unnecessary Motion) [4-6].

2. Problem Analysis

From the Current VSM, it is found that the assembly shop had DSTR of 4.24. From the production process time analysis, the bottlenecks include Set FR STRUT, Set RR SUSP, Set RR Shock and Mounting Engine. The flow process chart is analyzed and found that the waste comes from duplicated works between operators and in-line quality assurance officer (IQA), who is check man for recheck torque wrench. The duplicated process is the Tightening Torque which the worker needs to do it twice; therefore, these processes is the bottleneck and affects the next process since parts could not be sent according to the Takt time which is 2.28 minutes. For example, the station Set RR brakes Assembly RH and found that the assembly point has duplicated works between operators and IQA (Fig.4). The cycle time and man power of the bottleneck stations are summarized was shown in Table 2.

Item	Set RR brake Ass'y RH	Operation	Transportation	Inspection	Delay	Storage	VA	NVA
1	Check spec RR SUSP			●				0.05
2	Picking RR SUSP to jig fixture		●					0.15
3	Picking RR brake RH, LH to jig fixture		●					0.05
4	Assembly spindle to RR brake RH	●					0.44	
6	Temporary tightening SCR to PKB CABLE and PKB BRK'T RH (4 pcs)	●					0.16	
7	Tightening SCR (4 pcs)	●					0.1	
8	Assembly SENSANTISKID RR LH to RR brake ass'y RH	●					0.2	
9	Temporary tighten bolt to SENSANTISKID	●					0.15	
10	Tightening bolt to SENSANTISKID	●					0.1	
11	Assembly INSUL brake tube and press grommet to brake hose sensor harn BRK'T LH	●					0.15	
12	Assembly HSG RR AXEL to RR SUSP	●					0.2	
13	Assembly brake ass'y LH and HUB BRG wheel to RR SUSP	●					0.2	
14	Torque bolt FIX FLG to brake (4 pcs)	●					0.3	
15	Torque bolt SENSANTISKID	●					0.05	
16	Wiring brake tube to RR SUSP	●					0.05	
17	Check mark			●			0.05	
	Total	12	2	2	0	0	2.15	0.25
	Total C/T	2.4						

Item	Re-check torque RR brake RH	Operation	Transportation	Inspection	Delay	Storage	VA	NVA
1	Torque bolt FIX FLG to brake (4 pcs)	●					0.3	
2	Torque bolt SENSANTISKID	●					0.05	
3	Check mark			●			0.05	
	Total	2	0	1	0	0	0.4	
	Total C/T	0.4						

Figure 4. Flow Process Chart of RR SUSP (Set RR brake Ass'y)

Table 2. Cycle time and manpower for each process

Process	C/T *Remak Takt Time = 2.28 min	Man Power Operator	 IQA
1.RR SUSP			
1.1 Set RR brake ass'y RH	2.8	1	1
1.2 Set RR brake ass'y LH	2.8	1	1
1.3 Set RR Drum brake RH	2.8	1	1
1.4 Set RR Drum brake LH	2.8	1	1
2. Sub assembly FR STRUT			
2.1 Sub FR STRUT RH	2.3	1	1
2.2 Sub FR STRUT LH	2.3	1	1
3. RR Shock			
3.1 Set RR shock RH	2.6	1	1
3.2 Set RR shock LH	2.6	1	1
4. Mounting Engine			
4.1 Set mounting engine LH	2.38	1	
4.2 Set mounting engine RH	2.38	1	
4.3 Fix FR STRUT to FR Axel L	2.38	1	1
4.4 Fix FR STRUT to FR Axel L	2.38	1	1
4.5 Fix FR SUSP MBR	2.38	1	

3. Process Improvement

The standard quality assurance system of the factory has three levels of checking for tightening was shown in Table 3.

1) Alfa (α) level was a high level, i.e. checking without IQA but needs to bring the system wrench that records torque and the line stop system to help.

2) Beta (β) level was a medium level, i.e. checking by bringing shut-off wrench to help and needs to have IQA to check by touching.

3) Gamma (γ) level was a low level checking, i.e. system wrench or shut-off wrench are not used to help, but needs to be rechecked by operators and IQA.

Table 3. Standard assurance system

		Imp AB(tightening)	Imp C(tightening)
α	Built-in	HCT & Line stop	Shut-off wrench & Control the number of tightening
	Outflow prevention		
β	Built-in	HCT & Control the number of tightening	Shut-off wrench
	Outflow prevention	check by touching	check by touching (Same operator is possible for low JPH)
γ	Built-in	Torque wrench	Torque wrench
	Outflow prevention	Torque recheck	check by touching

IOP Conf. Series: Materials Science and Engineering **215** (2017) 012009 doi:10.1088/1757-899X/215/1/012009

Currently, the process of work of the factory has to check at the Grammar (γ) level; therefore, it should be improved the process by decreasing the waste from duplicated working of operators and IQA by bringing the system wrench which records torque point to help in α level as shown in Table 3. It will help to reduce in-line quality assurance as well (Fig.5)

Figure 5. System Wrench Nut Runner and Line Stop

4. Result and Discussion

After the improvement by adding the system wrench nut runner and line stop, it could decrease the waste from duplicated works between operators and IQA that needs to torque (Fig. 6) and reduce the bottleneck of the process as shown in Table 4.

Item	Set RR brake Ass'y RH	Operation	Transportation	Inspection	Delay	Storage	VA	NVA
1	Check spec RR SUSP			●				0.05
2	Picking RR SUSP to jig fixture		●					0.15
3	Picking RR brake RH, LH to jig fixture		●					0.05
4	Assembly spindle to RR brake RH	●					0.44	
6	Temporary tightening SCR to PKB CABLE and PKB BRK'T RH (4 pcs)	●					0.16	
7	Tightening SCR and torque(4 pcs)	●					0.2	
8	Assembly SENSANTISKID RR LH to RR brake ass'y RH	●					0.2	
9	Temporary tighten bolt to SENSANTISKID	●					0.15	
10	Tightening bolt to SENSANTISKID and torque	●					0.2	
11	Assembly INSUL brake tube and press grommet to brake hose sensor harn BRK'T LH	●					0.15	
12	Assembly HSG RR AXEL to RR SUSP	●					0.2	
13	Assembly brake ass'y LH and HUB BRG wheel to RR SUSP	●					0.2	
16	Wiring brake tube to RR SUSP	●					0.05	
17	Check mark			●			0.05	
	Total	12	2	2	0	0	2	0.25
	Total C/T			2.25				

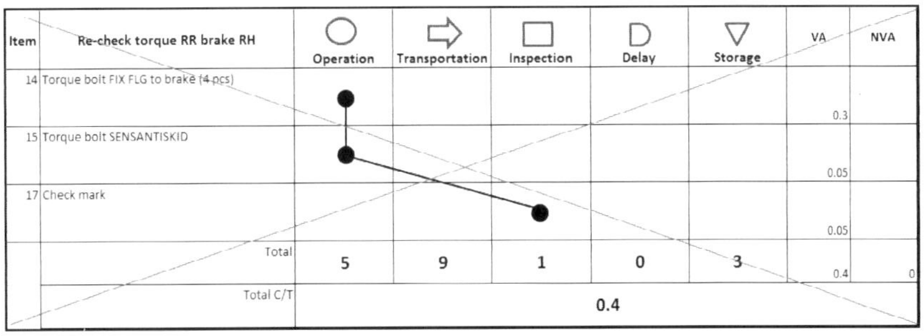

Figure 6. Flow Process Chart after Improvement

Moreover, it could reduce the number of IQA for 10 persons, from 216 persons to 206 persons, or 4.63% and the %JB increases from 73.20 to 75.57 or 3.24% as shown in Table 5. In addition, DSTR is reduced from 4.24 to 4.06 or 4.25%. Table 4 shows the new cycle time and manpower requirement after the improvement. The new Value Stream Mapping after improvement is shown in Fig. 7.

Table 4. Cycle time and manpower after improvement

Process	C/T *Remak Takt Time = 2.28 min	Man Power Operator	 IQA
1.RR SUSP			
1.1 Set RR brake ass'y RH	2.25	1	
1.2 Set RR brake ass'y LH	2.25	1	
1.3 Set RR Drum brake RH	2.25	1	
1.4 Set RR Drum brake LH	2.25	1	
2. Sub assembly FR STRUT			
2.1 Sub FR STRUT RH	1.74	1	
2.2 Sub FR STRUT LH	1.74	1	
3. RR Shock			
3.1 Set RR shock RH	2.15	1	
3.2 Set RR shock LH	2.15	1	
4. Mounting Engine			
4.1 Set mounting engine LH	2.08	1	
4.2 Set mounting engine RH	2.08	1	
4.3 Fix FR STRUT to FR Axel I	2.08	1	
4.4 Fix FR STRUT to FR Axel I	2.08	1	
4.5 Fix FR SUSP MBR	2.08	1	

Table 5. Percentage of job balance of assembly shop after improvement

Dept	Body Shop	Trim&Chassis	Paint Shop						
			PBS	Seal	Sursand	Surfacer	Top Coat	Touch up	Under Coat
Actual time	722.95	354.94	13.6	41.51	18.88	19.21	23.86	18.92	18.45
Tact Time	2.48	2.28	2.45	2.5	2.6	2.72	1.66	2.59	2.5
Man Power	325	206	6	18	8	8	16	8	8
Job Balancing Ratio	89.7	75.57	92.51	92.24	90.76	88.28	89.83	91.31	92.25
			91.03						
DSTR	1.08	4.06	0.81						

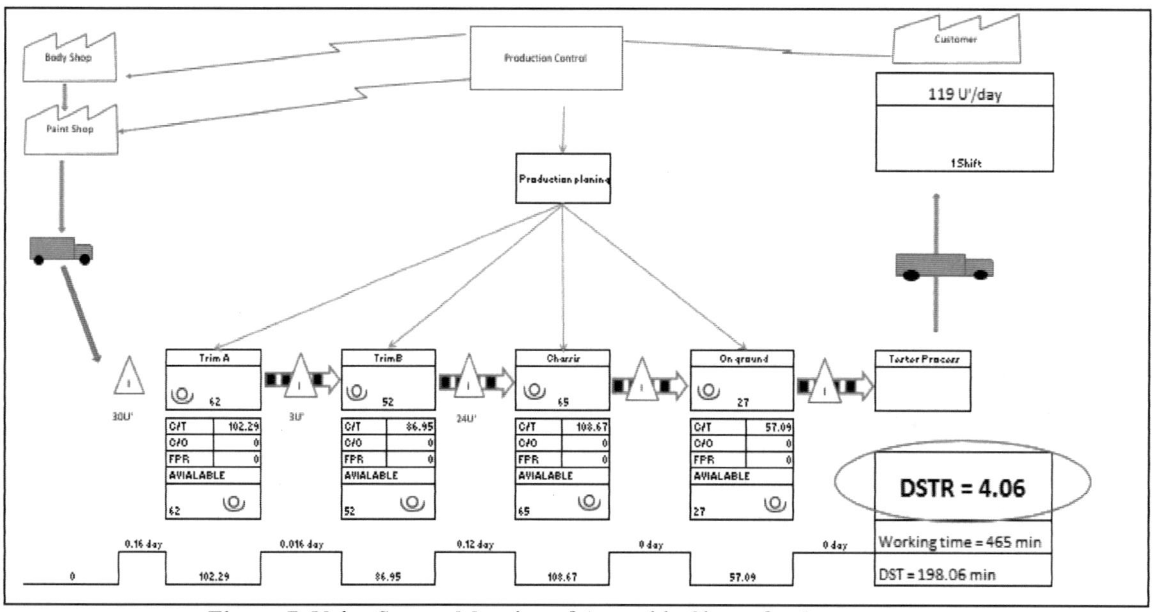

Figure 7. Value Stream Mapping of Assembly Shop after Improvement

From Figure 7, the value of DSTR is calculated as follows.

$$DSTR = \frac{Actual\ Working\ Time\ x\ \text{Man Power}}{\text{Design Standard Time}\ x\ \text{Production Volumn}}$$

$$DSRT = \frac{465\ x\ 206}{198.06\ x\ 119} = 4.06$$

5. Conclusion

This research aimed to use the Lean principle as a tool to improve the wasted process. The information is gathered from the Flow Process Chart to analyze jobs and classified them as valuable, wasted and duplicated jobs. The jobs are improved by adding the System Wrench Nut Runner and Line Stop. After the improvement, the waste from duplicated jobs between the operators and IQA is reduced. As a result, the value of DSRT is reduced significantly (i.e. 10 operators are removed from the assembly line) and the workload balance is improved.

References

[1] Feld, W.M. 2001 Lean Manufacturing : Tools, Techniques and How to them. Florida : St.Lucie.
[2] Markus L.Stamm, Thomas Noitzert. School of Engineering AUT University Auckland, New Zealand
[3] Sullivan, W.G., Mcdonold, T.N., Eileen, M. and Aken, V. 2002 Equipment Replacement Decision and Lean Manufacturing Elsevier Science Ltd.
[4] Hines, P. and Taylor, D. 2000 Going Lean. UK : Lean Enterprise Research Centre Cardiff Bussiness School.
[5] Ohno, T. 2002 Toyota Production System : Beyond Large Scale Production. Oregon : Productivity.
[6] Liker, J.K. 2004 The Toyota Way. New York : McGraw-Hill.

IOP Conf. Series: Materials Science and Engineering **215** (2017) 012010 doi:10.1088/1757-899X/215/1/012010

A Review of Theoretical Frameworks for Supply Chain Integration

AC Thoo[1], LC Tan[2], Z Sulaiman[1] and N Zakuan[1]

[1]Senior Lecturer, Faculty of Management, Universiti Teknologi Malaysia, Malaysia
[2]Senior Lecturer, Faculty of Geoinformation and Real Estate, Universiti Teknologi Malaysia, Malaysia

E-mail: tlchoon@utm.my

Abstract. In a world of fierce competition and business driven by speed to market, good quality and low costs, this environment requires firms to have a source of competitive advantage that is inimitable and non-substitutable. For a supply chain integration (SCI) strategy to achieve sustainable competitive advantage it must be non-substitutable, inimitable, path-dependent and developed over time. Also, an integrated supply chain framework is needed to tie the whole network together in order to reduce perennial supply chain challenges such as functional silos, poor transparency of knowledge and information and the inadequate formation of appropriate customer and supplier relationships. Therefore, this paper aims to evaluate the competitive impact of a SCI strategy on firm performance using the theory of Resource-based View (RBV) and relational view.

1. Introduction

The integrative philosophy of supply chain integration (SCI) involves internal, supplier and customer integration. The internal integration aims to eliminate traditional functional silos and integrate the functional departments of a company into a single entity (Flynn et al., 2010) in order to meet the requirements of customers at the lowest system-wide cost (Boon-itt and Wong, 2011). Brown (1983) suggested that inter-functional integration is based on the theory of interdependence, whereby the relations between two working units are described as individual or collection activities and behaviour of individuals or of a group (Topolsek et al., 2009). The internal integration can be optimal when the complementary cross-functional teams of a firm, including procurement, production, logistics, marketing, sales and distribution, act as a whole to coordinate the information flow, share resources and work as a team to achieve a mutual organisational goal. The internal integration can be operative or functional integration (Barki and Pinsonncault, 2005).

SCI should not just focus on a single entity and it should look at various sub-systems, activities, relationships and operations (Chandra and Kumar, 2000). SCI requires all the nodes in the network, whether inside or outside the firm, to communicate, exchange and share detailed and current information. The relational view is useful in understanding supply chain relationships with critical suppliers and customers. The arc of integration (outward facing) has helped organisations to achieve a higher level of business performance in terms of customer responsiveness, cost and time, as shown by a global study across industries (Frohlich and Westbrook, 2001).

This paper aims to evaluate the competitive impact of a SCI strategy on firm performance using the theory of Resource-based View (RBV) and relational view. The fundamental goal of the RBV is to explain why some firms outperform their counterparts, focusing on the notion that rare, values and inimitable resources lead to better business performance and higher profitability (Barney, 1991). This involves the acquiring of resources from internal and external parties, developing resources and divesting resources that are no longer useful for the strategy used by the firm (Hitt, 2011). Applied to SCI, the RBV supports the importance of managerial actions in managing the internal resources provided by different functional departments, and external resources provided by suppliers and customers to enhance firm performance. The resources must be deployed in ways that matched the strategies implemented by the firm in order to positively influence business performance.

In addition, the relationship between manufacturing firm and supply chain partners can be critical for the transfer of knowledge and other types of resources. Therefore, relational view represents another theoretical frame for the study of relationships between manufacturing firm and supply chain partners. This view focuses on the relational link between diverse functions and supply chain partners for better decision making and higher levels of coordination and collaboration (Sprague and Watson, 1979). In fact, RBV theory and structural network theory can be used to understand the acquisition of tacit and explicit knowledge from local suppliers through relational and contract mechanisms. The following sections review RBV and relational view in depth.

2. Resource-based View

The static view of the RBV posits that a firm's resource base is the antecedent to competitive advantage (Barney, 1991). Essentially, the RBV is based on the assumptions that firms have heterogeneous resources (Alchian and Demsetz, 1972; Coates and McDermott, 2002; Alvarez, 2005; Royer, 2005; DeSarbo et al., 2007) and the resources remain imperfectly mobile over time (Barney, 1991). Additionally, sustainable competitive advantage results when these four attributes of firm's resources are satisfied: valuable, rareness, non-substitutability and inimitability (Dierickx and Cool, 1989; Barney, 1991). First, the resources must be valuable to allow the firm to exploit opportunities or neutralise threats in its external environment. Second, the resources must be rare that are hard for rival firms to acquire. Third, the resources must be imperfectly imitable that other firms cannot easily develop. The firm resources that are imperfectly to imitate are created because of one reason or a combination of three reasons: (a) the ability of the firm to obtain resources is dependent upon unique historical conditions; (b) the relationship between the firm's resources and its competitive advantage is causally ambiguous; or (c) the firm's resources creating the competitive advantage is knowledge-based or socially complex. Finally, the resources are non-substitutable and do not have strategic equivalents. This means there must be no strategically equivalent valuable resources that are themselves either not rare or imitable (Barney, 1991; Hitt et al., 2010). The relationship between key attributes of firm resources and sustained competitive advantage is shown in Figure 1.

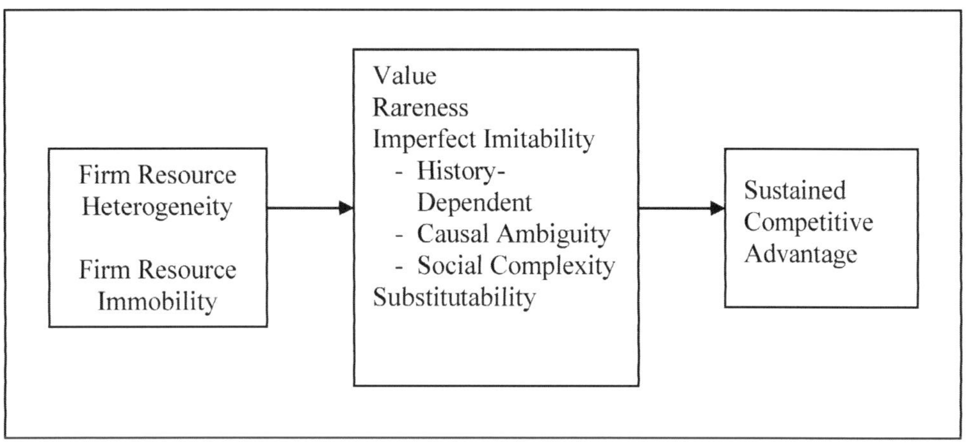

Figure 1 The relationship between firm resources and sustained competitive advantage
Source: Barney (1991, p. 112)

The firm resources can be categorised into physical capital resources, human capital resources and organisational capital resources. Physical capital resources consist of assets, technology equipment, a firm's plant, its geographic location and its access to raw materials. Human capital resources include the capabilities of managers and employees in terms of experience, judgement, intelligence, training and relationships. Also, organisational capital resources encompass a firm's formal and informal planning, coordinating systems and controlling and formal reporting structure, as well as its informal inter-organisational and intra-organisational relationships (Barney, 1991). Typically, some strategies often use a particular mix of physical capital, human capital and organisational capital resources to implement.

Consistent with the previous arguments, the empirical studies show that SCI has a positive and statistically significant effect on the firm performance in the manufacturing industry (Frohlich and Westbrook, 2001; Rosenzweig et al., 2003; Droge et al., 2004; Zailani and Rajagopal, 2005; Li et al., 2006; Swink et al., 2007; Flynn et al., 2010; Khang et al., 2010; Zolait et al., 2010; Chong et al., 2011; Salhieh, 2011; Sundram et al., 2011; Valmohammadi, 2013). Although the essence of the RBV is that firm consists of a bundled of valuable and rare resources to achieve a competitive advantage, however, it is insufficient to drive the competitive advantage. In fact, RBV has expanded to focus on how firms use value-oriented strategy to mobilise and utilise resources to maximise their competitive potential (Barney, 1991). In particular, managerial action to orchestrate resources is the key determinant to competitive advantage. The management capability and decision making of the firms are the dominant players in developing and configuring the resources in a way that enhance their sustained competitive advantage (Hitt, 2011).

The RBV can be used as a basis for the development of supply chain strategy taxonomy (McKone-Sweet and Lee, 2009). SCI requires a collaborative effort between a manufacturer and its suppliers and customers (Flynn et al., 2010). If a firm is completely vertically integrated, most of the prominent materials supplies are provided by its internal units. In fact, firms also acquire external resources from external parties such as suppliers and customers. Dyer and Singh (1998) highlighted that valuable resources are often provided by supply chain partners and argued that structuring the inter-organisational resources is more critical to achieve outstanding performance than a firm's own constrained resource base. Specifically, suppliers play an integral role in supplying essential resources to the focal firm as well as in the firm's implementation strategies. Accordingly, the external resources must be effectively managed and integrated with the internal resources of the firm in order to achieve superior performance (Hitt, 2011). The heterogeneous partner-specific resources therefore constitute the potential for a competitive advantage. Moreover, the sustainability of the competitive advantage can be achieved when the resources are immobile and difficult to imitate or appropriate (Morris et al., 2005). The RBV suggests that holding valuable, rare, inimitable and non-substitutable resources is important, however, it is more important for a firm to leverage and bundle its internal and external resources in order to create its competitive advantage.

3. Relational View

While the RBV focuses on the importance of managerial actions in managing resources flow in a supply chain in order to enhance firm performance, relational view theory can be used to understand a relationship between two or more supply chain partners. The relational view theory was first articulated by Dyer and Singh (1998) to suggest that idiosyncratic inter-organisational linkages can result in sustained competitive advantage. Four sources of relational rents are identified in the relational view: relation-specific asset investments, substantial knowledge sharing, complementary resources/capabilities and effective governance (Miguel and Brito, 2011). The relational rent is defined as "a supernormal profit jointly generated in an exchange relationship that cannot be generated by either firm in isolation and can only be created through the joint idiosyncratic contributions of the specific alliance partners" (Dyer and Singh, 1998, p. 662). The greater the alliance partners' investment is in relation-specific asset, inter-firm knowledge-sharing routines, complementary resources/capabilities and effective governance, the greater the potential will be for relational rents.

Williamson (1985) suggested that firms must have relation-specific assets that alliance partners will value and are fit for use. There are three types of relationship-specific investments: site specificity, physical asset specificity and human asset specificity. First, the firms may invest in site specificity by locating close to the alliance partners in order to reduce inventory and transportation costs. Second, investment in physical asset specificity such as customised machinery or equipment or tools allows the firm to improve its product quality and distinguish its products from others. Third, human asset specificity is the accumulation of knowledge and expertise that is specific to one alliance partner. Over time, the alliance partners work together and build up specialised information, language and know-how in order to communicate effectively and efficiently, which in turn leads to increased product quality and speed to market.

Dyer and Singh (1998) found that alliance partners can generate rents through substantial inter-organisational knowledge-sharing routines. The routines of knowledge-sharing are regular processes that create, transfer or recombine specialised knowledge between alliance partners. Also, firms can create relational rents through leveraging the benefits of complementary resource endowments of an alliance partner. The alliance partners pool their respective resources to eliminate deficiencies in each other's individual portfolios of resources, and create distinctive resource endowments that would have been difficult for either firm to achieve alone. Finally, the creation of relational rents also depends on the ability of alliance partners to align transactions with governance structures in a way that minimises transaction costs and maximises value-creation initiatives.

In general, the relational view provides insight into how a firm develops value-creating linkages with other firms to achieve high profit returns. Applied to SCI, the relational view suggests that collaborative relationships between manufacturing firm and its suppliers and customers can generate relational rents through relation-specific assets, knowledge-sharing routines, complementary resource endowments and "effective governance". These four activities create an idiosyncratic relationship that is difficult for competitors to imitate. Moreover, when the relational rents generated through effective inter-firm collaboration are causal ambiguity and time compression diseconomies (Barney, 1991). Therefore, the collaborative relationships could create differential advantage and confer supernormal rents to well-executed supply chain strategies.

The "relational capability" of supplier, manufacturer and customer, is a potential source of inter-firm competitive advantage and should be seen as the "winning" strategy (Gattorna, 2009). Hitt et al. (2006) found that relational capital in relationships between suppliers and buyers is positively related to firm performance. In this study, the researcher integrates the notions from the RBV and relational view theory to understand the relationships in strategic alliances when the both theories claim that the firm's resources creating the competitive advantage is socially complex interactions. Of course, the collaborative supply chain relationships are invariably based on trust, loyalty, a positive sum game, fairness in negotiations, goal and intent revelation, and commitment (Chandra and Kumar, 2000).

4. Conclusion

The theoretical perspectives of the RBV suggest that SCI is a non-substitutable, inimitable, path-dependent capability that a firm develops over time to achieve sustainable competitive advantage. Similarly, relational-specific capabilities can play a significant role in achieving sustained competitive advantage. As the theory of relational view becomes more fully developed, it will need to be based on the interaction between the internal functions of a firm, with valuable firm-specific resources and capabilities in place. The relational capabilities gained through supply chain members, such as suppliers, manufacturers and customers, will enable a unique combination of cross-organisational complementary resources; these can be a potential source of inter-firm competitive advantage. Lastly, this review found that for strategic SCI to be successful, practitioners must not focus on one particular inhibitor, but rather consider customer integration, supplier integration and internal functions in combination. In this regard, interdependence can be seen as an important part of the SCI.

Acknowledgments

This work was funded by Ministry of Higher Education Malaysia (MOHE) and Universiti Teknologi Malaysia GUP Tier 1 (PY/2016/06368 – Vot Number 15H61).

IOP Conf. Series: Materials Science and Engineering **215** (2017) 012010 doi:10.1088/1757-899X/215/1/012010

References

[1] Flynn, B. B., Huo, B. and Zhao, X. (2010). The Impact of Supply Chain Integration on Performance: A Contingency and Configuration Approach. Journal of Operations Management. 28: 58-71.
[2] Boon-itt, S. and Wong, C. W. (2011). The Moderating Effects of Technological and Demand Uncertainties on the Relationship between Supply Chain Integration and Customer Delivery Performance. International Journal of Physical Distribution & Logistics Management. 41(3): 253-276.
[3] Brown, D. (1983). Managing Conflict at Organizational Interfaces. New York: Addison-Wesley Publishing Company.
[4] Topolsek, D., Lipicnik, M. and Gajsek, B. (2009). The Importance of Internal Integration for a Successful External Integration of the Supply Chain. Business Logistics in Modern Management. 9: 45-54.
[5] Barki, H. and Pinsonncault, A. (2005). A Model of Organizational Integration, Implementation Effort and Performance. Organization Science. 16(2): 165-179.
[6] Chandra, K. and Kumar, S. (2000). Supply Chain Management in Theory and Practice: A Passing Fad or a Fundamental Change? Industrial Management & Data System. 100(3): 100-113.
[7] Frohlich, M. T. and Westbrook, R. (2001). Arc of Integration: An International Study of Supply Chain Strategies. Journal of Operations Management. 19: 185-200.
[8] Barney, J. B. (1991). Firm Resources and Sustained Competitive Advantage. Journal of Management. 17(1): 99-120.
[9] Hitt, M. A., Nixon, R. D., Clifford, P. G. and Coyne, K. (1999). The Development and Use of Strategic Resources. In Hitt, M. A., Clifford, P. G., Nixon, R. D. and Coyne, K. P. (Eds.) Dynamic Strategic Resources: Development, Diffusion, and Integration (pp. 1-15). New York: Wiley.
[10] Sprague, R. H., Jr. and H. J. Watson. (1979). Bit by Bit: Toward Decision Support Systems. California Management Review, XXII(1): 60-68.
[11] Alchian, A. A. and Demsetz, H. (1972). Production, Information Costs, and Economic Organization. *The American Economic Review*. 62(5): 777-795.
[12] Coates, T. T. and McDermott, C. M. (2002). An Exploratory Analysis of New Competencies: A Resource Based View Perspective. *Journal of Operations Management*. 20: 435-450.
[13] Alvarez, S. A. (2005). Resource and Hierarchies: Intersections between Entrepreneurship and Business Strategy. In Acs, Z. J. and Audretsch, D. B. (Ed.) *Handbook of Entrepreneurship Research: An Interdisciplinary Survey and Introduction* (pp. 247-266). United States of America: Springer Science+Business Media, Inc.
[14] Royer, S. (2005). Strategic Management and Online Selling: Creating Competitive Advantage with Intangible Web Goods. Oxon: Routledge.
[15] Desarbo, W. S., Benedetto, C. A. D. and Song, M. (2007). A Heterogeneous Resource Based View for Exploring Relationships between Firm Performance and Capabilities. Journal of Modelling in Management. 2(2): 103-130.
[16] Dierickx, I. and Cool, K. (1989). Asset Stock Accumulation and Sustainability of Competitive Advantage. Management Science. 35(12): 1504-1511.
[17] Rosenzweig, E. D., Roth, A. V. and Dean, Jr. J. W. (2003). The Influence of an Integration Strategy on Competitive Capabilities and Business Performance: An Exploratory Study of Consumer Products Manufacturers. Journal of Operations Management. 21: 437-456.
[18] Droge, C., Jayaram, J. and Vickery, S. K. (2004). The Effects of Internal versus External Integration Practices on Time-Based Performance and Overall Firm Performance. Journal of Operations Management. 22(6): 557-573.
[19] Zailani, S. and Rajagopal, P. (2005). Supply Chain Integration and Performance: US versus East Asian Companies. Supply Chain Management: An International Journal. 10(5): 379-393.
[20] Li, S., Ragu-Nathan, B., Ragu-Nathan, T. S. and Subba Rao, S. (2006). The Impact of Supply Chain Management Practices on Competitive Advantage and Organisational Performance. The International Journal of Management Science. Omega 34: 107-124.

[21] Swink, M., Narasimhan, R. and Wang, C. (2007). Managing Beyond the Factory Walls:Effects of Four Types of Strategic Integration on Manufacturing Plant Performance. Journal of Operations Management. 25(1): 148-164.
[22] Khang, T. S., Arumugam, V., Chong, A. Y. L. and Chan, F. T. S. (2010). Relationship between Supply Chain Management Practices and Organisational Performance: A Case Study in the Malaysian Service Industry. International Journal of Modelling in Operations Management. 1(1): 84-106.
[23] Zolait, A. H., Ibrahim, A. R., Chandran, V. G. R. and Sundram, V. P. K. (2010). Supply Chain Integration: An Empirical Study on Manufacturing Industry in Malaysia. Journal of Systems and Information Technology. 12(3): 210-221.
[24] Chong, A. Y. L., Chan, F. T. S., Ooi, K. B. and Sim, J. J. (2011). Can Malaysian Firms Improve Organizational/Innovation Performance via SCM? Industrial Management & Data Systems. 111(3): 410-431.
[25] Salhieh, L. (2011). An Exploratory Study of the Relationship between Supply Chain Management Practices and Technical Efficiency of Jordanian Manufacturing Companies. International Journal of Business and Management. 6(12): 126-134.
[26] Sundram, V. P. K., Ibrahim, A. R. and Govindaraju, V. G. R. C. (2011). Supply Chain Management Practices in Electronics Industry in Malaysia: Consequences for Supply Chain Performance. *Benchmarking: An International Journal.* 18(6): 834-855.
[27] Valmohammadi, C. (2013). Investigating Supply Chain Management Practices in Iranian Manufacturing Organizations. Operations and Supply Chain Management. 6(1): 36-42.
[28] McKone-Sweet, K. and Lee, Y-T. (2009). Development and Analysis of a Supply Chain Strategy Taxonomy. Journal of Supply Chain Management. 45(3): 3-24.
[29] Dyer, J. H. and Singh, H. (1998). The Relational View: Cooperative Strategy and Sources of Interorganizational Competitive Advantage. Academy of Management Review. 23: 660-679.
[30] Morris, S. S., Snell, S. A. and Wright, P. M. (2005). A Resource-based View of International Human Resources: Toward a Framework of Integrative and Creative Capabilities. Working paper,. Ithaca, NY: Cornell University, School of Industrial and Labor Relations, Center for Advanced Human Resource Studies.
[31] Miguel, P. L. S. and Brito, L. A. L. (2011). Supply Chain Management Measurement and its Influence on Operational Performance. Journal of Operations and Supply Chain Management. 4(2): 56-70.
[32] Williamson, O. E. (1985). The Economic Institutions of Capitalism. Free Press: New York, NY.
[33] Gattorna, J. (2009). Dynamic Supply Chain Alignment: A New Business Model for Peak Performance in Enterprise Supply Chains Across All Geographies. Great Britain: MPG Books Group, UK.

IOP Conf. Series: Materials Science and Engineering **215** (2017) 012011 doi:10.1088/1757-899X/215/1/012011

DESIGN PROCESS IMPROVEMENT FOR ELECTRIC CAR HARNESS

Thiwarat Sawatdee[1] and Parames Chutima[1,2*]

[1]Department of Industrial Engineering, Faculty of Engineering Chulalongkorn University, Bangkok, Thailand 10330

[2]Regional Centre for Manufacturing Systems Engineering, Faculty of Engineering Chulalongkorn University, Bangkok, Thailand 10330

[1]Thiwarat.Sa@student.chula.ac.th, [2]cparames@chula.ac.th

Abstract. In an automobile parts design company, the customer satisfaction is one of the most important factors for product design. Therefore, the company employs all means to focus its product design process based on the various requirements of customers resulting in high number of design changes. The objective of this research is to improve the design process of the electric car harness that effects the production scheduling by using Fault Tree Analysis (FTA) and Failure Mode and Effect Analysis (FMEA) as the main tools. FTA is employed for root cause analysis and FMEA is used to ranking a High Risk Priority Number (RPN) which is shows the priority of factors in the electric car harness that have high impact to the design of the electric car harness. After the implementation, the improvements are realized significantly since the number of design change is reduced from 0.26% to 0.08%.

1. INTRODUCTION

Electric car harness is an equipment that connects to the automobile central nervous system. It consists of various wires bundled together, with which each individual wire is responsible for carrying the electric signals and energy used in the vehicles such as engine, lights, meters, navigation systems, power windows, doors, etc. This research is focuses on the problems occurred in the electric car harness design company in Thailand which is the designer of the electric car harness for Japanese and European passenger cars. The product named "electric car harness model 17M type T1" as shown in Fig. 1 is selected in this study since its designs are changed most frequent. At present, the company faces downward trend of the market share. Two key reasons come from customers are not satisfied with the products (measured by the customer satisfaction survey) and the design of the products is not match with customer needs. These two reasons are the key factors that lead to high number of design changes [1]. Currently, the percentage of design changes per year is 0.26% which is higher than the 0.10% target set by the company. As a part history, the company has to improve the design process and eliminate the root causes to reduce the number of design changes so as to meet the target set by the company.

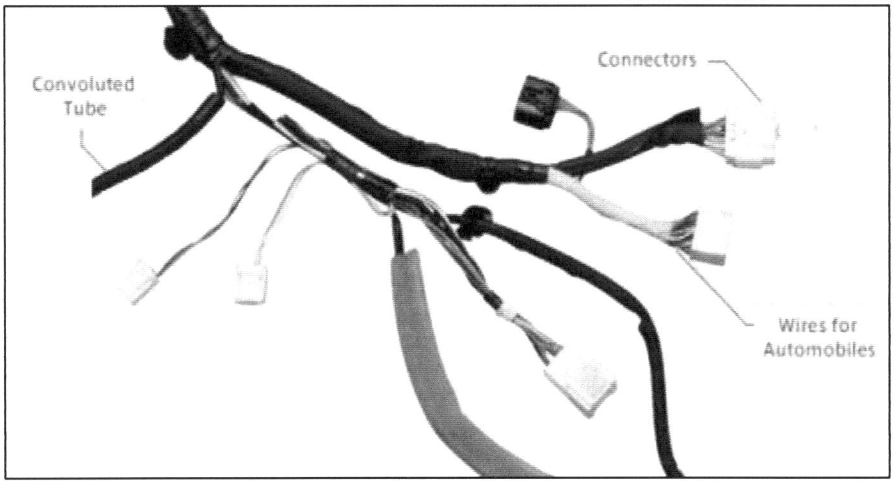

Figure 1. An example of the electric car harness designed by the company
(Source: http://www.sws.co.jp/en/product/wireharness)

The historical design defect data from customer satisfaction survey reported between April 2015 to March 2016 illustrated by the Pareto chart (Fig. 2) shows the major proportions of defects that effects to the high number of design change, including wrong wire size 15.29%, wrong electrical connection 10.59%, wrong type of wire and broken conductor 9.41%, protection part mistake 8.24%, protection part broken 7.06% and vibrating surface testing 7.06%. These problems are subjected to the improvement. The appropriate tools that are employed in this research are FTA (Fault Tree Analysis) [2] and FMEA (Failure Mode and Effect Analysis) [3] because they are the tools that can be easily used and cautiously analysed the problems.

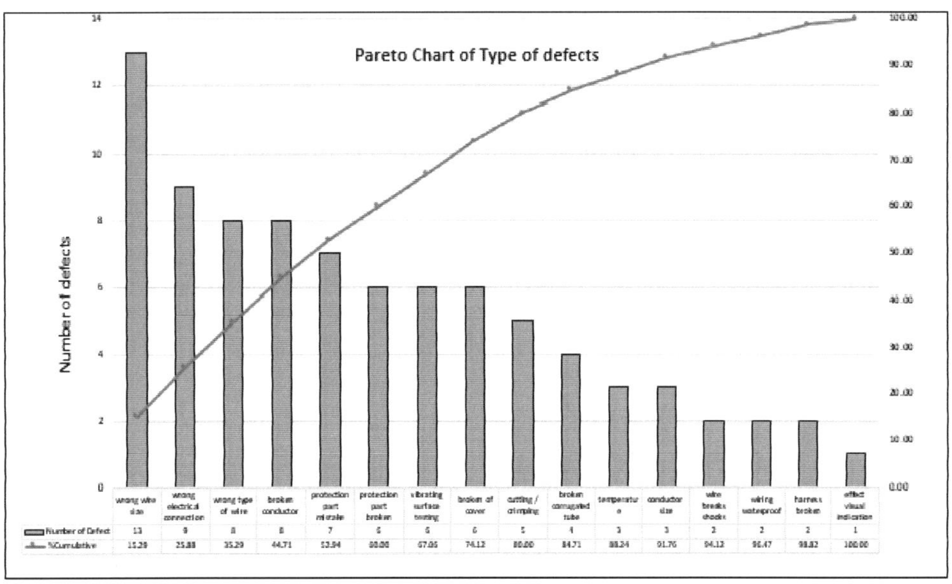

Figure 2. Pareto chart of defects

Upon reviewing literature, FTA and FMEA techniques are two approaches that could be used to discover and prioritize root causes of failures. The analysis of the causes and modes of failures of the collector electro motors, used for cooling systems of motor vehicles, is presented in the paper. Vehicle cooling system is one of most important system for internal combustion engine security and safety. It provides that engine's working temperature is in permit limits and without breakdown. Based on a detailed review of the structure and operation modes of the observed object and other relevant data, a fault tree for collector electromotor is formed. Thus, a logical relation between the peak event and the basic initiating events from the fault tree is established. In conclusion, the paper presents possible applications of the achieved results [4].

Now a days Reliability of any mechanical system is the most important factor of the product design, so the need for reliability estimation & prediction of critical modes of failure for mechanical system became the talk of the town. Lead acid battery which has been in use for different applications for over 13 decades. Their areas of application have transcended the traditional areas of automotive vehicle and have spread to newly developed area such as in traction of hybrid-electric vehicles, in un-interruptible power supply units and in telecommunication system for standby duties. The objective of this paper is to present the failure mode, its severity and effects on the service life of Lead Acid Battery. The Risk Priority Number (RPN) of the automobile battery based upon the failure data from a battery. This is very useful to understand Battery faults & maintenance [5].

2. IMPROVEMENT METHOD

To reduce the number of design changes, FTA and FMEA are applied for this research study. The purpose of applying FTA is to analyze in details for root-cause analysis and developing appropriate measures to prevent the recurrent of the problems. FTA uses a graphical representation to show the information about the faults that could be occurred in terms of logical probability. It enables to find causes-and-consequences connections of the breakdown elements. With the help of FTA, it is possible to analyze reliability for improving the defects related to design changes. The FTA of the aluminium electric car harness model 17M type T1 is shown in Fig. 3 which illustrates the causes of design changes found in the trial prototype of the car assembly process. From the causes of faults, we can determine the appropriate approach to improve the design of the products.

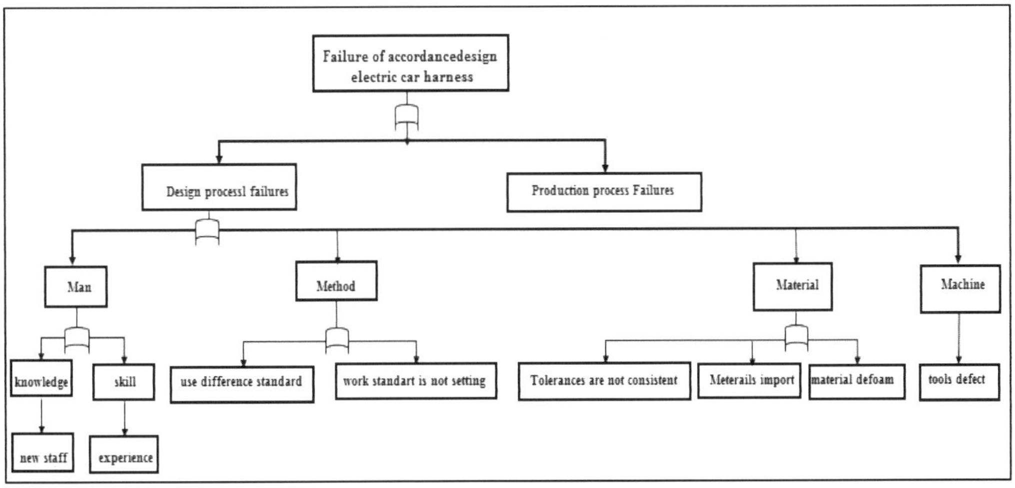

Figure 3. FTA of the design changes in the electric car harness

FMEA is a proactive analysis tool, allowing engineers to define, identify and eliminate known and/or potential failures, problems, errors and so on from the design process. Furthermore, FMEA is an inductive approach to support risk assessment studies and correction of defects. The risk priority number (RPN) is a mathematical product of the seriousness of a group of effects (severity), the likelihood that a cause will create the failure associated with those effects (occurrence), and an ability to detect the failure before it gets to the customer (detection). In the equation form, RPN = S*O*D, where S = Severity; O = Occurrence; and D = Detection. Severity is an assessment of the seriousness of the effect of the potential failure mode after it has occurred. The severity of the failure should be predicted and recorded on a 1 to 10 scale, which is called severity rating. Since the severity rating is based solely on the effect, and not the cause of the failure, it always remains the same, regardless of its possible causes. The "occurrence of failure" is related to the risk evaluation. In other words, occurrence is the probability that the failure mode will occur as a result of a specific cause. The estimate of the probability of the failure occurrence must be recorded by using all available knowledge and it would be ranked on a 1 to 10 scale. Detection is the probability that a potential failure will be detected before it reaches the customer. The detection rate should also be estimated based on a scale of 1 to 10. The meaning of different ranking scales used in FMEA is shown in Table 1 [6]. The working team of the company including head engineer, engineers and project leader discusses the rating scales of failure modes. With the approval of the section manager, the RPN with scoring higher 100 must be improved. The results of the FMEA meeting is shown in Table 2. The recommendations for corrective actions to improve the design process of the electric car harness are also discussed in the meetings.

Table 1. Severity Ranking, Occurrence Ranking & Detection Ranking

Rank	Severity Ranking		Occurrence Ranking		Occurrence Ranking	
	Effect	**Criteria**	**Effect**	**Criteria**	**Effect**	**Criteria**
1	No	No effect	Almost never	Failures unlikely	Almost Certain	Proven detection methods available in concept stage
2	Very Slight	Customer Not annoyed	Remote	Very few Nos. of failures likely	Very High	Proven computer analysis available in early design stage
3	Slight	Customer slightly annoyed	Very Slight	Few Nos of failures likely	High	Simulation and/or modelling in early stage
4	Minor	Customer experiences minor nuisance	Slight	Occasional High Nos of failures likely	Moderately High	Test on early prototype system element
5	Moderate	Customer experiences	Low	Medium Nos of	Medium	Test on preproduction
6	Significant	Customer experiences discomfort	Medium	Moderately High Nos of failures likely	Low	Test on similar system components
7	Major	Customer dissatisfied	Moderately High	Moderately High Nos of failures likely	Slight	Test on product with prototypes and system components installed
8	Extreme	Customer very dissatisfied	High	High Nos of failures likely	Very Slight	Proving durability test on Test on product with system components installed
9	Serious	Potential hazardous effect	Very High	Very high Nos of failures likely	Remote	Only unproven or unreliable techniques available
10	Hazardous	Hazardous effect	Almost Certain	Failure almost certain	Almost impossible	No known techniques available.

Table 2. FMEA of design process for electric car harness.

Failure Modes	Effects	Causes	S	O	P	RPN	Recommendations
Use difference standard	Method	broken conductor	8	6	5	240	Use documented design standard for work
Work standard is not setting	Method	wrong type of wire	4	9	7	252	Use documented design standard for work
Tolerances are not consistent	Method	protection part broken	5	6	7	210	Use documented design standard for work
Materials import	Material	protection part mistake	8	4	7	224	Requiring the use of raw materials in a system control plan
Knowledge	Man	wrong electrical connection	8	8	4	256	Use form of quality control to check by work flow
Skill	Man	wrong wire size	8	8	5	320	Focusing on training of employees to comply with design standard
Tools defect	Machine	vibrating surface testing	5	6	5	180	Check the computer program daily and control by annual maintenance plans

3. RESULT AND DISCUSSION

The working team use FTA and FMEA are applied to reduce the number of design changes of electric car harness model 17M type T1. The results show that is very helpful to find the significant factors and analysis of the failures occurred in the design process of the electric car harness. The results of FTA indicate that the "defects" such as Man, Method, Machine and Material are the most critical components [7]. The RPN determined by FMEA is used to prioritize the failure modes that are urgently needed for corrective and preventive actions [8, 9]. In this case study, those failure modes with RPNs more than 100 are subjected to take actions immediately. The improvements could be performed by developing a new design working standard, control the procurement process of the raw materials from abroad, quality control of the design process to comply with the standard work flow, provide appropriate training to the designers, check the computer program daily and perform annual maintenance of the computers. After the improvement, the design change data are collected for 4 months. It is found that the design changes are reduced substantially from 0.26% to 0.08 % (Fig. 4 and 5). In addition, the designers understands the importance of controlling these processes to ensure that the performance is maintained or further improved in the future.

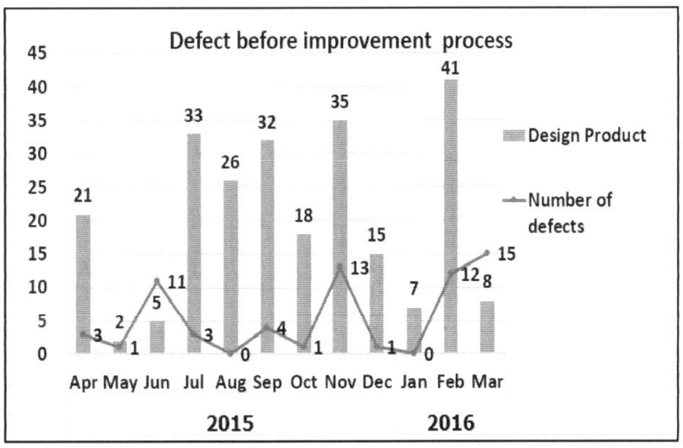

Figure 4. The number of design changes before the improvement

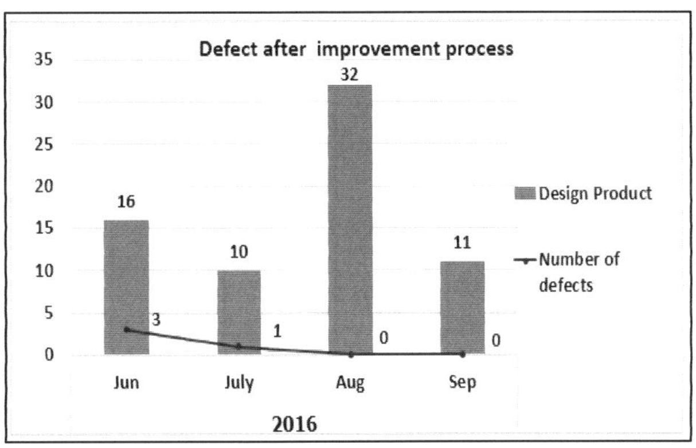

Figure 5. The number of design changes after the improvement

IOP Conf. Series: Materials Science and Engineering **215** (2017) 012011 doi:10.1088/1757-899X/215/1/012011

4. CONCLUSION

The objective of this research is to apply FTA and FMEA to reduce the number of design change occurred in the design process of the electric car harness. It is found that FTA and FMEA are effective tools to identify the significant factors and prioritize the root causes of the failure modes, as well as being a useful media to facilitate the recommendations for the improvements. The results indicate that the number of design changes is reduced from 0.28% to 0.08% which is lower than 0.10% target set by the company. In addition, this results could be assessed in terms of the market share improvement of 5% increasing.

References

[1] J.W. Priest, 1988, "Engineering design for productibility and reliability", New York: Marcel Dekker Inc.

[2] M. Vujosevic, 1983, " Tree fault analyze; view of basic concept and technique", Tehnika 38, 11, 1546-1555.

[3] D.H. Stamatis, 1995, "Failure mode and effect analysis FMEA from theory to Execution", ASQC Quality Press.

[4] G. Ivanovic, 1991, "Tree fault analyze - basic and use in vehicle projecting", Masinstvo 40, 5-6, 373-381.

[5] E.S. Kumar and B. Sarkar, 2012, "Improvement of Life Time and Reliability of Battery", International Journal of Engineering Sceince & Advanced Technology, 2-5, 1210-1217.

[6] E. Povolotskaya and P. Maach, 2012, "FMEA and FTA Analyses of the Adhesive Joining Process using Electrically Conductive Adhesives", Acta Polytechnica, 52.

[7] J.P. Bentley, 1993, "An introdution to reliability and quality engineering" , New York: John Wiley & Son Inc.

[8] E. J. Henley and H. Kumamoto, 1992, "Probablilistic risk assesment", IEEE Press, USA.

[9] L.S. Lipol and J. Haq, "Risk analysis method: FMEA/FMECA in the organizations", International Journal of Basic & Applied Sciences, 11, 5.

Real Option in Capital Budgeting for SMEs: *Insight from Steel Company*

F M Muharam[1,2] and M A Tarrazon[2]

[1]Faculty of Management, Universiti Teknologi Malaysia, Skudai, Johor, Malaysia
[2]Departamiento de Empresa y Empresariales, Universitat Autonoma de Barcelona, Barcelona,Spain

Email: merlinda@utm.my

Abstract. Complex components of investment projects can only be analysed accurately if flexibility and comprehensive consideration of uncertainty are incorporated into valuation. Discounted cash flow (DCF) analysis has failed to cope with strategic future alternatives that affect the right value of investment projects. Real option valuation (ROV) proves to be the right tool for this purpose since it enables to calculate the enlarged or strategic Net Present Value (*ENPV*). This study attempts to provide an insight of the usage of ROV in capital budgeting and investment decision-making processes of SMEs. Exploring into the first stage processing of steel industry, analysis of alternatives to cancel, to expand, to defer or to abandon is performed. Completed with multiple options interaction and a sensitivity analysis, our findings prove that the application of ROV is beneficial for complex investment projects independently from the size of the company and particularly suitable in scenarios with scarce resources. The application of Real Option Valuation (ROV) is plausible and beneficial for SMEs to be incorporated in the strategic decision making process.

Keywords: Real Option Valuation, Capital Budgeting, Flexibility, Strategic Decision Making and SMEs

1. Introduction

Option theory in real investment analysis, real option valuation (ROV) has received great attention in natural resources related projects, land and real estate development, electric sector, construction of large-scale infrastructures and energy-generating plants, pharmaceutical and biotechnological industries, research and development strategies, start-up ventures, Internet companies, capital intensive industries (such as airlines and railroads) and other industries subject to volatile demand (consumer electronics, toys, machine parts) or supply (oil and electric power facilities, chemicals, crop switching), among others, due to its ability to value flexibility[1-4]. The approach allows managers to price different alternatives throughout the project's life before formulating their path of strategy in investment activities.

Real option valuation (ROV) extends the valuation of the Net Present Value (*NPV*) based on discounted cash flows (DCF) to the Enlarged (or Strategic) Net Present Value (*ENPV*). Although many decisions in capital budgeting are dominated by the *NPV*, it is obvious that this method is insufficient to capture the whole complex nature of investment projects[5], let alone to provide tools to value managerial flexibility. Real option valuation takes care of these flaws by adding into the traditional *NPV* the value of flexibility calculated as the value of one or several real options.

$$\text{ENPV} = [\text{NPV} + \text{Real Option(s)}]$$

As managerial flexibility is important and complex, several issues are being highlighted especially in the valuation methodology. Pindyck[6] emphasizes the issues of irreversibility and uncertainty into investment valuation while McDonald and Siege[7] incorporate microeconomic theory of valuation.

When real option valuation has been applied to price strategic components of complex projects, these were specifically large-scale investments undertaken by multi-national enterprises (MNEs). The same approach then is simulated in the first stage processing of steel industry, an activity undertaken by SMEs worldwide, a sector heavily connected with natural resources. The unique characteristic of this sector is that both input and output prices follow random walks, which implies that firms do not have (total) control over them. Furthermore, more firms are managing their operations in small scale according to mini-mill systems. More and more small firms are participating in the supply chain networks, hence creating a significant percentage growth of the whole SMEs involvement in the industry.

Therefore, besides an attempt to apply real option valuation into a specific sector of industry that has not attracted the interest of researchers (and practitioners) up to now, this study is also tackling ROV for small scale sector, might it be a business unit, subsidiary or SME. The case study approach of the research applies a valuation methodology that does not differ from ROV applied in the analysis of large-scale investments when discrete-time treatment is used. The objective is to show that real option analysis is also feasible for SMEs managers. Thus the convenience to apply ROV does not depend on the size of the company undertaking the investment project but on the degree of complexity of the project itself by answering three basic questions. Firstly, in the evaluation of projects undertaken by SMEs, how can the high degree of uncertainty about input and output prices as well as reserves quantity be incorporated appropriately? Secondly, why does *NPV* fail to cope with strategic decision-making valuation of small-scale investment projects in the steel industry? Finally, how do real option valuation (ROV) and the *enlarged NPV* (*ENPV*) contribute to access the optimal policy approach to continue, expand, defer or abandon natural resource investments.

The case used in this study is on a mini-mill iron smelting project, potentially being carried out by small businesses in the steel industry, has been chosen as a subject of study.

2. Literature review: real option and capital budgeting

The discounted cash flow (DCF) method that leads to the calculation of the traditional *NPV* is the approach that dominates capital budgeting analysis [5]. However, due to the inability of DCF to capture the whole nature of complex projects, often these calculations lead to undervaluation of investments in different economic sectors. New proposals to deal with these problems started to appear in the 1980s, once option pricing of financial assets had been consolidated during the previous decade.

The key year for the inception of option theory is 1973, with the publication of the Black-Scholes model to value *European options* and the Merton model to price *American options*. Afterwards, more researchers enriched option theory. The introduction of fundamental economic principles of option pricing by arbitrage methods and simple numerical procedures for valuing options have been made by Cox and Ross[8] and Cox, Ross and Rubenstein[9]. The design of more complex option contracts, known as non-standard options or *exotic options*, enlarged the scope. Some exotic options such as i) *compound options* – which are options on options [10] ii) *Asian options* – where the payoff depends on the average price of the underlying asset during at least some part of the life of the option [11] or iii) options to exchange one asset for another, also referred to as *exchange options*[12], do not only play an important role in financial option strategies but have also proved to be extremely useful in real option valuation (ROV).

ROV incorporates the financial assets valuation approach into the evaluation of investments in capital budgeting analysis. As stated in equation (1), the enlarged *NPV* (*ENPV*) – calculated as the sum of the traditional *NPV* discounted at the risk-adjusted rate of return plus the value of the real option(s) involved in the project's nature – becomes the right measure to be taken into account by managers.

Besides being applied in investments related to natural resources, many other economic fields also profit from ROV methodology. Just to quote some contributions, among many others: Kellogg and Charnes[13] in valuing a biotechnology company, Schwartz and Moon[14] pricing an Internet company, Grenadier and Weiss[15] valuing investments in technological innovations, Alonso, Azofra and de la Fuente[16] studying an electric company, and Willigers and Hansen[17] analysing a pharmaceutical industry and still expanding.

All the above suggest the possibility of applying real option analysis in the first stage processing sector of the steel industry. However, few studies have been conducted in this economic field. Cortazar, Schwartz and Salinas[18] have conducted a very close case study of options available for steel industry. Their study in a smelter plant identified options to expand, to close and open a plant based on maximum capacity, to temporary closing as well as to contract. In order to identify the possible suitable options available for this particular sector, it is important to first analyse the nature and characteristics of a business before incorporating real option valuation, which has been the base of the study as illustrated in detail under the introduction chapter as Research Setting A.

3. Methodology

The research conducted in this study follows a stylized fact case study approach as exploratory research following Cooper and Slagmulder [19]. The approach is similar to other studies in the field of real options, such as Brennan and Schwartz [20-21] Dixit and Pindyck[3] and Trigeorgis[4], in the case of natural resource activities. This approach requires of the construction of a base case with various sources of information representative in a worldwide scenario.

The project chosen for the evaluation is an investment in first stage steel processing, a mini-mill smelting project as per **Research Setting A** in Chapter 1. To sum up, there is a new proposal of building up an iron smelting plan based on new process innovation, mini-mill iron smelting. The investment requires €10 million, €6 million in t_0 and €4 in t_1. By investing this amount, the firm will have a mini-mill plant with capacity of producing 182000 tons per year. However, due to Kyoto Protocol, the plant is allowed to produce only up to 75% of its capacity in order to maintain emission and effluent at minimum level.

The investment has an expected useful life of 10 years. 2 years are dedicated for construction and the rest 8 years are operational. Volatility is forecasted to be σ = 30 %. Two discount rates are employed which are 5% of risk-free rates and 12 % of adjusted-risk rate. Holding to this information, discounted investment cost is I, is €9.81 million while discounted net cash flows, V is €9.25 million resulting in NPV of €-0.56 million (*negative NPV*).

Analyses conducted have added 4 individual options to the case. Instead of investing €9.81 million and getting return of NPV € -0.56 million (*negative NPV*), the investment is now embedded with option to defer, option to cancel during construction, option to expand and option to abandon. The scenario is illustrated as per Figure 1.

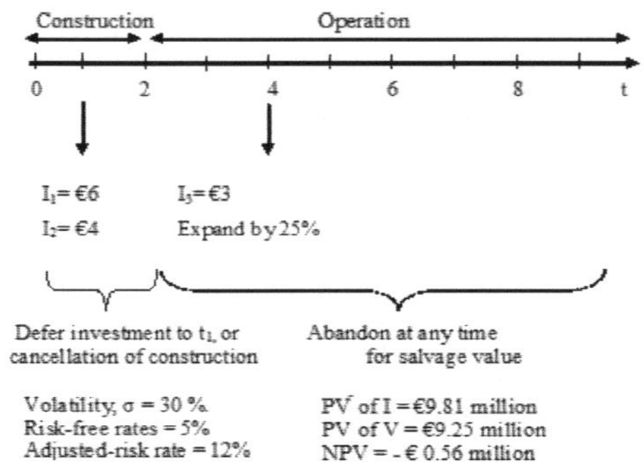

Figure 1: Investment Project with Expected Option of Deferring Investment, Cancellation during Construction, Expansion and Abandonment. (Source: Own design)

The analysis and calculation done follows the log-transformed binomial lattice approach developed by Trigeorgis[22]. Calculation is performed by software *DerivaGem.*

4. Analysis and Results

4.1. Individual Options

The real options involved in our case study are built up based on the characteristics and nature of a standard mini-mill iron smelting plant. The four options are developed are as follows:

Option to defer[23-25], is valued as an American call, where the strike price is the required outlay if the investment is made in the future. This option allows project initiation to be delayed to the next year, t_1. The projected net cash flows will remain static but the cost of investment will increase by 5%. Real option value of this option is €1.485 million, representing about 16% of the gross project value. The *ENPV* of the project is €0.925 million.

Option to cancel during construction, which is valued as a compound option of a call on a put [26]. In the case where the commodity price is unfavourable, construction can be cancelled at any time without any penalty. The firm might earn any invested amount, being discounted at the adjusted-risk rate of 12%. Once cancelled, the project cannot be deferred, expanded or abandoned (i.e., the other three options considered). The amount disinvested is taken as the strike price since it represents the saving due to discontinuation of the project. Real option value of this option is €0.17 million, representing nearly 2% of the gross project value. The *ENPV* of the project is €-0.39 million (negative *ENPV*).

Option to expand[20-21,7,27-28] is valued as a European call. This is the option to make further investment and increase production output if future conditions are favourable. The strike price of the call option is the cost of creating the additional capacity discounted over to the time of exercising it. Two years after operation, i.e. at t_4, the firm has the opportunity to increase capacity by 25% without being penalized for polluting the environment according to the Kyoto protocol. Real option value of this option is €3.22 million, representing about 35% of the gross project value. The *ENPV* of the project is €2.66 million.

Option to abandon is valued as an American dividend paying put [28].This is an option to sell down or close a project where the put value is on the project´s value. Meanwhile, the strike value is the liquidation (or resale) value of the project after deducting all closing down costs. At any time during operational years, the project can be abandoned for alternative use and enjoy a salvage value of, in principle, 50% of accumulated capital outlays net of 10% average annual depreciation, being discounted at adjusted-risk rate of 12%. Real option value of this option is €1.17 million, representing about 13% of the gross project value. The *ENPV* of the project is €0.61 million.

4.2. Multiple Interacting Options

Besides the four individual options above, there is possibility of two or more options being combined together. Therefore, it is important to analyse the sequence of options in order to identify any possibility of combining several real options embedded in the nature of an investment project. Smit [29], for example, has identified the stages in a petroleum offshore concession project before valuing it.

In our case study, four possible options have been identified for the purpose of strategic investment valuation. Based on this base case, first, the firm has an option whether to invest now or defer investment. By choosing an option to invest, it opens up to more options. After investment is undertaken, the firm may choose to cancel it during construction. However, by exercising this option, it closes the opportunity of incorporating other options later. If construction proceeds, investment has the options to expand and/or abandon in later time. Figure 2 shows possible option combinations for the investment according to the sequence proposed.

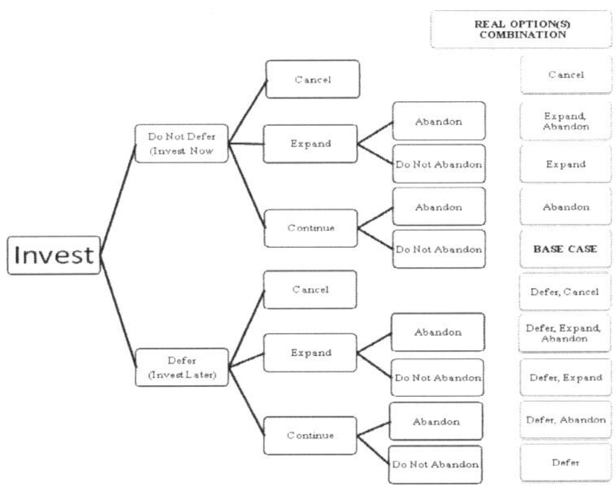

Figure 2: Combination of Options Embedded in the Investment According to the Proposed Sequence - Viewed in a Tree Diagram.

To comment briefly, the option to expand and abandon is valued at €1.81 million, resulting in *ENPV* of € 1.25 million. The options combined together are negatively interacting with individual options to expand and to abandon by €4.14 million.

The next combined option is to defer and cancel, valued at €0.23 million, resulting in an *ENPV* of €-0.33 million (negative *ENPV*). The options combined together are negatively interacting with individual options to defer and to cancel by €-1.42 million.

The third combination joins together the most possible options available for the project, which are the options to defer, to expand and to abandon. This option is valued at €2.37 million, resulting in an *ENPV* of €1.81 million. The interaction among individual options is negative €-4.14 million.

The fourth option combination is to defer and to expand, valued at €2.12 million, resulting in *ENPV* of € 1.56 million. The options combined together are negatively interacting with individual options to defer and to expand by €-2.58 million.

Finally, is the option combination to defer and to abandon, valued at €1.63 million, resulting in an *ENPV* of € 1.07 million. The options combined together are negatively interacting with individual options to defer and to expand by €-1.58 million. Table 1 summarizes the results of valuation for both individual options and multiple interacting options.

Table 1: Values of Options, ENPV and Interactions in € millions.

Item	Value of Option	*ENPV*	Interaction [a]
1 option			
Defer	1.48	0.92	
Cancel	0.17	-0.39	Not
	3.22	2.66	Applicable
Expand	1.17	0.61	
Abandon			
2			
options	0.23	-0.33	-1.42
D&C	2.12	1.56	-2.58
D&E	1.63	1.07	-1.58
D&A	1.81	1.25	-2.58
E&A			
3			
options	2.37	1.81	-4.14
D,E&A			

[a] The value of the combined options minus the sum of separate values

5. Discussions

5.1. Impact of ROV into Project Valuation

Referring to Table 1, on individual options, cancellation during construction option illustrates an example where adding up the value of real option does not necessarily make an investment attractive. In this case, if the investment is cancelled during construction once is has been undertaken, the *ENPV* of the project ends up being negative -€0.39 million. Yet, having an option to cancel reduces the loss.

For the rest of options, individually embedded in the project, the value of *ENPV* is positive. The most significant difference between *NPV* and *ENPV* appears when an expansion plan is considered. The expansion option gives the highest option value available for this project, which is €3.22 million and results in an *ENPV* of €2.66 million.

Combinations of multiple options may result in additive or non-additive value [31]. Based on the case study, it is found out that the interaction among multiple options is all non-additive. The biggest interaction as stated in Table 1 is noted when deferral, expansion and abandonment options are considered simultaneously causing a negative effect of €-4.14 million. It is identified that the cause of these negative interaction among all groups of multiple options is due to the different types of option, the period of exercising the options and the types of option being exercised earlier.

5.2. Sensitivity Analysis

Sensitivity analysis is a powerful complementary management tool in strategic decision-making. In this subsection we study the effect of changes in volatility and risk-free interest rate towards each option. Volatility is based on the average annual standard deviation of price fluctuation of outputs that are being traded in terminal commodity exchanges. Volatility being 30% in the base case is assumed in the sensitivity analysis to fluctuate by +/- 5% to 25% and 35%.

The second parameter, the risk-free interest rate, is based on the future prevailing yield on two- to four-year Treasury bonds. In the base case, the risk-free interest rate is 5%. Changes analysed consider the risk-free interest rate sinking to 3% (2.95588% continuously compounded) or rising to 7% (6.765865% continuously compounded).

For individual options, higher volatility results in higher option values while effects of changes in the risk-free interest rate depend on the type of option. Increases in the risk-free interest rate result in higher values of deferral and expansion options, denoting that both are call options. On the contrary, option values fall in cancellation during construction and abandonment reflecting the put option nature of these two real options.

For multiple options, the sensitivity analysis considered must reflect all the factors interacting during valuation. These factors depend on the types of options and the degree of overlapping of exercise regions. It has been identified that the cause of these negative interactions among all groups of multiple options is due to the different types of option, the period of exercising the options and the types of option being exercise earlier which are consistent with finding of Trigeorgis[30].

The whole sensitivity analysis is performed with other variables held constant, and only considering only changes in volatility and the risk-free interest rate. The result is presented in Table 2. From Table 2, it is found that the movement of option values depends on the final nature of options being exercised. For example, in deferral and cancellation, it can be seen that the values are increasing with increment of volatility rate but decrease when risk-free interest rate falls.

Table 2: Sensitivity Analysis on Individual Strategic Options to Reflect Changes of Volatility and the Risk-free Interest Rate. (€ in millions)

Risk.-Free Interest Rate	Volatility 25%	Volatility 30%	Volatility 35%
Cancellation			
3%	0.13	0.21	0.29
5%	0.10	0.17	0.25
7%	0.08	0.14	0.21
Deferral			
3%	1.11	1.35	1.59
5%	1.24	1.48	1.72
7%	1.38	1.62	1.86
Expansion			
3%	3.05	3.08	3.11
5%	3.20	3.22	3.25
7%	3.34	3.36	3.38
Abandonment			
3%	1.18	1.41	1.64
5%	0.98	1.17	1.39
7%	0.86	1.04	1.21
Deferral and Cancellation			
3%	0.19	0.30	0.40
5%	0.14	0.23	0.33
7%	1.10	0.18	0.27
Deferral and Expansion			
3%	1.91	1.98	2.05
5%	2.07	2.12	2.18
7%	2.31	2.26	2.31
Deferral and Abandonment			
3%	1.64	1.81	2.00
5%	1.46	1.63	1.80
7%	1.43	1.47	1.64
Expansion and Abandonment			
3%	1.79	2.05	2.33
5%	1.58	1.81	2.03
7%	1.39	1.63	1.85
Deferral, Expansion and Abandonment			
3%	2.41	2.62	2.81
5%	2.29	2.37	2.58
7%	2.29	2.29	2.35

It should be noted that the changes of option value towards volatility and the risk-free rate indicate its minimum value regardless of any changes in volatility or the risk-free interest rate. In the combination of deferral, expansion and abandonment options, when the risk-free interest rate is at 7%, any volatility below 32% will not affect the option value. The option value is affected when volatility grows above 32%. When risk-free rate is 5%, option value starts to be constant at €2.29 million as volatility falls below 28%. Meanwhile, the option value started to be constant at €2.29 million as risk-free rate is 3% once the volatility falls below 23%. Figure 3 illustrates this condition, where the asymptotic behaviour of the real option value functions becomes evident.

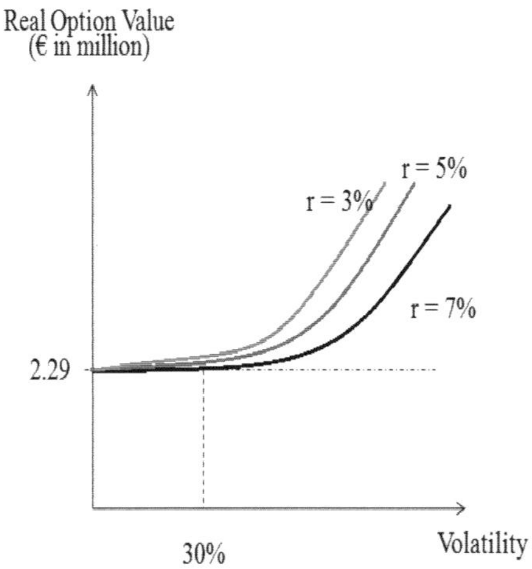

Figure 3: Effect on Option Value to Defer-Expand-Abandon When Changes in the Risk-free Interest Rate and Volatility Occur.

6. Conclusion

The complexity of real option valuation may intimidate managers, in particular those of SMEs, and deter them from applying it. This complexity may give the impression that this method is suitable only for large firms with complex activity. Steel industry also bears the same complexity compared to other economic fields, such as mining [18, 20-21], oil and gas concession [29], Internet companies [14] or pharmaceutical research and development activities [31], just to quote some examples.

Bearing the advantage for SMEs that activities can be staged out and carried by different smaller entities, strategic investment decision making becomes crucial in the steel industry under various circumstances. However, it is more crucial during recession because resources are scarce for all companies, particularly for SMEs especially in their attempts to have access to new financing funds. The attractiveness of steel industry arouses the interest of additional SMEs [32] to invest, but such investment requires careful and meticulous analysis before decision is made, as SMEs do not have the same capacity to absorb losses as large firms. Furthermore, a collapse of one member in the supply chain will most probably trigger the fall of other as this industry is capital intensive and industry specific [33]. Therefore, measures have to be taken to prevent as much as possible wrong investment decisions. In other words, the complex characteristics of the production activity in the first stage processing of the steel industry, mainly undertaken by SMEs worldwide, offers the opportunity to prove how convenient (if not necessary) ROV happens to be for the investment analysis in this industry.

Embedding options into the enlarged *NPV (ENPV)* through real option valuation (ROV) contributes interestingly to strategic management of any company and, as shown in this chapter through a case study analysis in the steel industry, of SMEs too. Uncertainty and flexibility of steel industry have to be incorporated into investment decision-making, a goal that this methodology achieves. By being capable of reflecting the complex nature of steel industry, ROV becomes a highly valuable tool to support managers in their strategic investment decisions. Contribution of ROV does not only reflect the market value of the project comprehensively. It also helps managers of small businesses in steel industry to decide upon the convenience of undertaking an investment, deferring it, altering the production conditions or abandon the business activity, i.e. a wide array of situations, since this methodology is capable of pricing flexibility and strategic resource allocation accurately.

It is important for managers, especially in SMEs, to understand that flexibility has its value when it comes to investment analysis. ROV values this flexibility as options embedded in the investment project before combining them to result in *ENPV*. Yet, the traditional valuation method through discounted cash flows is not totally discarded, in fact it is being enhanced. The stylized approach applied in this chapter should be a stepping-stone for SMEs' managers to realize the scope of the benefits of ROV. Furthermore, the existence of different software programmes easily available in the market – like *DerivaGem* used in this chapter – turn applying ROV into a feasible decision making tool for SMEs' managers.

ROV provides solutions to managers' dilemmas about capital budgeting decision-making and helps them efficiently in the evaluation of alternative future scenarios in any conditions – growth or contraction, market boom or recession, even at any point of the firm's learning curve. The advantages are not linked to the size of companies but to the complexity of the investment projects that are being analysed, making it more attractive, competitive and meaningful compared to the traditional *NPV*.

References

[1] Brennan M and Trigeorgis L (Eds.) 2000 *Project Flexibility, Agency and Competition:New Development in Theory and Application of Real Options* (New York: Oxford Press)

[2] Copeland T, Antikarov V and Copeland T 2001 *Real Options: A Practitioner's Guide* (London: Texere)

[3] Dixit A and Pindyck R 1994 *Investment under Uncertainty* (Princeton University Press)

[4] Trigeorgis L 1996 *Real Options, Managerial Flexibility and Strategy in Resource Allocation* (MIT Press)

[5] Ross S 1995 Uses, abuses, and alternatives to the Net-Present-Value rule *Financial Management* **24**(3): 96-102

[6] Pindick R 1991 Irreversibility, Uncertainty and Investment *Journal of Business*, **65** (1):1-29

[7] McDonald R and Siegel D 1985 Investment of the firms when there is an option to shut down *International Economic Review* **26**(2): 331-49

[8] Cox J and Ross S 1976 The valuation of options for alternative stochastic processes. *Journal of Financial Economics* **3**: 145-66

[9] Cox J Ross S and Rubenstein M 1979 Option pricing: A simplified approach *Journal of Financial Economics* **7**: 229-63.

[10] Geske R 1979 The valuation of compound options. *Journal of Financial Economics* **7**: 63-81.

[11] Kemna A G and Vorst A C F 1990 A pricing method for options based on average asset values *Journal of Banking & Finance* **14** (1) :113-29.

[12] Margrabe W 1978 The value of an option to exchange one asset for another *The journal of finance* **33**(1) 177-86

[13] Kellogg D and Charnes J M 2000 Real-options valuation for a biotechnology company *Financial analysts journal* **56**(3) 76-84

[14] Schwartz E and Moon M 2000 Rational pricing of internet companies *Financial Analyst Journal* May/June: 62-75

[15] Grenadier S R and Weiss A M 1997 Investment in technological innovations: An option pricing approach *Journal of financial Economics* **44**(3) 397-416.
[16] Alonso S, Azofra V and De la Fuente G 2009 The value of Real Options: The case of Endesa's takeover of Enersis *Journal of Finance Case and Research* **10**(1) 1- 26
[17] Willigers B J and Hansen T L 2008 Project valuation in the pharmaceutical industry: a comparison of least-squares Monte Carlo real option valuation and conventional approaches *R&d Management* **38**(5):520-37.
[18] Cortazar G, Schwartz E and Salinas M 1998 Evaluating environmental investments: A real options approach *Management Science* **44**(8):1059-70
[19] Cooper R and Slagmulder R 2004 Inter-organisational cost management and relating context. *Accounting, Organizations and Society* **29**(1): 1-26
[20] Brennan M and Schwartz E 1985a Evaluating natural resource investments *Journal of Business* **58**(2): 135-57
[21] Brennan M and Schwartz E 1985b A new approach to evaluating natural resource investments *Midland Corporate Finance Journal* **3**: 37-47.
[22] Trigeorgis L 1991 A log-transformed binomial numerical analysis method for valuing complex multi-option investment *Journal of Financial and Qualitative Analysis* **26**(03), 309-26
[23] Tourinho O A (1979). *The option value of reserves of natural resources* (No. 94). University of California at Berkeley.
[24] McDonald R and Siegel D 1986 The value of waiting to invest *Quarterly Journal of Economics* **101**(4): 707-27
[25] Paddock J Siegel D and Smith J 1988 Option valuation of claims on real assets: The case of offshore petroleum leases *The Quarterly Journal of Economics* **103**(3): 479-508
[26] Cortazar G and Schwartz E 1993 A compound option model of production and intermediate inventories. *Journal of Business* **66**(4): 517- 40
[27] Pindyck R 1988 Irreversible investment capacity choice and the value of the firm *American Economic Review* **78**: 969-85.
[28] Myers S and Majd S 1990 Abandonment and project life *Advances in Futures and Options Research* **4**: 1-21
[29] Smit H 1997 Investment analysis of offshore concessions in the Netherlands *Financial Management* **26**(2): 5-17
[30] Trigeorgis L 1993 The nature of option interactions and the valuation of investments with multiple real options *Journal of Financial and Quantitative Analysis* **28**(1): 1–20
[31] Nichols N 1994 Scientific management at Merck: An interview with CFO Judy Lewent *Harvard Business Review* **72**(1): 88-99
[32] Crone M and Watts D 2000 Supply linkages and the local SME sector: Evidence from Yorkshire and Humberside *Local Economy* **15**(4): 325-337
[33] Sato H 2009 The iron and steel industry in Asia: Development and restructuring. *IDE Discussion Paper* No. 210. Japan: Institute of Developing Economies

Factors Influencing Consumers Intention for Online Grocery Shopping – A Proposed Framework

SFF Pauzi[1], AC Thoo[1], LC Tan*[2], FM Muharam[1] and NA Talib[1]

[1] Faculty of Management, Universiti Teknologi Malaysia, Malaysia
[3] Faculty of Geoinformation and Real Estate, Universiti Teknologi Malaysia, Malaysia

*E-mail: tlchoon@utm.my

Abstract. Nowadays, Internet is one of the most popular platforms for people to do online shopping including grocery items. Many studies have been conducted to investigate the determinants of customer intentions for online grocery shopping. Till now, there is no consensus on what are the factors that actually influencing people to shop grocery items through Internet. This paper aims to explore the factors such as social influences, facilitating conditions, hedonic motivations, perceived risk and perceived trust that influence the consumer intention to purchase grocery online. Questionnaires will be the main instrument of the study and they will be distributed to target respondents using Internet survey. Respondents of the study will be selected using convenience sampling. After data collection, Statistical Package for the Social Sciences (SPSS) will be employed for data analysis. Overall, the result of the study is important to retailers to identify the important factors in increasing their customers' intention to purchase grocery online.

1. Introduction

Nowadays, the Internet has dramatically changed our everyday lives. It has revolutionized to become one of the most popular platforms for people to do online shopping [1]. Online shopping allows consumers to purchase products or services over the Internet from a seller. Online shopping platform requires consumers to transform their traditional ways of shopping since information finding, transaction and purchasing processes as well as the issues in logistics are absolutely different as compared to the existing groceries shopping experience using brick-and-mortar concept [2]. Many online stores will provide extra information about their grocery products. In fact, consumers' purchase intention is usually based on the quality and quantity of the product information [3]. Hedonic motivated consumers will find more information about online grocery shopping as they feel overwhelmed and happy in what they do [4].

Many studies have been conducted to investigate the determinants of customer intentions for online grocery shopping. Social influences play direct impacts on consumer online shopping behavior [5]. Internet self-efficacy or facilitating condition is important to encourage consumers to adopt the online shopping technology processes [6]. However, some consumers have some doubt when purchasing grocery online. The major hindrances such as risk in term of payment that using credit card or debit card and lack of trust in term of quality of grocery products might impede consumers to shop grocery online. Till now, there is no consensus on what are the factors that actually influencing people to shop grocery items through Internet. Therefore, this paper aims to investigate the factors including social influences, facilitating conditions, hedonic motivations, perceived risk and perceived trust which influence the consumer intention to purchase grocery online.

2. Literature Review

Online Grocery Purchase Intention

Purchase intention is defined as "the probability that customers will aim or be disposed to buy any product and service later" [7] and it is considered as the standard step to reward the real purchasing behavior [8]. According to Huang and Su [9], consumers' purchase intention can be classified as a part of a consumer's cognitive behaviour that discloses the way of a person is expected to purchase any specific brand. Online purchase intention is a situation which customers show their eagerness to attempt an online transaction [10]. Moreover, measurement of purchase intention actually indicates future purchasing behavior [11]. The measurements of purchase intention should be low cost and easy to understand [12]. According to Morwitz et al. [13], the findings of the purchase intention is used to predict demand of recent products for operation, production schedule management, advertising and costing policy.

Bai et al. [14] stated that final purchasing behavior can be derived from consumer intention, so it is important to understand the purchase intention. Based on the research outcome, consumers are the ones that make decision on whether they want to proceed with the purchase [15]. Furthermore, consumers' purchase intention is usually based on the quality and quantity of the information that they have [3]. Currently, sellers are not only focusing on convincing consumers to use the websites that sell their goods, but also influencing the consumers to repeat purchasing their products through the channels [16]. However, trust, information and recommendations from friends or parents will affect the final buying decision when risks are involved [17]. Therefore, the reduction of production and monetary risks will lead to an increase probability of a future purchase [18].

According to He et al. [19], the main barrier to grow online business is the lacking in consumer intention to shop online. According to the theory of planned behavior (TPB), it indicates that the intention to do online shopping is mostly affected by the consumers' behavior itself and from the people around them [20]. These two factors play important roles to influence consumers' purchase intention and directly affect their behavior to shop online. Moreover, Jamil and Mat [21] also presented that consumers' purchase intention positively affects the expected online purchasing response. As such, Kim and Hong [22] stated that an online website also should understand the behavior of customers' purchase in order to create and retain the excellent relationship with consumers.

Moreover, there is a study about acceptance and the intention of online groceries shopping in Malaysia by Yunus et al. [23] which stated that attitude, subjective norm and perceived behavioural control are the factors that influence groceries purchase intention. The result shows that attitudes and subjective norms are the factors that commonly involves when consumers make a decision to purchase groceries online. In addition, an empirical study has been conducted to investigate the consumer purchase intention of Zalora and it has been concluded that Zalora's consumer purchase intention is determined by hedonic motivation, trust and attitude [24]. In that case, it can be stated that consumer have the intention to shop online when they feel safe, trust and enjoy when shopping online. Therefore, the study suggested an implication for sellers in the B2C e-commerce, the sellers should create an effective listing template with product information in order to meet the customer's need. This is because the amount of product information listed by sellers in the market is proven to be the most crucial predictor to determine consumers' purchase intentions.

Based on extensive literature review, there are many factors affect online grocery purchase intention. However, in this study, online grocery purchase intention is affected by social influences, facilitating conditions, hedonic motivation, perceived risk and perceived trust. Also, online purchase intention is measured using 5-item in terms of consumers probability to shop groceries online in the near future, consumers think that Internet as a medium for them to do their groceries shopping and consumers have intention to buy a grocery items through Internet soon, consumers like to increase their frequency of purchase groceries online compared to physical store and consumers think online retailers website will be a good decision for them to make transaction [25].

Social Influences

According to Venkatesh et al. [26], social influence is defined as the level of an individual perception to use a new system based on the influences of his or her important persons. In a previous research, it has been observed that social influence is built on the merger of individual norm and persona attributes [27]. These attributes are the key points of prominent societal forces that might influence customers' awareness and actions. Moreover, Rashotte [28] defined social influence as how other individuals indirectly or directly change the behavior, attitude, feelings and thoughts of an individual. In other words, it is a process by which individuals develop their real feelings and behaviors as a result of the interaction with other people who are perceived as similar, desirable or expert changes. People adjust their beliefs about others according to the psychological principles to keep their social life balance. People are influenced most when the social group concerning an individual expects them to maintain a particular attitude, and the individual is likely to adopt it as well. The other individuals include parents and peers who interact with the individual frequently will also impact the belief and behaviour of the individual. In addition, the individual can also be influenced by the media like Facebook, Twitter and Instagram because this social media has become a significant online platform for individuals to develop their social network [29].

Moreover, Chen et al. [30] emphasized that online word-of-mouth have influence on purchase intention. According to Jalilvanda et al. [31], reviews of online consumer play two roles in social influence which is informative and suggestive. Informative provide an additional user-focused information whereas suggestive give the product popularity positive or negative signals. Consumers who feel associated to a social network consume the social media by seeking for more online activities, and this will lead to a virtual type of friend pressure [32]. Wang et al. [33] stated that the tendency of consumers to communicate with their friends about consumption positively influences their perspective toward any products and services, which results in either they will buy the same brand or avoid the brands in order to be different with their friends. Power and Philips-Wren [34] also agreed that friend pressure on social media is faster and more complete than face-to-face experience.

Hansen et al. [35] stated that the Internet is a social communication and shopping is a cultural practice, therefore it is expected that the influence of subjective norm is significant in the process of adoption of Internet shopping. Therefore, in the context of a developing country, any attempt to quantify the forces responsible for the determining subjective standard and adoption of Internet channel for purchase should be included. Many empirical studies have confirmed that the subjective norm positively give an effect toward intention of buying behavior through the Internet channel [36]. In addition, an empirical study has been done on university academic staff members in Jordan and it is concluded that subjective norm has a significant positive influence on consumer intentions to purchase online [37]. This result also proves by Ranadive [38] in his research about online grocery shopping intentions where subjective norm also be the one of the factors that influences consumers to purchase groceries through the Internet. Moreover, there are other opinions that can impact a consumer's purchase decision such as normative intent and informational social influence [26]. Normative intent means that when an individual follows the behavior of his or her peers and work partners in order to fulfill a self-defining relationship. On the other hand, informational social influence is viewed as how individuals accept opinions and information from others without questioning for problem solving.

Facilitating Conditions

Venkatesh et al. [26] defined that facilitating conditions are the level of an individual perception on the organizational and technical infrastructure for a new system. Besides, people who are labelled as 'digital natives' or individuals who have been exposed to information technology (IT) since their childhood is the type of people that the organisations target for online purchase. Therefore, the individual can be presumed to be rather assertive with using different types of IT applications and connecting in web-enabled social environments [39]. However, if the users are younger generations who mostly live with their parents, their use of the Internet and computer may be influenced, observed and perhaps managed by their parents. As a consequence, the facilitating condition is important predictor of the purchasing intention. Based on social cognitive theory (SCT) [40], self-efficacy is exerted as the facilitating conditions.

According to Bandura [41], social cognitive theory (SCT) postulates a triadic reciprocity between the individual's behavior that intends to fulfill his/her cognitive impression and the environment. Moreover, SCT asserts that the individual tend to be more preferable in accomplishing the behaviors that they believe that will lead to favorable outcome instead of behaviors that will result in negative ones. Furthermore, belief about one's capability to conduct a specific behavior that is known as self-efficacy and it influences their choices about what behaviors to be implemented [40]. Compeau and Higgins [42] developed an evaluation scale for analyzing computer self-efficacy in ascertaining IT adoption. In addition, the literature has also constructed an equivalent dissimilarity and evaluates the relationship between specific and general computer self-efficacy which is specific self-efficacy is associated with a particular software and application whereas general self-efficacy describes self-efficacy with the computers [43].

Additionally, Internet self-efficacy (ISE) refers to the judgment of one's ability to organize and execute actions related to Internet in accomplishing online tasks [44] or simply self-evaluations consumption of his/her capacities to shop online. In fact, online shopping transactions is a long and complex process that requires a certain degree of competence to complete it successfully. Therefore, individuals with low ISE are likely evolve less favorable attitudes toward the adoption of purchases through the Internet channel. Indeed, it has also been confirmed that consumers with low self-efficacy of Internet are doubtful and are uncomfortable with online purchases [45]. Apparently, the acquisition of ISE is potentially a crucial level in achieving the adoption of process in online shopping by potential consumers [46].

In fact, many research studies have empirically showed that Internet self-efficacy is a crucial step in the adoption of online shopping technology processes [6]. Besides, empirical research that addresses the online shopping context has also concluded that internet self-efficacy will positively affect consumers' behavioral intention to use this technology to shop online [47]. From the discussion above, it is clearly stated that self-efficacy with the computer and Internet is an important thing that consumers need to know in order to purchase grocery through the Internet and the consumers need to know how to use the computer and Internet as well. That is the reason why people who are not so skilful in using computer and Internet will less likely to shop grocery online than people can master the skill.

Hedonic Motivations

According to Babin et al. [48], hedonic shopping motivations are described as the benefits such as fun and satisfaction gained from the buyer's purchasing decision based on their occurrence. Therefore, the appearance of hedonic on purchases consists of pleasure and entertainment based on the shopping experience. In online shopping, hedonic motivations are related to the different types of purchases, for example, commercial experience, idea, shopping satisfaction, and value purchases [49]. Moreover, hedonic motivation definitely affects the attitude of the consumers toward online shopping [50] and effective shopping technology [51]. Apart from that, based on To et al. [52], hedonic buyers are seeking natural experiences based on physical, psychological stimulation, value, and aesthetics which increase the contentment and pleasure of online shopping. Since hedonic purchasers increase in number, providing hedonic value also became a significant source of income in business online [53].

There are different reasons for shopping compared to the buying needs in the past [54]. In the past, consumers do shopping because of their need but nowadays, many consumers are driven by hedonic motivation when doing their shopping as they will achieve joy and entertainment in the process [55]. Sometimes, for some consumers, they can get the feel of enjoyment when they do online shopping using information technology [56]. Some researchers have also claimed that hedonic motivation plays an important role in forming a positive behavior to online shopping [57]. Consumers with strong hedonistic motivation will enjoy the process of finding information about the product than the usefulness of the purchased product [4]. Moreover, hedonic motivated consumers will also spend more time to search, compare and select the online store because they enjoy doing the actions. Therefore, online grocery structure provides consumer with an option to search, compare, and access information about the grocery items much more convenient and at a profound levels than within the bricks and mortar groceries store [58]. Thus, hedonic motivated consumers will find more information about online groceries shopping as they feel overwhelmed and happy in what they do.

From the past research, hedonic motivation construct has also been concluded to be an influential factor of the information technology uses in the context of consumer behavior [50]. Furthermore, To et al. [52] also emphasize that hedonic motivation construct is constantly significant due to the recognizable motivations that attract consumers to visit websites of a sellers' online. From previous study that conducted in Taiwan, Liao, Fei, and Chen [59] those motivations of online shopping of adults from hedonic motivation which it is not only influence the search intention, but indirectly influence the intention to shop online. Therefore, Chiu et al. [60] showed that hedonic motivation has positive relationship with consumer's purchase intention in online shopping.

In addition Mathwick et al. [61] evaluated that the experimental value in online shopping which is the intrinsic experimental value comprising pleasure and aesthetics. Such experimental value should be considered as hedonic value. Furthermore, Kim and Shim [62] claimed that customers who shop online are not only seeking for the excitement in gathering information and purchase products but are also meeting their demands of experience and sentiment, which proves that online buyers are like buyers in the actual world. Parsons [63] affirmed that online buyers are influenced by hedonic motivations to shop online. Hence, it can be assume that hedonic motivations in online grocery shopping do exist.

Perceived Risk

Perceived risk is extremely important for online buyers [64] particularly when the buyers have no information about seller [65]. Also, perceived risk is subjective which differentiates the users individually because every person has his or her own thought on matter and experiences. According to Featherman and Pavlou [66], perceived risk is also known as consumers' degree of doubt concerning the outcome of their decision in online purchasing. Nevertheless, if the online purchasers are able to know and comprehend the information regarding the product that they want to purchase, the risk can be minimized [67]. Perceived risks consist of different types which are monetary risk, time risk, social risk, psychological risk, source risk and general risk. Bhatnagar et al. [68] described that risks are linked to not gaining what is anticipated which in turns triggers the dissatisfaction.

Based on previous research, perceived risk is becoming the key element when consumers want to make the decision in the online shopping process [69]. In the context of customers' perspective, purchases that happen in a conventional store are less risky compared to the purchases that take place in online store [70]. From a managerial sentiment, managers who understand the consumer's risks and consumer reaction are able to lessen the risks and help them to improve their business possibility and strategies [71]. According to Miyazaki and Fernandez [72], buyers who perceived high risks in online shopping are more probably to make lesser purchases than buyers who perceived low risks.

In the e-commerce research, perceived risk has always been one of the vital barriers to the development of multichannel shopping [73]. Consequently, buyers could see different types of risks like their incapacity to examine a product before buying it [74], fright of giving information of their credit card to an online seller, loss of seclusion indirect channels and expending too much time on purchase transaction or the waiting for the product to be delivered [75]. Ranganathan and Ganapathy [76] found that concerns about security had the greatest effect on the consumers' intention to purchase groceries online. Results acquired by Odekerken-Schro¨der and Wetzels [77] state that online consumers worry about the products quality. These results also prove by Geuens et al. [78] where most online consumers worry about the quality of fresh food products that are delivered to them from an online grocery store. Hansen [79] conceptualizes perceived risk as the extent to which consumer believes it is risky to use the Internet for grocery shopping.

Besides, Internet trustworthiness is also related to the customers' concerns about privacy when they want to shop online. These concerns involve the unauthorized obtaining of personal information during the usage of Internet or the supplying of personal information collected by the vendors [80]. Similarly, the study found that the belief in the personal information privacy was related with negative perspective toward online purchasing. The available literatures on online purchasing and privacy stated that a large number of Internet shoppers do not have enough trust in the vendors in order to provide their personal information [81]. The study suggested that as the privacy concerns increases, the likelihood of online purchasing will decreases.

From my point of view, perceived risk will give a big impact toward intention to purchase grocery online. Customers take this as an important thing before making any decision to do an Internet purchasing. However, there are some customers that have less concern about this issue. This is due to customers who like to purchase more on the Internet are more concerned about the creation of privacy laws. So, it can be concluded that customers that place importance on the perceived risk of online shopping are less likely to purchase grocery online.

Perceived Trust

According to Chung and Kwon [82], trust is defined as a safe feeling to others and something that individuals can depend on. Fam et al. [83] stated that trust and satisfaction of the customer can be built in certain specific time. Consequently, trust can also be described as an assumption that people will not behave unscrupulously and that the sellers have to provide what has been promised to the customers [84]. Regardless of the variety of definitions, trust is commonly considered crucial in online atmosphere because of different types of risks that people will face during the buying process [85]. Perceived trust is classified into two processes: trust as a way of thinking, faith, viewpoint and trust as a perceived likelihood that including liability and unreliability [86]. In addition, McCole [87] summarized ten most quoted elements of trust that are regularly used in the literature such as opportunity, capability, constancy, distinctiveness, justice, morality, faithfulness, acceptance, agreement and accomplishment. Therefore, consumer's perceived trust in the electronic commerce situation of purchasing has an effect on the transactions and personal information privacy [88].

Kim et al. [15] stated that the characteristics of trust have been examined in many studies in numerous divergent scopes, like fiscal, administration, technology, civil, customer actions and psychology. Trust is a form of the purchaser's assumption that the dealer does not have an opportunistic stance and will not take benefit of the condition, yet will act in a reliable, behavioral and socially responsible attitude and fulfilling the responsibility in spite of the purchaser's subjection [84]. Consequently, the purchasers' trustworthiness is probable to decide the ultimate purchase decision between a purchaser and a dealer [89]. According to Li et al. [90], trust plays more crucial role for online than for offline as customers perceive greater risk in online shopping due to their incapacity to visit a store and examine the product that they are interested in personally. In addition, trust plays a pivotal role in determining intentions of online purchase [91] and purchasing decisions [92]. Trust is also being the main point for consumer loyalty and long term relationship between purchasers and sellers [93].

In the online groceries shopping, trust is a pivotal factor that influences consumers in their decision to buy or not especially for fresh products. Most of customers lack trusts in employees' choice and product quality without seeing it themselves like they do in groceries stores [94]. It is because putting trust into the employees who aren't necessarily as observant isn't easy. Example of consumers purchases groceries in Amazon, consumers cannot choose by themselves between "ripe" and "not ripe" avocados because it chooses by their employees. So, it can assume that everything in online groceries shopping is a gamble [95]. Based on the previous studies, it shows that people tend to shop online if they discern high trust in online shopping [96]. On the other hand, Wang and Emurian [97] proposed that trust is becoming an important factor of online shopping in the future. Wen [98] also stated that purchasers' trust has a positive effect toward online purchase intention. In that case, it can be said that trust influences consumers' intention to shop grocery online. It is because people tend to purchase and repurchase if the vendor are able to provide them the feeling of trust about their grocery products. In addition, the retailers who are able to deliver what they promise and avoid any mistake that will disappoint the customers can build stronger trust among the consumers. This is due to customers have the power of word of mouth that can give negative or positive effect to the retailers.

3. Research Framework

Conceptual framework guides the researchers in determining the result and the statistical relationship that will be examined between the dependent variable and independent variables. As indicated in Figure 1, factors including social influences, facilitating conditions, hedonic motivations, perceived trust and perceived risk are used to examine their effects on the online grocery purchase intention.

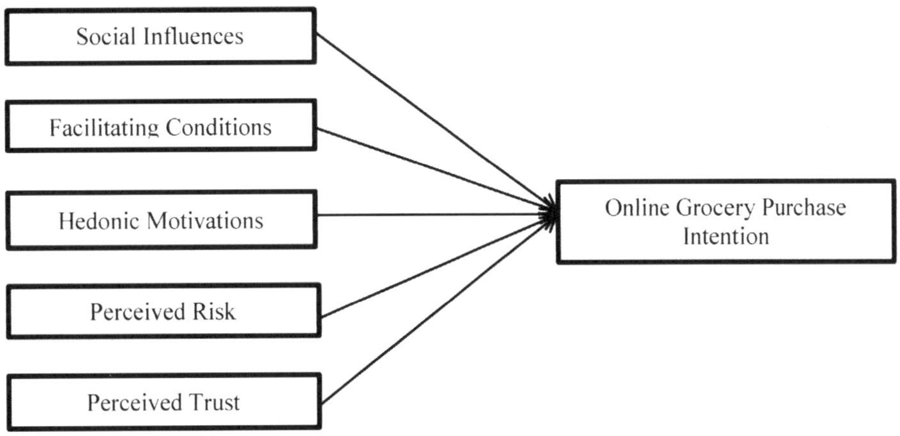

4. Research Methodology

Descriptive study will be used in this study and it is employed to investigate a representative sample of online consumers in order to identify the relationships between social influences, facilitating conditions, hedonic motivations, perceived risk, perceived trust and online grocery purchase intention. Online survey is the major instrument of the study. Convenience sampling technique will be used to select the respondents. Since convenience sampling provides easy access to choose its respondent, a link of the questionnaire via private message and email will be sent to the respondents. The data of the study will be analysed using normality test, reliability test, validity test and multiple regression analysis. Multiple regression will be used to investigate the relationships between social influences, facilitating conditions, hedonic motivations, perceived trust, perceived risk and online grocery purchase intention. The data collected from the respondents will be analysed using Software Package used for Statistical Analysis (SPSS).

5. Conclusion

Future empirical studies are needed to test the effect of social influences, facilitating conditions, hedonic motivations, perceived risk and perceived trust on consumer intention to purchase grocery online. The result is important to online retailers as it can help them to understand what factors that drive Malaysian consumers for online grocery purchase intention. Additionally, the findings will assist them in generating necessary and effective actions to attract the attention of Malaysian consumers towards online grocery shopping since purchasing groceries through Internet is not extensively used in Malaysia. Finally, this paper contributes to the literature of factors influencing consumers in online shopping, particularly in Malaysia.

Acknowledgments

This work was funded by Ministry of Higher Education Malaysia (MOHE) and Universiti Teknologi Malaysia GUP Tier 1 (PY/2016/06368 – Vot Number 15H61).

References

[1] Bourlakis, M., Papagiannidis, S. and Fox, H. (2008). E-consumer behaviour: Past, present and future trajectories of an evolving retail revolution. *International Journal of E-Business Research*. 4: 64-76.

[2] Seitz, C. (2013). E-grocery as new innovative distribution channel in the German food retailing. *In Proceedings of the MakeLearn International Conference*. 125-133.

[3] Park, D.H., Lee, J., Han, I. (2007). The effect of on-line consumer reviews on consumer purchasing intention: the moderating role of involvement. *Int. J. Electron. Commer.* 11(4): 125–148.

[4] Mikalef, P., Giannakos, M. and Pateli, A. (2013). Shopping and Word-of-Mouth Intentions on Social Media. *Journal of Theoretical and Applied Electronic Commerce Research.* 8(1): 17-34.

[5] Cetina, I., Munthiu, M-C., and Radulescu, V. (2012). Psychological and Social Factors that Influence Online Consumer Behavior. *Procedia - Social and Behavioral Sciences.* 62: 184-188.

[6] Hsu, M.H., Chiu, C.M. (2004). Internet self-efficacy and electronic service acceptance. *Decision Support Syst*em. 38(3): 369-381.

[7] Wu, P.C.S., Yeh, G.Y.Y., Hsiao, C.R. (2011). The effect of store image and service quality on brand image and purchase intention for private label brands. *Australasian Marketing Journal.* 19(1): 30-39.

[8] De Magistris, T. and Gracia, A. (2008). The decision to buy organic food products in Southern Italy. *British Food Journal.* 110(9): 929-947.

[9] Huang, X. and Su, D. (2011). Research on Online Shopping Intention of Undergraduate Consumer in China-Based on the Theory of Planned Behavior. *International Business Research.* 4(1): 86-92.

[10] Ling, K.C., Chai, L.T., & Piew, T.H. (2010). The Effects of Shopping Orientations, Online Trust and Prior Online Purchase Experience Toward Customers' Online Purchase Intention. *International Business Research.* 3: 63-76.

[11] Grewal, D., Krishnan R., Baker J., & Borin N. (1998). The effect of store name, brand name and price discounts on consumers' evaluations and purchase intentions. *Journal of Retailing.* 74(3): 331-352.

[12] Armstrong, J. S., Morwitz, V.G. and Kumar, V. (2000). Sales forecasts for existing consumer products and services: Do purchase intentions contribute to accuracy?. *International Journal of Forecasting.* 16: 383-397.

[13] Morwitz, V.G., Steckel, J.H., & Gupta, A. (2007). When do purchase intentions predict sales?. *International Journal of Forecasting.* 23(3): 347-364.

[14] Bai, B., Law, R., & Wen, I. (2008). The impact of website quality on customer satisfaction and purchase intentions: evidence from Chinese online visitors. *Int. J. Hosp. Manag.* 27: 391-402.

[15] Kim, D.J., Donald, L.F., & Rao, H.R. (2008). A trust-based consumer decision-making model in electronic commerce: the role of trust, perceived risk and their antecedents. *Decis. Support. System,* 44(2): 544-564.

[16] Chiu, C.M., Hsu, M.S., Lai, H., & Chang, C.M. (2012). Re-examining the influence of trust on online repeat purchase intention: the moderating role of habit and its antecedents. *Decis. Support. Syst*em. 53(4): 835-845.

[17] Wang, J.C., & Chang, C.H. (2013). How online social ties and product-related risks influence purchase intentions: a Facebook experiment. *Electron. Commer. Res. Appl.* 12: 337-346.

[18] Suwelack, T., Hogreve, J. and Hoyer, W.D. (2011). Understanding money-back guarantees: Cognitive, affective, and behavioral outcomes. *Journal of Retailing.* 87(4): 462-478.

[19] He, D., Lu, Y., & Zhou, D. (2008). Empirical Study of Consumers' Purchase Intentions in C2C Electronic Commerce. 13: 287-292.

[20] Laohapensang, O. (2009). Factors influencing internet shopping behaviour: a survey of consumers in Thailand. *Journal of Fashion Marketing and Management: An International Journal.* 13(4): 501-513.

[21] Jamil, N. A., & Mat, N. K. (2011). To Investigate the Drivers of Online Purchasing Behavioral In Malaysia Based on the Theory of Planned Behavior (TPB): A Structural Equation Modeling (SEM) Approach. *International Conference On Management.* 453-460.

[22] Kim, E., & Hong, T. (2010). Segmentating Customers in Online Stores from Factors that Affect the Customer's Intention to Purchase. 383-388.

[23] Muhamad Yunus, N. S. N., Abdul Ghani, N. M., & Wan Rashid, W. E. (2016). The Acceptance and the Intention of Online Groceries Shopping in Malaysia. *Journal of Applied Environmental and Biological Sciences*. 6(6): 29-34.
[24] Shierly, L. and Sihombing, S.O. (2015). Predicting Online Purchase Intention: An Empirical Study. *Management Dynamics Conference*. 3-13.
[25] Moon, J. W. and Kim, Y. G. (2001). Extending the TAM for the World-Wide-Web context. *Information and Management*. 38(4): 217-230.
[26] V. Venkatesh, M.G., Morris, G.B. and Davis, F.D. (2003). User acceptance of information technology: toward a unified view. *MIS Quarterly*. 27: 425-478.
[27] Johnston, A.C., & Warkentin, M. (2010). Fear appeals and information security behaviors: an empirical study. *MIS Q*. 34(3): 549-566.
[28] Rashotte, L. (2007). "Social influence." *The blackwell encyclopedia of social psychology*. 9: 562-563.
[29] Nelson, M. R., & McLeod, L.E. (2005). Adolescents brand consciousness and product placement: Awareness, liking and perceived effects on self and others. *International Journal of Consumer Studies*. 25(1): 1-13.
[30] Chen, Y., Fay, S., & Wang, Q. (2011). The Role of Marketing in Social Media: How Online Consumer Reviews Evolve. *Journal of Interactive Marketing*. 25(2): 85-94.
[31] Jalilvanda, M. R., Esfahani, S. S., & Samiei, N. (2011). Electronic Word-of-Mouth: Challenges and Opportunities. *Procedia Computer Science*. 3: 42-46.
[32] O'Brien, C. (2011). The Emergence of the Social Media Empowered Consumer. *Irish Marketing Review*. 21: 32-40.
[33] Wang, X., Yu, C., & Wei, Y. (2012). Social Media Peer Communication and Impacts on Purchase Intentions: A Consumer Socialization Framework. *Journal of Interactive Marketing*. 26(4): 198-208.
[34] Power, D. J., & Philips-Wren, G. (2011). Impact of Social Media and Web 2.0 on Decision-Making. *Journal of Decision Systems*. 20(3): 249-261.
[35] Hansen, T., Jensen, J.M., & Solgaard, H.S. (2004). Predicting online grocery buying intention: a comparison of the theory of reasoned action and the theory of planned behavior. *Int. J. Inf. Manag*. 24(6): 539-550.
[36] Choi, J. and Geistfeld, L.V. (2004). A cross-cultural investigation of consumer e-shopping adoption. *Journal of Economic Psychology*. 25(6): 821-838.
[37] Al-Jabari, M.A., Othman, S.N., & Mat, N.K.N. (2012). Actual online shopping behavior among Jordanian customers. *Am. J. Econ*. 125–129.
[38] Ranadive, A. (2015). An Empirical Study on the Online Grocery Shopping Intentions of Consumers in Vadodara City. *International Journal of Management and Social Sciences Research*. 4(3): 8-14.
[39] Palfrey, J., & Gasser, U. (2008). Born digital: Understanding the first generation of digital natives. NY: Basic Books, New York.
[40] Bandura, A. (1977). Self-efficacy: Toward a unifying theory of behavioural change. *Psychological Review*. 84(2): 191-215.
[41] Bandura, A. (1986). Social foundations of thought and action: A social cognitive theory. NJ: Prentice Hall, Englewood Cliffs.
[42] Compeau, D. R., & Higgins, C. A. (1995). Computer self-efficacy: Development of a measure and initial test. *MIS Quarterly*. 19(2): 189-211.
[43] Agarwal, R., Sambamurthy, V., & Stair, R. M. (2000). The evolving relationship between general and specific computer self-efficacy – An empirical assessment. *Information Systems Research*. 11(4): 418-430.
[44] Tsai, M.J., Tsai, C.C. (2003). Information searching strategies in web-based science learning: the role of Internet self-efficacy. *Innov. Educ. Teach. Int*. 40(1): 43-50.
[45] Monsuwe, T.P.Y., Dellaert, B.G.C. and Ruyter, K.D. (2004). What drives consumers to shop online? A literature review. *International Journal of Service Industry Management*. 15(1): 102-121.
[46] LaRose, R., & Rifon, N.J. (2007). Promoting i-safety: effects of privacy warnings and

privacy seals on risk assessment and online privacy behavior. *Journal Consumer Affair.* 41(1): 127-149.

[47] Delafrooz, N., Paim, L.H., & Khatibi, A. (2010). Students' online shopping behavior: an empirical study. *J. Am. Sci.* 6(1): 137-147.

[48] Babin, J., Darden, R. and Griffin, M. (1994). Work and/or fun: measuring hedonic and utilitarian shopping value. *Journal of Consumer Research*. 20(4): 644-657.

[49] Arnold, M.J., & Reynolds, K.E. (2003). Hedonic shopping motivations. *Journal of Retailing*. 79(1): 77-95.

[50] Childers, T.L., Carr, C.L., Peck, J., & Carson, S. (2001). Hedonic and utilitarian motivations for online retail shopping behavior. *Journal of Retailing*. 77: 511-535.

[51] Kim, J. and Forsythe, S. (2007). Hedonic usage of product virtualization technologies in online apparel shopping. *Journal of Retail Distribution. Management*. 35(6): 502-514.

[52] To, P., Liao, C. and Lin, T. (2007). Shopping motivations on Internet: a study based on utilitarian and hedonic value. *Technovation*. 27: 774-784.

[53] Sorce, P., Perotti, V. and Widrick, S. (2005). Attitude and age differences in online buying. *International Journal of Retail and Distribution Management*. 33(2): 122-132.

[54] Dhaundiyal, M., & Coughlan, J. (2009). The Effect of Hedonic Motivations, Socialibility and Shyness on the Implusive Buying Tendencies of the Irish Consumer. *Irish Academy of Management*. 1-29.

[55] Kazakeviciute, A., & Banyte, J. (2012). The Relationship of Consumers Perceived Hedonic Value and Behavior. *Inzinerine Ekonomika-Engineering Economics*. 532-540.

[56] Zaidi, S. D., Gondal, B. J., Yasmin, A., & Rizwan, M. (2014). Antecedents of Online Shopping Intention: A Study held in Pakistan. *Journal of Sociological Research*. 5(1): 231-247.

[57] Sarkar, A. (2011). Impact of Utilitarian and Hedonic Shopping Values on Individual's Perceived Benefits and Risks in Online Shopping. *International Management Review*. 7(1): 58-65.

[58] Lynch Jr, J. G., & Ariely, D. (2000). Wine online: search costs affect competition on price, quality, and distribution. *Marketing Science*. 19(1): 83–103.

[59] Liao, S. H., Fei, W. C., & Chen, C. C. (2007). Knowledge sharing, absorptive capacity, and innovation capability: An empirical study of Taiwan's knowledge/intensive industries. *Journal of Information Science*. 33(3): 340-359.

[60] Chiu, C. M., Wang, E. T., Fang, Y. H., & Huang, H. Y. (2014). Understanding customers' repeat purchase intentions in B2C e-commerce: The roles of utilitarian value, hedonic value and perceived risk. *Information Systems Journal*. 24(1): 85-114.

[61] Mathwick, C., Malhotra, N., & Rigdon, E. (2001). Experiential value: conceptualization, measurement and application in the catalog and internet shopping environment. *Journal of Retailing*. 77(1): 39-56.

[62] Kim, Y.M. and Shim, K.Y. (2002). The influence of intent shopping mall characteristics and user traits on purchase intent. *Iris Marketing Review*. 15(2): 25-34.

[63] Parsons, A.G. (2002). Non-functional motives for online shoppers: why we click. *The Journal of Consumer Marketing*. 19(5): 380-392.

[64] Doolin, B., Dillon, S., Thompson, F., & Corner, J.L. (2005). Perceived risk, the internet shopping experience and online purchasing behaviour: a New Zealand perspective. *J. Glob. Inf. Manage*. 13: 66-88.

[65] Finch, B.J. (2007). Customer expectations in online auction environments: an exploratory study of customer feedback and risk. *Journal of Operation Management.* 25(5): 985-997.

[66] Featherman, M. S., & Pavlou, P. A. (2002). Predicting e-Services Adoption: A Perceived Risk Facets Perspective. *In Eigth American Conference on Information System.* 9: 1034-1046.

[67] Gefen, D., & Straub, D.W. (2004). Consumer trust in B2C e-Commerce and the importance of social presence: Experiments in e-Products and e-Services. *Omega: The International Journal of Management Science*. 32(6): 407-424.

[68] Bhatnagar, A., Misra, S., & Rao, H.R. (2000). On risk, convenience, and Internet shopping behavior. *Commun. ACM*. 43(11): 98-105.
[69] Liebermann, Y. and Stashevsky, S. (2002). Perceived risks as barriers to Internet and e-commerce usage. *Qualitative Market Research*. 5(4): 291.
[70] Suki, N. (2007). Online buying innovativeness: effects of perceived value, perceived risk and perceived enjoyment. *International Journal of Business and Society*. 8(2): 81.
[71] Comegys, C., Hannula, M., & Vaisanen, J. (2006). Longitudinal comparison of Finnish and US online shopping behavior among university students: the five-stage buying decision process. *Journal of Targeting, Measurement and Analysis for Marketing*. 14(4): 336.
[72] Miyazaki, A.D., & Fernandez, A. (2001). Consumer perceptions of privacy and security risks for online shopping. *The Journal of Consumer Affairs*. 35(1): 27.
[73] Dholakia, R.R., Zhao, M. and Dholakia, N. (2005). Multichannel retailing: a case study of early experiences. *Journal of Interactive Marketing*. 19: 63-74.
[74] Konuş, U., Verhoef, P.C., & Neslin, S.A. (2008). Multichannel shopper segments and their covariates. *Journal of Retailing*. 84(4): 398-413.
[75] Forsythe, S., Liu, C., Shannon, D., & Gardner, L.C. (2006). Development of a scale to measure the perceived benefits and risks of online shopping. *J. Interact. Marketing*. 20: 55-75.
[76] Ranganathan, C., & Ganapathy, S. (2002). Key dimensions of business-to-consumer web sites. *Information and Management*. 39(6): 457-465.
[77] Odekerken-Schro¨der, G., & Wetzels, M. (2003). Trade-offs in online purchase decisions: two empirical studies in Europe. *European Management Journal*. 21(6): 731-739.
[78] Geuens, M., Brengman, M., & S'Jergers, R. (2003). "Food retailing, now and in the future: a consumer perspective". *Journal of Retailing and Consumer Services*. 10(4): 241-251.
[79] Hansen, T. (2005). "Consumer adoption of online grocery buying: a discriminant analysis". *International Journal of Retail & Distribution Management*. 33(2): 101–121.
[80] George, J.F. (2002). Influences on the intent to make Internet purchases. *Internet Research*. 12(2): 165-180.
[81] Hoffman, D.L., Novak, T.P., & Peralta, M. (1999). Building Consumer Trust Online. *Communications of the ACM*. 42(4): 80-85.
[82] Chung, N., & Kwon, S. J. (2009). Effect of trust level on mobile banking satisfaction: a multi-group analysis of information system success instruments. *Behaviour & Information Technology*. 28(6): 549-562.
[83] Fam, K. S., Foscht, T., & Collins, R. D. (2004). Trust and the online relationship-an exploratory study from New Zealand. *Tourism Management*. 25(2): 195-207.
[84] Gefen, D., Karahanna, E., & Straub, D.W. (2003). Trust and TAM in online shopping: an integrated model. *MIS Q*. 27(1): 51-90.
[85] Van der Heijden, H., Verhagen, T. and Creemers, M. (2003). Understanding online purchase intentions: contributions from technology and trust perspectives, *European Journal of Information Systems*. 12: 41-48.
[86] Chen, C. (2006). Identifying significant factors influencing consumer trust in an online travel site. *Information Technology & Tourism*. 8: 197-214.
[87] McCole, P. (2002). The role of trust for electronic commerce in services. *International Journal of Contemporary Hospitality Management*. 14(2): 81-87.
[88] Kim, H., Kim, T., & Shin, S.W. (2009). Modeling roles of subjective norms and eTrust in customers' acceptance of airline B2C eCommerce websites. *Tourism Management*. 30(2): 266-277.

[89] Gupta, P., Yadav, M.S., & Varadarajan, R. (2009). How task-facilitative interactive tools foster buyers' trust in online retailers: a process view of trust development in the electronic market place. *Journal of Retailing*. 85(2): 159-176.
[90] Lee, Y. (2014). Impacts of decision-making biases on eWOM retrust and risk-reducing strategies. *Comput. Hum. Behavior*. 40: 101-110.
[91] Hong, I.B. and Cho, H. (2011). The impact of consumer trust on attitudinal loyalty and

purchase intentions in B2C e-market places: intermediary trust vs seller trust. *Int. J. Inf. Manag*. 31(5): 469-479.
[92] Buttner, O.B., & Goritz, A.S. (2008). Perceived trustworthiness of online shops. *Journal of Consumer Behavior*. 7(1): 35-50.
[93] Santos, C.P. and Fernandes, D.V.H. (2008). Antecedents and consequences of consumer trust in the context of service recovery. *Braz. Adm. Rev*. 5(3): 225-244.
[94] Galante, N., López, E. G., & Monroe, S. (2013). The future of online grocery in europe. *Perspect. Retail Consum. Goods*. 1: 22-31.
[95] Vaknin, S. (2013). 'AmazonFresh vs. supermarket: A hands-on shopping test'. Retrieved from https://www.cnet.com/news/amazonfresh-vs-supermarket-a-hands-on-shopping-test/ on 9 October 2016.
[96] Corbitt, B. J., Thanasankit, T., & Yi, H. (2003). Trust and e-commerce: a study of consumer perceptions. *Electronic Commerce Research and Applications*. 2(3): 203-215.
[97] Wang, Y. D., & Emurian, H. H. (2005). An overview of online trust: concepts, elements, and implications. *Computers in Human Behavior*. 21(1): 105-125.
[98] Wen, I. (2010). Online travelers' decision makings: a new equation model to evaluate impacts of website, search intention, and trust. *Information Technology & Tourism*. 12(2): 153-173.

IOP Conf. Series: Materials Science and Engineering **215** (2017) 012014 doi:10.1088/1757-899X/215/1/012014

Understanding Supply Chain Management Practices for Small and Medium-Sized Enterprises

AC Thoo[1], Z Sulaiman[1], SL Choi[2] and UHA Kohar[1]

[1]Senior Lecturer, Faculty of Management, Universiti Teknologi Malaysia, Malaysia
[2]Associate Professor, Raffles University Iskandar, Johor Bahru, Johor, Malaysia.

E-mail: zuraidahs@utm.my

Abstract. Small and medium enterprises (SMEs) are a major source of dynamism, innovation and flexibility for emerging and developing countries, as well as for the economies of the most industrialised nations. However, the survival and growth of SMEs can be difficult in the current competitive business environment and global marketplace. It can be a real challenge to deliver the right product and service at the most opportune time and at the lowest possible cost to the right customer. The challenge stresses the importance of managing cross-boundary relationships between business partners. For gaining a competitive advantage, supply chain management (SCM) is an effective tool to SMEs. Therefore, this paper aims to review the tenet of SCM, its benefits and practices to SMEs.

1. Introduction

In the past decades, there has been a massive surge of interest in supply chain management (SCM) (Burgess et al., 2006; Zhou and Benton, 2007; Fawcett et al., 2008) due to its innovative approach to business (Tracey et al., 2004; Fawcett et al., 2008) and competitive advantage (Koh et al., 2007). The popularity of SCM concept can be seen from the remarkable rise in conferences, academic research and publications, professional development programs and courses in university (Burgess et al., 2006) since the 1980s (Stock and Boyer, 2009). Many fields are devoted to the proliferation of SCM literature, including logistics and transportation, supply and purchasing, operations management, organisational theory, strategic management and marketing, hoping to fully understand the tenets of SCM (Chen and Paulraj, 2004). The fundamental concept of SCM revolves around the efficient flow of goods, service, information and money to the business with aim to provide right product to the right customers at the right cost, right time, right quality and right quantity. SCM can deliver a number of benefits to small and medium sized-enterprises (SMEs); therefore, adoption of SCM can help SMEs to achieve competitive advantages.

SMEs are now considered to be a major source of dynamism, innovation and flexibility for emerging and developing countries, as well as for the economies of the most industrialised nations. SMEs contribute substantially to the nation through economic development and employment generation (Koh et al., 2007). In addition, SMEs form the potential economic back-bone of many regions and make a greater contribution to employment than large firms (Peng, 2009).

2. Tenet of Supply Chain

A supply chain results from the efforts of organisations to produce and deliver a finished good from suppliers' supplier to customers' customer. The efforts include all activities involved in these five basic processes – plan, source, make, deliver and return, which encompass matching supply with demand, sourcing components and raw materials, producing final products, delivering to end customers and providing post-manufacturing services such as return, repair and warranty (Supply-Chain Council, 2016).

As defined by Mentzer et al. (2001), supply chain is a network consists of all parties involved (e.g. manufacturer, supplier, retailer, customer, etc.), directly or indirectly, in manufacturing and delivering products or services to ultimate consumers – both in upstream and downstream sides through physical distribution, flow of information and finances. As stated by Chopra and Meindl (2010), a typical supply chain includes the following five stages: component/raw material suppliers, manufacturers, retailers, wholesalers/distributors and customers. These five stages are connected through flows of products, information and money. As pointed out by Chandra and Kumar (2000), managing a supply chain network is complex and difficult, since it involves various sub-systems, activities, relationships and operations. Therefore, an integrated supply chain framework is needed to tie a whole network together with the goal of reducing perennial supply chain challenges.

SCM is about managing the supply chain (Mentzer et al., 2001). SCM is a watchword in business (Li, 2002), Wall Street, and in the news media (Stephens, 2001). As defined by the Council of Supply Chain Management Professionals (CSCMP), SCM is the management and planning process that involve all activities from sourcing and procurement, transformation of raw materials to finished goods, as well as logistics activities. The supply chain incorporates all the collaboration and coordination within and outside an organisation with its channel partners (e.g. suppliers, intermediaries, customers, and third party service providers) with the hope of integrating supply and demand management.

As proposed by Mentzer et al. (2001), there are three categories of SCM definitions: a philosophy of management, a set of activities to implement a philosophy of management and series of management practices. These definitions are under the category of management philosophy, which views SCM as a whole system including cooperative efforts for managing and distributing a finished product from supplier to end customer (Monczka et al., 2005), as well as information and funds coordination (Handfield and Nichols, 1999); SCM philosophy aims to integrate supply chain partners to create customer value and satisfaction.

To embrace a successful SCM philosophy, various activities are needed, such as development of long-term relationships (Min and Mentzer, 2004; Chen and Paulraj, 2004), mutually sharing information, integrated behaviour (Tan, 2002), cooperation, risks and rewards, integration of processes, agreed goals, a focus on serving customers, (Min and Mentzer, 2004) and partnership with supply chain members. In addition, SCM is about the process of managing materials flow, information and relationships between companies in fulfilling customer requests. This category emphasises the importance of all supply chain functions as key business processes (Ho et al., 2002), such as customer service management, order fulfilment, customer relationship management, product development, procurement and commercialisation.

By the same token, Li (2002) developed three SCM categories, including purchasing and supply management, integrated logistics management, and integrated SCM. Purchasing and supply management involve partnering and integrating with suppliers to better management of the supply and purchasing functions. Integrated logistics management deals with physical distribution and logistics activities within and beyond organisations, including suppliers and customers. Definitions under an integrated SCM focus on strategic nature of coordination between trading partners within a supply chain network, with hopes to enhance performance of supply chains and organisations.

There have been an enormous number of definitions of SCM when the tenet has gained tremendous popularity since 1980s – approximately 173 definitions of SCM have been proposed in total (Stock and Boyer, 2009). As SCM covers a wide range of disciplines, the researchers have continued to suggest the SCM definitions as according to their own research directions. The lack of an overarching SCM definition (Mentzer et al., 2001; Burgess et al., 2006) may affect the development and adoption of SCM theory. Without a single consensus SCM definition, researchers are unable to further develop the theory and practice (Stock and Boyer, 2009). Therefore, an encompassing and inclusive definition of SCM is of paramount importance to help scholars and practitioners, such as supply chain executives to develop sound SCM strategies (Mentzer et al., 2001; Burgess et al., 2006).

The definition given by Stock and Boyer (2009) provides an all-encompassing and complete definition of SCM. The researcher concurs with this definition, which posits that SCM involves network relationships management within and across companies. These entities include marketing, logistics, production facilities, purchasing, material suppliers and related systems are connected through forwards and reverse flows of information, materials, services and finances for the purpose of delivering the final products from original producer to end customer, thus enhancing customer satisfaction, bringing value-enhancing and maximising profitability through efficiencies.

3. Supply Chain Management Practices

SCM is one of the competitive priorities that have evolved from a one-dimensional subject with a rather narrow focus on logistics and physical aspects of material flow, into a multifaceted theory encompassing a broad range of subjects. For example, Bidgoli (2010) contended that modern SCM simultaneously incorporates strategic differentiation and value enhancement as well as operational efficiency improvements and cost reduction. According to Sweeney (2007), SCM has become the business model of the 21st century. This view was echoed by Ketchen et al. (2008), who supported the idea of a contemporary SCM approach – the best value supply chains are designed to deliver superior total value to the customer in terms of speed, cost, quality and flexibility and becomes a key competitive weapon for the 21st century.

The concept of SCM begins to gain popularity in the past thirty years (Fredendall and Hill, 2001), with an emphasis on supplier collaborative relationship or partnership (Lummus and Vokurka, 1999; Tan, 2001). The SCM practices play a critical role in sustainable organisational performance enhancement. SCM practices include a set of approaches and activities utilised by an organisation to effectively integrate supply and demand for improving the management of its supply chain (Li et al., 2005; Wong et al., 2005; Koh et al., 2007). SCM practice is a multi-dimensional concept which can be measured from different perspectives such as stated in Sundram et al. (2011). Different dimensions of SCM practice have been highlighted in past literature (Koh et al., 2007).

Alvarado and Kotzab (2001) emphasised the importance of inter-organisational system such as electronic data interchange (EDI), concentration on core competencies and elimination of excess inventory levels in SCM practices. Tan (2002) also suggested that a seamless supply chain includes shared IT and information sharing, managing materials and supply issues, customer service, extending all the way to internal operations. Therefore the author affirmed that SCM practices need to cover information sharing, customer service management, SCI, supply chain characteristics, just-in-time (JIT) capability and strategic location. Sahay and Mohan (2003) used aligned supply chain and business strategies, information technologies, partnerships in inventory management and SCI to measure SCM practices. Chen and Paulraj (2004) defined SCM practices to include long-term relationships, communication, supplier involvement, supply base reduction, and cross-functional teams. Min and Mentzer (2004) noted that SCM practices are composed of information sharing, agreed customer focus and vision, cooperation, sharing of risk and rewards, accepted supply chain leadership, process integration and long-term relationships. This model is grounded on strategic partnerships, with the purpose of achieving mutual benefits. Similarly, Tracey et al. (2004) proposed a construct of SCM assimilation consisting of technology utilisation (internal and supply chain technology usage), internal relationships (participative leadership, manufacturing and logistics participation in strategy), external relationships (supplier selection, evaluation and management), transportation, product development and inventory management (inventory control, warehousing and packaging).

The literature largely addresses only downstream customers, upstream suppliers of the supply chain or internal functions. Hence, Li et al. (2005) developed and validated a measurement instrument for six SCM practices constructs which cover supply chain information flows (quality and sharing of information), downstream (relationship with customer), upstream (strategic supplier partnership) and internal supply chain processes (postponement and internal lean practices). Even though the proposed SCM practices framework portrays the major perspectives of SCM practices, however, it does reveal a number of limitations. These limitations include the use of a single manufacturing sector in the study – the constructs like postponement and waste reduction practices are not suitable for retailers and distributors.

Wong et al. (2005) defined SCM practices to include supply chain performance, SCM initiatives, product uniqueness, inventory-cost and lead-time management, postponement and customisation, information sharing and coordination, supplier and customer relationships, retail strategy, coordination and sharing of information, distribution and logistics. Burgess et al. (2006) proposed a SCM construct with seven practices including the relationships of inter- and intra- organisational, leadership, information system, process improvement orientation, logistics, business outcomes and results. Koh et al. (2007) explored the relationships between SCM practices, operational performance and SCM-related organisational performance in Turkish SMEs. A list of twelve SCM practices was identified in the study. The SCM practices involve building dense relationships with suppliers and customers, implementing inventory management including JIT and safety stock approaches, using inter-organisational systems such as e-procurement, engaging third-party services through 3PL (third-party logistics), outsourcing and subcontracting, and managing internal side of the supply chain via strategic planning, supply chain benchmarking and managing the numbers of suppliers (with just a few or many suppliers). Through factor analysis, such SCM practices are classified on the following two dimensions: outsourcing and multi-suppliers (OMS), and strategic collaboration and lean practices (SCLP).

Zhou and Benton (2007) used JIT, supply chain planning, delivery practices, and production to represent SCM practices. Also, Chow et al. (2008) compared supply chain practices between Taiwan and the US. The framework of supply chain practices for US comprises supply chain features, supplier and customer management, information sharing, speed and communication. On the other hand, SCM practices for Taiwan consist of customer service management and SCI and features. Robb et al. (2008) developed a supply chain practice to focus on customer and supplier relationships, e-commerce and enterprise software. Hsu et al. (2009) studied SCM practice as a mediator between operations capability and firm performance. The SCM practices include increased suppliers' JIT capabilities, participating in sourcing decisions, geographical proximity of suppliers and formal information-sharing agreements. Similarly, Singh et al. (2010) identified five factors such as use of technology, customer satisfaction, inventory management, SCI and supply chain speed as SCM practices in relationship of competitive advantage and organisational performance.

Chong et al. (2011), basing their surveys from Malaysian manufacturers and service firms, suggested that SCM practices should cover the following six dimensions: strategic supplier partnership, customer relationship, information technology, information sharing, internal operation and training. Cook et al. (2011) used the following practices to represent SCM: RELATION – long-term relationships with supply chain members; SHARE – information sharing among supply chain members; SOURCE – supplier network design; INTERNET – internet-based technology leverage; PLAN – usage of advanced planning systems; and DISTRIBUTION – distribution network design. Miguel and Brito (2011) conceptualised SCM practices as including information sharing, long term relationships, cooperation, and process integration.

Kushwaha (2012) proposed that SCM practices include information and communication technology practices, strategic sourcing and supplier relationship practices, supply chain manufacturing practices, inventory and warehousing management system, transportation and distribution management system, and customer relationship management. Sukati et al. (2012) targeted three impacts of SCM practices (strategic supplier partnership, customer relationship and information sharing) on supply chain performance. Toyin (2012) used five dimensions to measure SCM practices including strategic supplier partnership, customer relationship, postponement, level of information sharing and quality of information sharing. Besides, Sukati et al. (2013) suggested three relationships from perspectives of internal-firm, firm-supplier and firm-customer to implement SCM practices.

While many efforts proposing a model of SCM constructs have been made, there has been no clear convergence on a single unifying construct to incorporate all facets of SCM. Despite lacking common set of constructs, these features do possess some similarities.

4. Why SCM is important to SMEs

The main focus of SCM is to provide right product to the right customers at the right cost, right time, right quality and right quantity. Meanwhile, the short-term strategic goal of SCM is to reduce cycle time and inventory and thus increasing productivity, whereas the long-term goal is to enhance profits through market share and customer satisfaction (Tan, 2002). According to Mohanty and Deshmukh (2005), the quantifiable benefits of SCM include lower supply chain costs, overall productivity, inventory reduction, forecast accuracy, delivery performance, as well as fulfilment cycle time and fill rates. SCM delivers improvement up to 60%, with ranges between 10% and 60%. Fulfilment cycle time records the highest improvement, jumping from 30% to 60%.

A review of extent literature reveals that adoption of SCM may deliver a number of potential benefits to the organisations. Different criterion have been used to measure organisation performance such as financial, non-financial, innovation performance, market share and customer satisfaction (Talib et al., 2010). For example, the benefits include operation costs reduction, customer satisfaction, order fulfilment lead times (Chin et al., 2004; Tumala et al., 2006; Fawcett et al., 2008), inventory improvement (Chin et al., 2004; Tumala et al., 2006; Talib et al., 2010), responsiveness to customer requests (Fawcett et al., 2008; Talib et al., 2010), remain competitive (Chin et al., 2004; Tumala et al., 2006); on-time delivery (Fawcett et al., 2008), increased flexibility (Tumala et al., 2006), increased sales, increased internal coordination between departments, increased supplier and customer coordination, improved supply chain communication, a reduction in risk and a reduction in the duplication of inter-organisational processes (Talib et al., 2010).

In the context of SMEs, cost-effective SCM is critical for survival and growth, as purchasing cost makes up the largest share in terms of sales revenue – approximately 80% (Quayle, 2003). Meehan and Muir (2008) conducted a study in SMEs in Merseyside, UK to reveal the perceived benefits of SCM to SMEs. The potential benefits include increases in customer service and responsiveness, improved supply chain communication, risk reduction, a reduction in product development cycle time processes, a reduction in duplication of inter-organisational processes, inventory reductions, and improvement in electronic trading. Another study by Koh et al. (2007) involving SMEs manufacturing companies in Turkey found that the execution of SCM practices could deliver benefits to SMEs in terms of reduced inventory level, reduced lead time in production, increased flexibility, forecasting accuracy, cost saving and accurate resource planning.

While there is voluminous research literature available on SCM potential benefits but till now there is limited study to reveal the actual benefits of SCM (Lambert et al., 2005; Stock et al., 2010). Explicit SCM does not seem to occur often enough in practice. This is mainly because respective functional managers view the focus of SCM differently; this disagreement in terms of SCM goal provides difficulty to supply chain managers (Fawcett et al., 2008). This dilemma often happens in big companies. However, in comparison to large enterprises, SMEs usually have a small management team and the owner as the top manager (Jaidee and Beaumont, 2003; Schatz, 2006; Woolderink, 2010). An owner-manager can develop SCM roadmap more easily in order to drive for superior supply chain performance (Thakkar et al., 2008). As the review has indicated, there is a lack of empirical research confirming these significant benefits of SCM. Therefore, it is important to explore the benefits of SCM to SMEs.

5. Conclusion

The survival and growth of SMEs can be difficult in the current competitive business environment and global marketplace; customers are more demanding of better and cheaper products, higher service levels, more product varieties and faster delivery (Tachizawa and Thomsen, 2007; Chow et al., 2008; Ketchen et al., 2008). It can be a real challenge to deliver the right product and service at the most opportune time and at the lowest possible cost to the right customer (Chin et al., 2004; Li et al., 2005).

SMEs have financial constraints, so a high priority is placed on inventory management and control, because capital tied up in inventory can be huge (Meehan and Muir, 2008). Therefore, SMEs must strike a balance between responsiveness and efficiency. A higher level of inventory may increase supply chain responsiveness; however, it increases inventory holding costs (Chopra and Meindl, 2010). There are a variety of partnering initiatives and inventory solutions are being employed to encourage collaborative inventory management that involves inter-dependent trading partners, such as vendor-managed inventory (VMI), JIT and collaborative planning, forecasting and replenishment (CPFR) (Sari, 2007) with the hopes of mitigating bullwhip effect, minimising demand uncertainty, reducing costs and in the end improving profitability.

References

[1] Burgess, K., Singh, P. J. and Koroglu, R. (2006). Supply Chain Management: A Structured Literature Review and Implications for Future Research. International Journal of Operations and Production Management. 26(7): 703-729.
[2] Zhou, H. and Benton, Jr. W. C. (2007). Supply Chain Practices and Information Sharing. Journal of Operations Management. 25: 1348-1365.
[3] Fawcett, S. E., Magnan, G. M. and Mccarter, M. W. (2008). Benefits, Barriers, and Bridges to Effective Supply Chain Management. Supply Chain Management: An International Journal. 13(1): 35-48.
[4] Tracey, M., Fite, R. W. and Sutton, M. J. (2004). An Explanatory Model and Measurement Instrument: A Guide to Supply Chain Management Research and Applications. American Journal of Business. 19(2): 53-69.
[5] Koh, S. C. L., Demirbag, M., Bayraktar, E., Tatoglu, E. and Zaim, S. (2007). The Impact of Supply Chain Management Practices on Performance of SMEs. Industrial Management & Data Systems. 107(1): 103-124.
[6] Burgess, K., Singh, P. J. and Koroglu, R. (2006). Supply Chain Management: A Structured Literature Review and Implications for Future Research. International Journal of Operations and Production Management. 26(7): 703-729.
[7] Stock, J. R. and Boyer, S. L. (2009). Developing a Consensus Definition of a Supply Chain Management: A Qualitative Study. International Journal of Physical Distribution & Logistics Management. 39(8): 690-711.
[8] Chen, I. J. and Paulraj, A. (2004). Towards a Theory of Supply Chain Management: The Constructs and Measurements. Journal of Operations Management. 22: 119-150.
[9] Peng, M. W. (2009). Global Business. South-Western Cengage Learning, Canada.
[10] Supply-Chain Council (2016). Definition of Supply Chain. Retrieved November 1, 2016, from http://supply-chain.org/
[11] Mentzer, J. T., Dewitt, W. and Keebler, J. S. (2001). Defining Supply Chain Management. Journal of Business Logistics. 22(2): 1-25.
[12] Chopra, S. and Meindl, P. (2010). Supply Chain Management: Strategy, Planning, and Operation. (4th ed.) Upper Saddle River, New Jersey: Pearson Education, Inc.
[13] Chandra, K. and Kumar, S. (2000). Supply Chain Management in Theory and Practice: A Passing Fad or a Fundamental Change? Industrial Management & Data System. 100(3): 100-113.
[14] Li, S. (2002). An Integrated Model for Supply Chain Management Practice, Performance and Competitive Advantage. Doctor of Philosophy Degree in Manufacturing Management, The University of Toledo, USA.
[15] Stephens, S. (2001). Supply Chain Operations Reference Model Version 5.0: A New Tool to Improve Supply Chain Efficiency and Achieve Best Practice. Information Systems Frontiers. 3(4): 471-476.
[16] Monczka, R. M., Trent, R. J. and Handfield, R. B. (2005). Purchasing and Supply Chain Management. (3rd ed.) Ohio, USA: Thomson South-Western.
[17] Handfield, R. B. and Nichols, E. L. (1999). Introduction to Supply Chain Management. Upper Saddle River, NJ: Prentice Hall.
[18] Min, S. and Mentzer, J. T. (2004). Developing and Measuring Supply Chain Concepts. Journal of Business Logistics. 25(1): 63-99.
[19] Tan, K. C. (2002). Supply Chain Management: Practices, Concerns, and Performance Issues. Journal of Supply Chain Management. 38(1): 42-53.
[20] Ho, D. C. K., Au, K. F. and Newton, E. (2002). Empirical Research on Supply Chain Management: A Critical Review and Recommendations. International Journal of Production Research. 40(17): 2215-4430.
[21] Bidgoli, H. (2010). The Handbook of Technology Management: Supply Chain Management, Marketing and Advertising, and Global Management: Volume 2. United States of America, New Jersey: John Wiley and Sons.
[22] Ketchen, Jr. D. J., Rebarick, W., Hult, G. T. M. and Meyer, D. (2008). Best Value Supply

Chains: A Key Competitive Weapon for the 21st Century. Business Horizons. 51: 235-243.
[23] Fredendall, L. D. and Hill, E. (2001). Basics of Supply Chain Management. United States of America: CRC Press LLC.
[24] Lummus, R. R. and Vokurka, R. J. (1999). Defining Supply Chain Management: A Historical Perspective and Practical Guidelines. Industrial Management & Data Systems. 99(1): 11-17.
[25] Li, S., Subba Rao, S., Ragu-Nathan, T. S. and Ragu-Nathan, B. (2005). Development and Validation of a Measurement Instrument for Studying Supply Chain Management Practices. Journal of Operations Management. 23: 618-641.
[26] Wong, C. Y., Arlbjorn, J. S. and Johansen, J. (2005). Supply Chain Management Practices in Toy Supply Chains. Supply Chain Management: An International Journal. 10(5): 367-378.
[27] Sundram, V. P. K., Ibrahim, A. R. and Govindaraju, V. G. R. C. (2011). Supply Chain Management Practices in Electronics Industry in Malaysia: Consequences for Supply Chain Performance. Benchmarking: An International Journal. 18(6): 834-855.
[28] Alvarado, U. Y. and Kotzab, H. (2001). Supply Chain Management: The Integration of Logistics in Marketing. Industrial Marketing Management. 30(2): 183-198.
[29] Sahay, B. and Mohan, R. (2003). Supply Chain Management Practices in Indian Industry. International Journal of Physical Distribution & Logistics Management. 33(7): 582-606.
[30] Tracey, M., Fite, R. W. and Sutton, M. J. (2004). An Explanatory Model and Measurement Instrument: A Guide to Supply Chain Management Research and Applications. American Journal of Business. 19(2): 53-69.
[31] Chow, W. S., Madu, C. N., Kuei, C-H., Lu, M. H., Lin, C. and Tseng, H. (2008). Supply Chain Management in the US and Taiwan: An Empirical Study. Omega: The International Journal of Management Science. 36: 665-679.
[32] Robb, D. J., Xie, B. and Arthanari, T. (2008). Supply Chain and Operations Practice and Performance in Chinese Furniture Manufacturing. International Journal of Production Economics. 112: 683-699.
[33] Hsu, C. –C., Tan, K. C., Kannan, V. R. and Leong, K. (2009). Supply Chain Management Practices as Mediator of the Relationship between Operations Capability and Firm Performance. International Journal of Production Research. 47(3): 835-855.
[34] Singh, R., Sandhu, H. S., Metri, B. A. and Kaur, R. (2010). Relating Organised Retail Supply Chain Management Practices, Competitive Advantage and Organisational Performance. The Journal of Business Perspective. 14(3): 173-190.
[35] Chong, A. Y. L., Chan, F. T. S., Ooi, K. B. and Sim, J. J. (2011). Can Malaysian Firms Improve Organizational/Innovation Performance via SCM? Industrial Management & Data Systems. 111(3): 410-431.
[36] Cook, L. S., Heiser, D. R. and Sengupta, K. (2011). The Moderating Effect of Supply Chain Role on the Relationship between Supply Chain Practices and Performance: An Empirical Analysis. International Journal of Physical Distribution & Logistics Management. 41(2): 104-134.
[37] Kushwaha, G. S. (2012). Operational Performance through Supply Chain Management Practices. International Journal of Business and Social Science. 3(2): 222-232.
[38] Sukati, I., Hamid, A. B. A., Baharun, R. and Yusoff, R. M. (2012). The Study of Supply Chain Management Strategy and Practices on Supply Chain Performance. Procedia Social and Behavioral Sciences. 40: 225-233.
[39] Toyin, A. I. (2012). Supply Chain Management (SCM) Practices in Nigeria Today: Impact on SCM Performance. European Journal of Business and Social Sciences. 1(6): 107-115.
[40] Sukati, I., Hamid, A. B. and Baharun, R. (2013). Testing the Effect of the Supply Chain Management Implementation on Business Performance: An Empirical Study. International Business Research. 6(1): 63-75.
[41] Mohanty, R. P. and Deshmukh, S. G. (2005). Supply Chain Management: Theories and Practices. Delhi: Biztantra Publication.
[42] Talib, F., Rahman, Z. and Qureshi, M. N. (2010). Integrating Total Quality Management and Supply Chain Management: Similarities and Benefits. Journal of Information Technology and Economic Development. 1(1): 53-85.

[43] Chin, K-S., Rao Tumala, V. M., Leung, J. P. F. and Tang, X. (2004). A Study on Supply Chain Management: The Hong Kong Manufacturing Perspectives. International Journal of Physical Distribution & Logistics Management. 30(6): 505-524.

[44] Tumala, V. M. R., Philips, C. L. M. and Johnson, M. (2006). Assessing Supply Chain Management Success Factors, A Case Study. Supply Chain Management: An International Journal. 11(2): 179-192.

[45] Fawcett, S. E., Magnan, G. M. and Mccarter, M. W. (2008). Benefits, Barriers, and Bridges to Effective Supply Chain Management. Supply Chain Management: An International Journal. 13(1): 35-48.

[46] Quayle, M. (2003). A Study of Supply Chain Management Practice in UK Industrial SMEs. Supply Chain Management: An International Journal. 8(1): 79-86.

[47] Meehan, J. and Muir, L. (2008). SCM in Merseyside SMEs: Benefits and Barriers. The TQM Journal. 20(3): 223-232.

[48] Lambert, D., Garcia-Dastugue, S. and Croxton, K. (2005). An Evaluation of Process-Oriented Supply Chain Management Frameworks. Journal of Business Logistics. 26(1): 25-51.

[49] Stock, J., Stefanie, L., Boyer, S. and Harmon, T. (2010). Research Opportunities in Supply Chain Management. Journal of the Academy of Marketing Science. 38(1): 32-41.

[50] Jaidee, S. and Beaumont, N. (2003). Factors Affecting SMEs' Owner/Managers in Adoption of Business-To-Business Techniques: A Research Framework. Working Paper 47/03, Monash University, Melbourne, Australia.

[51] Schatz, C. (2006). A Methodology for Production Development: The Body of Knowledge Approach. Norwegian University of Science and Technology, Faculty of Engineering Science and Technology, Institute for Production and Quality Engineering. NTNU.

[52] Woolderink, M. (2010). "Walking the Extra Mile". Analysis of the Effect of Leadership Behaviour on the Organizational Citizenship Behaviour Shown by Employees in Small and Medium Sized Enterprises. Master Thesis, University of Twente, Netherlands.

[53] Thakkar, J., Kanda, A. and Deshmukh, S. G. (2008). Supply Chain Management in SMEs: Development of Constructs and Propositions. Asia Pacific Journal of Marketing and Logistics. 20(1): 97-131.

[54] Tachizawa, E. M. and Thomsen, C. G. (2007). Drivers and Sources of Supply Flexibility: An Exploratory Study. International Journal of Operations & Production Management. 27(10): 1115-1136.

[55] Sari, K. (2007). Exploring the Benefits of Vendor Managed Inventory. International Journal of Physical Distribution & Logistics Management. 37(7): 529-545.

DEFECTIVE REDUCTION IN FROZEN PIE MANUFACTURING PROCESS

Oranuch Nooted[1] and Somkiat Tangjitsitcharoen[2]

Department of Industrial Engineering, Chulalongkorn University, Bangkok 10330, Thailand

E-Mail: nczh_17@hotmail.com[1], somkiat.t@eng.chula.ac.th[2]

Abstract. The frozen pie production has a lot of defects resulting in high production cost. Failure mode and effect analysis (FMEA) technique has been applied to improve the frozen pie process. Pareto chart is also used to determine the major defects of frozen pie. There are 3 main processes that cause the defects which are the 1st freezing to glazing process, the forming process, and the folding process. The Risk Priority Number (RPN) obtained from FMEA is analyzed to reduce the defects. If RPN of each cause exceeds 45, the process will be considered to be improved and selected for the corrective and preventive actions. The results showed that RPN values decreased after the correction. Therefore, the implementation of FMEA technique can help to improve the performance of frozen pie process and reduce the defects approximately 51.9%.

1. Introduction

Currently, the bakery manufacturing is growing rapidly with an highly competitive rate from domestic enterprises and foreign investors. Key success factors of production and quality are the customer satisfaction including cost reduction for maximum profit and higher competency. Cost reduction by controlling the quality of product can help to reduce the production time and increase the production rate without defects.

Hence, the case study of frozen pie production is conducted and the Failure Mode and Effect Analysis (FMEA) is applied to reduce the defects and wastes. In bakery company which manufacture various kinds of bakery to domestic and international customers, the frozen pie production is focused in this study. Since the highest number of defects (123,373 pieces) when comparing with other productions about 1,480,476 Baht per year.

Failure Mode and Effect Analysis (FMEA) technique is employed to define, identify, and eliminate the potential product failures from the process [1]. Hence, FMEA technique is utilized to analyze potential failure modes and prioritized according to how serious their consequences are, how frequently they occur and how simply they can be detected. The purpose of the FMEA is to take actions to reduce failures. Pareto analysis is used to determine the operations where appear the majority of failure modes in the frozen pie production [2].

2. Define phase of defect types in frozen pies production

This phase can help to define the rank of main defects according to the relative impact on the defective rate significantly. The historical data from April 2015 to March 2016 showed that the proportions of defects are classified as follows: cracked pies 27.10%, pleated pies 17.3%, twisted pies 14.12%, unfolded pies 11.20%, fissure pies 10.45% and other 19.83% shown in Fig. 1.

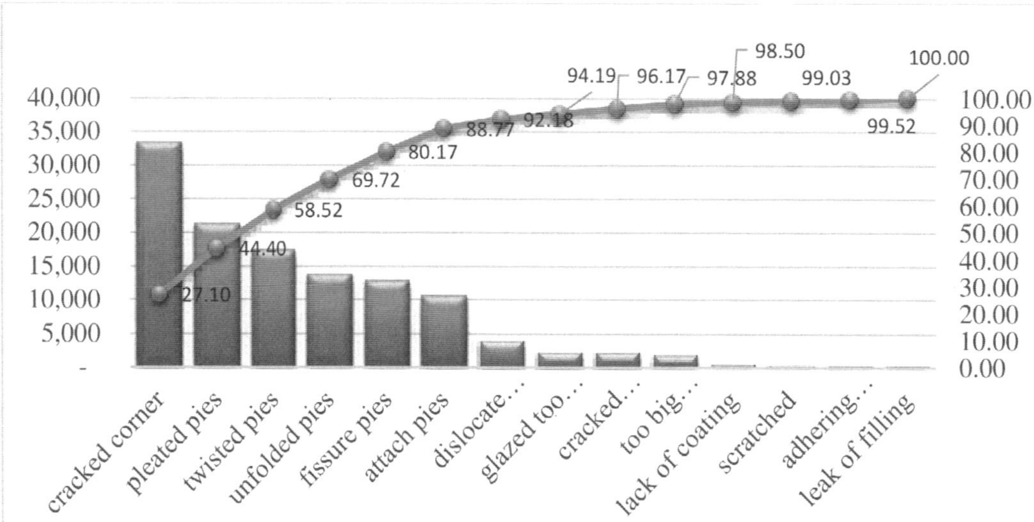

Figure 1. Pareto chart of defect (source: Production data of frozen pie from April 2015 to March 2016)

3. Root cause analysis of defects

Fishbone diagrams or the Cause and Effect diagrams are diagrams with 4M (Manpower, Machine, Material, and Method), to identify the factors causing an overall effect that may be part of the problem. These may be systems, equipment, materials, external forces, people involved with the problem. Fishbone diagrams or the Cause and Effect is applied to this process to indicate the root-causes of the defects [3]. Root causes of each type of defects can be summarized as shown in Table 4. There are 4 potential causes which are the lack of skills and expertise, the failure to check and control during process, the inappropriate equipment and improper method.

A brainstorming meeting by expert team who specializes in the frozen pie production is set to analyze and discuss the root-cause of the defects [4]. Hence, the failure modes of defects are determined by the cause-and-effect diagram as shown in Table 4.

4. FMEA in frozen pie production

FMEA is utilized to identify the potential process failure modes by ranking failures and helping to establish priorities according to the relative impact on the frozen pie. Implementing FMEA will help to identify the potential causes in order to reduce the occurrence and detect failure in frozen pie production.

The relative risk of a failure and its effects is determined by 3 factors which are Severity (S), Occurrence (O), Detection (D) [5] by multiplying the ranking for the three factors (S x O x D). A risk priority number (RPN) will be determined for each potential failure mode and effect [6]. Criteria for ranking of severity(S) occurrence (O) and detection (D) are selected and determined suitably by specialist in the frozen pies production as shown in Tables 1 to 3 [7]. The score of each parameter was recorded in FMEA table as shown in Table 4.

Table 1. Ranking of severity

Effect	Severity of effect ranking	Ranking
Very high	Failure causes a high degree of customer dissatisfaction	5
High	Failure result in a primary function and becomes impacts the quality of the product	4
Moderate	Failure results in a partial malfunction of the product	3
Low	Failure creates enough of an effect to cause the customer to complain	2
Minor	No effect	1

Table 2. Ranking for possible failure rates (Occurrence)

Possibility of failure	Possible Failure Rates (Occurrence)	Ranking
Very high	Extremely high; bound to occur (> 1 in 2)	5
High	High; occur frequently (1 in 10)	4
Moderate	Intermediate; occur occasionally (1 in 100)	3
Low	Low; occur rarely (1 in 1,000)	2
Minor	Zero; never occur (< 1 in 100,000)	1

Table 3. Ranking of detection

Detection	Criteria	Ranking
Very high	Uncontrolled process	5
High	Very low likelihood of detecting potential cause, failure mechanism	4
Moderate	Intermediate possibility of detecting potential cause, failure mechanism	3
Low	High possibility of detecting potential cause, failure mechanism	2
Minor	Virtual certainty of detecting potential cause, failure mechanism	1

Table 4. Failure Mode and Effect Analysis in frozen pie production

NO.	Failure Mode	Potential effects of Failure	S	Potential Cause(s) of Failure	O	Current Control, Prevention	D	RPN	Action Results				
									Action	S	O	D	RPN
1	Operators fold dough at inappropriate temperature	cracked pies	3	Failure to check temperature before folded	3	Observation of the operator usage	5	45	Dough temperature checking form was created	3	2	3	18
2	Folding Machine be in disrepair	fissure pies	5	Failure to check machine before folded	2	Observation and Checking of the machine	5	50	Maintenance , Upgrading and machine inspection form was created	5	3	1	15
3	Delay time of forming process	twisted pies	5	Failure to control delay time	5	Uncontrolled	3	75	Design new the conveyor process from convey on basket into belt directly	5	1	2	10
4	Operators trim edges over the edges	unfolded pies	5	Lack of skills and expertise	3	Observation of the operators	3	45	- Training and Informing of staff - Created specified the defects on work Instruction and place on site	5	2	3	30
5	Operators ruffle the sheet while forming	pleated pies	5	Lack of skills and expertise	3	Observation of the operators	3	45					
6	Operators spread sheet too tight	fissure pies	5	Lack of skills and expertise	3	Observation of the operators	3	45					
7	Gap between fringes of conveyor	cracked pies	5	Inappropriate equipment	4	Uncontrolled	5	100	Design new fringe of conveyor	5	2	2	20
8	Improper Method to bring pies out of basket	cracked pies	5	Not implemented for method	5	Uncontrolled	5	125	Design new the conveyor process from convey on basket into belt directly	5	2	2	20

The Risk Priority Number (RPN) of different processes are found out. Eight causes of failure modes are selected and classified by Pareto diagram. The RPN values before and after corrective action are shown in Table 4. After applying all new corrective and preventive actions to the process, the RPN value is reduced approximately 60.08 %. Finally, FMEA analysis showed that the percentage of defects in the frozen pies production decreased by 51.9%.

5. Conclusions

In this study, FMEA is utilized to reduce the defects during the frozen pies production. The risk priority number (RPN) for the prioritization of failure modes of frozen pies is calculated. The results showed that RPN values of the 1st freezing to the glazing process, the forming and the folding process decreased by 33% to 87%. Furthermore, Pareto chart is applied firstly to elucidate which can help to find the main cause, the frequency and importance of the failures for each defects that may affect quality of frozen pie. An integration of PFMEA and Pareto chart to reduce the defects in the frozen pie production is proposed as these could be useful quality tools for improving the quality and the performance of the bakery manufacturing processes.

Acknowledgement

We would like to express our acknowledgement to the Bakery Company for the financial support in this research, and especially to the expert team who specializes in the frozen pies production.

References

[1] D.H. Stamatis 2003 *Failure Mode and Effect Analysis: FMEA From Theory to Execution* (Milwaukee: ASQ, Quality Press, Inc)

[2] Andre Segismundo, Paulo Augusto and Cauchick Miguel, 2008 Failure mode and effects analysis (FMEA) in the context of risk management in new product development: A case study in an Automotive company, International Journal of Quality & Reliability Management, Volume 25 Issue 9, pp.89 – 912, 2008

[3] R. Tague 2004 *Seven Basic Quality Tools the Quality toolbox* (Milwaukee Wisconsin: American Society for Quality, Inc)

[4] M. Clute 2009 *Food industry Quality Control Systems* (New York: CRC Press Taylor & Francis Group)

[5] J.T. Jones and M.F. Berard 1972 *Ceramics: Industrial Processing and Testing,* Iowa State University Press

[6] R.J. Mikulak and M.R. Beauregard 1996 *The Basics of FMEA* (New York: Productivity, Inc)

[7] P.H. Tsarouhas and D. Arampatzaki 2010 Application of Failure Modes and Effects Analysis (FMEA) of a Ceramic Tiles Manufacturing Plant, Technological Educational Institute of Thessaloniki

[8] I.S. Arvanitoyannis and T.H. Varzakas 2009 Application of failure mode and effect analysis (FMEA) and cause and effect analysis for industrial processing of common octopus (Octopus vulgaris), International Journal of Food Science and Technology, Volume 44, Issue 1, pp. 79-92

REDUCTION OF DEFECTS IN JEWELRY MANUFACTURING

Phitchaya Phanomwan na Ayudhya[1] and Somkiat Tangjitsitcharoen [2]
Department Industrial Engineering, Chulalongkorn University, Bangkok 10330,Thailand

E-Mail: p.phitchaya@gmail.com[1],somkiat.t@eng.chula.ac.th[2]

Abstract. The aim of this research was to reduce the defects of gem bracelet found during manufacturing process at a jewelry company. It was found that gem bracelet product has the highest rejects compared to the rejects found in ring, earring, and pendant products. Types of defect were classified by using Pareto Diagram consisting of gem falling, seam, unclean casting, impinge, and deformation. The causes of defect were analyzed by Cause and Effect Diagram and applied Failure Mode and Effects Analysis (FMEA) was applied during manufacturing processes. This research found that the improvement of manufacturing process could reduce the Risk Priority Number (RPN) and total of all defects by 48.70% and 48.89%, respectively.

1. Introduction

Currently, Thailand jewelry export is top 15 countries creating a great economic value to the country. Consumers in the ASEAN countries are interested in jewelry products. This results in overall jewelry business become active and jewelry export of Thailand is leap forward. Therefore, it is a great opportunity of Thailand to expand its jewelry export and supports the revenue of its population and tourism business. The study of Kasikorn Research Center found that export forecast in the year 2016 is expected to increase by 4% from the year 2015 with a total export value of approximately 286,000 million Baht and once AEC (Asean Economic Community) is opened, the competition will increase. It is necessary to improve the performance to get high quality products according to customer specifications of which will lower the cost and the defective products and subsequently result in high revenue.

The Failure Mode Analysis and Effect (FMEA) technique is a step by step approach for identifying all possible failures in a design, a manufacturing or assembly process, or a product or service [1]. The FMEA is a technique used to identify the associated something potentially going wrong such as creating a defect or out of specification in the production of the product. The FMEA identifies controls that are placed in the production process to detect any defect at various stages on the processing [2].

A case study of a jewelry company which manufacture various kinds of jewelry was carried out. The gem bracelet product was focused in this study since the highest number of defects (3.05% or 30,547 PPM) were found when comparing to other products. Therefore, the research aims to study the reduction of defective parts in gem bracelet because management level give the priority to reduction of defective parts in gem bracelet and has get a goal to reduce the defects to halve or as much as possible

This research has focused on the revision of manufacturing process of bracelets, especially those inlaid with gemstone around the work piece. The principles of FMEA was used for the reason that this technique is useful to identify and eliminate failures and mistakes during design and manufacturing processes. It can select defects that should be fixed first and it is a technique widely used in the automobile industry. It is also can be applied to many other industries. The researcher therefore sees the possibility of applying FMEA techniques along with studying and analyzing the real root cause of the problems in order to approach the solution permanently and to prevent the problem repeatedly occur; and to serve as a benchmark in terms of production and quality control in the jewelry industry in order to minimize the defects.

2. Methods and Result

2.1 Defining phase

The historical manufacturing data of gem bracelets from January 2015 to October 2015 was collected in order to determine types of defect by using Pareto chart. Five defect types account for 80 percent of all defects were then taken into consideration to find the cause. A Pareto chart is a bar graph. The lengths of the bars represent frequency or cost (time or money), and are arranged with longest bars on the left and shortest to the right. In this way, the chart visually depicts which situations are more significant [3].

Five defect types account for 80 percent of all defects were determined as follows gem falling off 25.32%, seam 18.68%, unclean casting 12.09%, impinge 11.47% and deformation 10.63% as shown in Fig. 1.

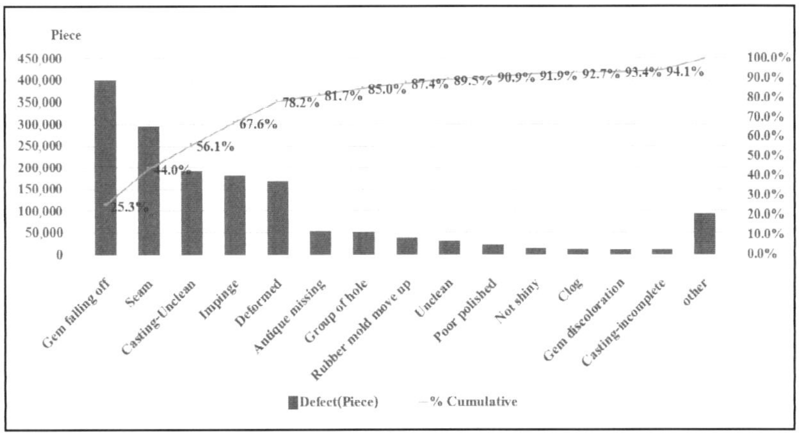

Figure 1. Pareto chart of defect (Source: Production data of bracelet from January 2015– October 2015)

2.2 Root-cause analysis

The Cause and Effect Diagram (Fishbone Diagram) developed by Ishikawa in 1968, help users to think through problem causes [4]. For that particular reason it is also termed as "Cause-Effect analysis". In a typical fishbone diagram the main problem which is required to be resolved has been put on the head of the diagram and the causes are put as the bones and then smaller bones are created as the resemblances of the sub-causes. Ultimately after completion of the diagram it is a comprehensive evaluation of the causes of the main problems and also reveals the root causes as well [5].

The case study was brainstormed by expert team who are specialists in jewelry production process. The Cause and Effect Diagram or fishbone diagram was applied to this process to find the root causes of the problems.

Ten root causes of defect were determined by the Cause and Effect Diagram as shown in Table 1.

Table 1. Root causes of defects

Defect type	Root cause
Gem falling off	1. Water pressure used to clean the product is too high. 2. Operators make a wax-setting to repeat bezel
Seam	1. There is no equipment to help operators while pressing the rubber mold during wax injection 2. incomplete rubber mold
Unclean Casting	1. Water pressure used to clean the casting tree is too low.
Impinge	1. Improper equipment to place the products

	2. Operators stack the products 3. Operators accidently drop the product
Deformation	1. Operators do not have enough skill to remove product from the flask. 2. Operators remove the wax from the rubber mold too soon

3. Analysis of defect by FMEA technique

From Table 1, it was found that the causes of problem occurred only during wax process comprising of wax injection, wax cleaning, wax setting, treeing and casting process, investment, burn out, casting, cleaning, black oxidation, cutting treeing processes. These are very important in the manufacture of jewelry. If defects can be reduced during these processes, a great number of defects will decrease also wax injection and casting process as shown is Fig.2. This research therefore aims to study those 2 processes only.

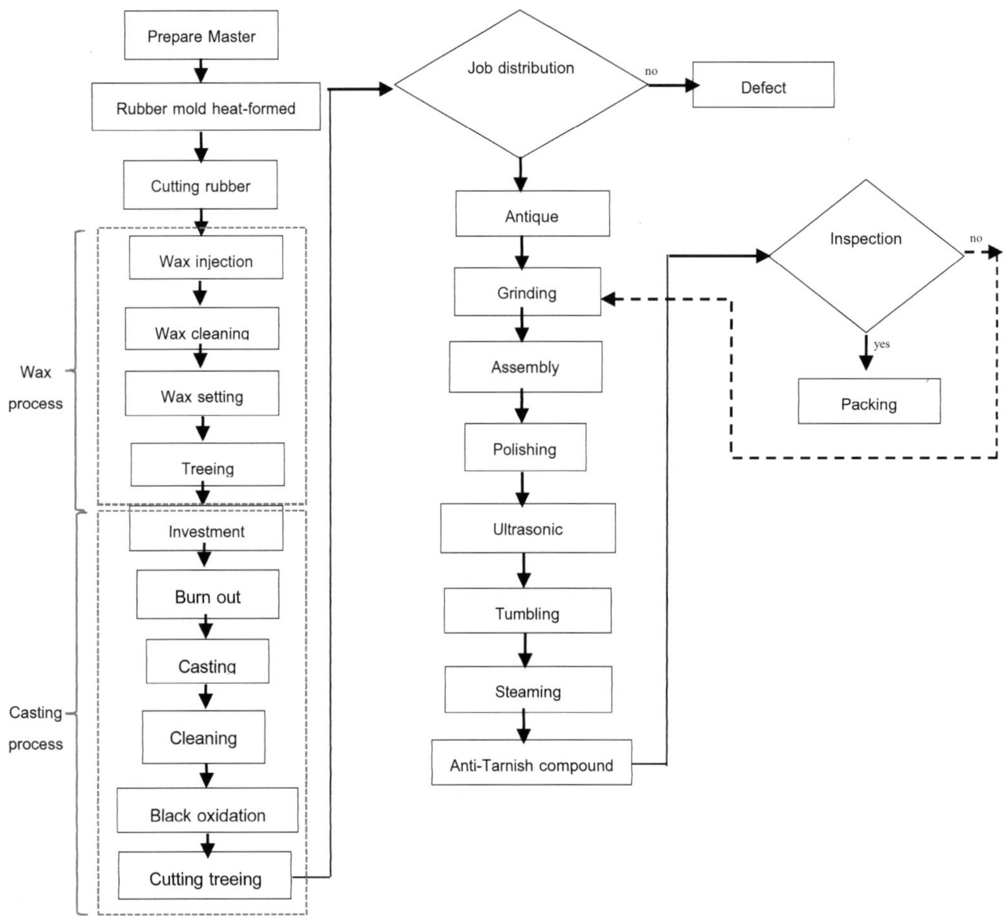

Figure 2. Flow chart of case study jewelry process

FMEA method is used all over industry for a sort of applications and this flexible method can be executed at diverse steps in the product life cycle. FMEA method can be used to carry design, development, manufacturing, service and other activities to get better reliability and enlarge efficiency. As an example, there is extensive use of both design and process FMEAs inside the automotive industry and documentation of this investigation is a general requisite for automotive suppliers. This technique is also generally used in the aerospace, medical. Nuclear and other manufacturing industries [6]. The FMEA table was applied to the jewelry industry by adjusting severity (S), occurrence (O) and detection (D) Score is determined during the meeting between Production Department, Quality Assurance, Maintenance Departmnet and workers to determine the criteria of scoring according to the level of effects as shown in Table 2. The score of each effect was recorded in the FMEA as shown in Table 3.

Table 2. Scores of severity Level (S), occurrence (O), detection (D) in jewelry manufacturing process

Effect	Severity Level (S)	Possible failure rate (O)	Detection failure (D)	Score
Serious	Tendency of the defect severely impacts the quality of the product resulting in demolishing all the product	1 in 20	Uncontrolled process	5
Major	Tendency of the defect severely impacts the quality of the products. Correction some of the products out of the production line may be required before acceptance.	1 in 100	Not easy detection (i.e. random)	4
Moderate	Tendency of the defect severely impacts the quality of the products. Correction all the products at the point of operation may be required before acceptance.	1 in 1,000	Detection after process by operator	3
Low	Little problems occur to some manufacturing process at the point of operation	1 in 10,000	Detection after process auto	2
Minor	No impact felt	<1 in 100,000	Detection in process	1

The risk priority number (RPN) can be calculated according to the equation below.

$$RPN = S \cdot O \cdot D \qquad (1)$$

Table 3. FMEA table before and after improvement

No.	Defect	Root cause	S	O	D	RPN Before	Action	S	O	D	RPN After
1	Incomplete of rubber mold	Failure to check the rubber mold before the trial wax-injection	5	5	5	125	Rubber mold inspection line was established	5	3	3	45
2	Improper Equipment to place the products	Use of plastic bags to transport the items to the next process will cause impinge	4	5	5	100	A new tray to place the product was designed	4	3	3	36
3	No equipment to help operators while pressing rubber mold during wax injection	Operators use his/her hands to press the rubber mold causing the stiffness	4	5	5	100	Equipment to hold rubber mold was designed	4	3	5	60
4	Operators repeat the setting	Bezel and pave are deformed due to operators repeat the setting without fixing it first	4	5	5	100	Q-Point and pictures of defect were posted at the point of operation	4	3	3	36
5	operators stage products	operators stage the products in basket	4	5	4	80	Design new basket to place the casing tree	4	3	4	48
6	Operator remove wax from the rubber mold too soon	There is no schedule to remove the wax from rubber mold in the work instruction	5	5	3	75	Details of schedule to remove wax from the rubber mold in work	5	5	3	45

							instruction was added.				
7	Water pressure used to clean casting plaster is too high	Failure to check water pressure before the operation	4	4	4	64	Water pressure checking form was added	4	3	3	36
8	Water pressure used to clean casting plaster is too low	Failure to check water pressure before the operation	4	4	4	64	Water pressure checking form was added	4	3	3	36
9	Operators incorrectly remove casting tree from the flask	Operators clean the flask at the edge of the flask first, therefore the casting tree hits the cleaning cabinet	3	5	4	60	1.Working instruction to remove casting tree from the flask was added. 2.Training staff to follow WI	3	3	3	27
10	Operators accidently drop the casting tree	Operator rush to clean the casting tree	4	3	4	48	Training staff to follow WI	4	2	3	24

4. Conclusion

This research analyzes the manufacturing process of bracelets, pendants, earrings, and rings. The study found that there was 5 categories of defect of gem bracelets. Root causes of the problem were analyzed by the application of Pareto Charts, Cause and Effect Diagram, and Failure Mode and Effect Analysis (FMEA) together with the experts of the plant in order to find improvements and solutions. The study found 10 root causes of which has been improved by 1. Amendment of machines 2. Improvement of methodology 3. Staff training. After the correction and improvement it was found that RPN was decrease by 48.70% and resulting in a total number of defects decrease by 48.89%. This research can be used as a guide to improve processes, and to reduce defects during in manufacturing of jewelry and can be applied to other product of jewelry industry.

5. Reference

[1] Nancy R. Tague's 2004 *The quality Toolbox* (Second Edition) p 236-240
[2] Bester field Dale 2003 *Total Quality Management* p 449-453
[3] Nancy R. Tague's 2005 *The quality Toolbox* (Second Edition) p 376-378
[4] Ishikawa K (1986). Guide to Quality Control, Tokyo: Asian Productivity Organization.
[5] Balanced Scorecard Institute (2007), Basic tools for process improvement, Module 5 – Cause and Effect diagram, Accessed From: http://www.balancedscorecard.org/files/c-ediag.pdf, Accessed by: March 2, 2011.
[6] Stamatis, D. H. (2003). Failure Mode and Effect Analysis: FMEA from Theory to Execution. (2nd ed.). Milwaukee: American Quality Press, (Chapter 2).

Technology Assessment Need: *Review on Attractiveness and Competitiveness*

Siti Salwa Sait, Farrah Merlinda Muharam, Thoo Ai Chin and Zuraidah Sulaiman

Faculty of Management, Universiti Teknologi Malaysia, Skudai Johor, Malaysia

Email: merlinda@utm.my

Abstract. Technology assessment is crucial in managing technology for the purpose of technology exploitation. With business environment continuously changing, firms have to address this issue critically as technology is considered one of the important elements to evaluate performance and gain competitive advantage. Missteps in deciding the best technology to be developed, employed or maintained would cost the firm overall value. To fulfil the need of finding the appropriate scale to assess suitable technology, this paper summarizes that technology assessment should cover two main aspects, namely technology attractiveness and competitiveness. These components are seen capable to link the scale suggested towards evaluation of financial and non-financial performance towards competitive advantage.

Keywords: technology assessment, technology attractiveness, technology competitiveness, firms' performance, competitive advantage.

1. Introduction

Technology is known as knowledge or science that needs to be applied to a definite purpose. Assessment of technology is one of a crucial tool that all organizations need to focus on to define the suitability of current technology used within a system of organization to achieve better performance and simultaneously creating competitive advantage[1]. The assessment evaluated the needs and competitiveness of current technology and its management process. This will lead to better performance in the future. According to Stewart[2],assessments involve organization's members shared the same mission even though working in different area of expertise. Technology assessment (TA) is also an input for decision making and it is answered some uncertainties which to be more efficient and effective towards improving firm's performance[3]. Besides, other issues concerned about technology assessment were circulated on the issues of quality, scarcity of resources, the viability of the technology, strategy development, and benchmarking[2] and its relations towards organization performance, regardless operation, market or financial[4].

However, to find an appropriate assessment measures which are feasible to be linked to performance for the purpose of achieving competitive advantage is crucial[5]. Normally, problem occurs when managers failed to assess the implementation of technology that eventually will be affected an organization in daily operation and will put the organization left behind the other competitor competing in the same industry. Thus, it leads to the next problem – is there any relationship between performance measurement to firm performance.

The problem above struck debate on what is the effective measurement for organizational to decide on the technology exploitation[4]. Most measurement circulated on financial indicator yet non-financial indicators also important because it lead to be more action and future oriented in order to improve organization's technology capabilities in terms of planning and strategy implementation [5].

The purpose of this review is to identify and propose a suitable measurement of technology that able to access the relevancy of technology towards financial and non-financial performance towards creation of competitive advantage. It is essential as this field is newly emerging discipline that gained focus recently [6]. Besides giving insights to technology based firms on the measurement criteria of technology, the review aims to provide framework for researchers to handle the issue of TA towards firms' performance and gaining competitive advantage.

2. Technology assessment towards firms' performance and competitive advantage

2.1. *Performance Indicators*

Basically, performance measurement is a tool to quantify the efficiency of resources used and effectiveness of the process that will impact on product or service offered [7]. The indicators assist organization's objectives in finding the actual condition that happened in the company in the certain period of time and pushed the company to come out the variety of solutions for the purpose of surviving and sustaining competition from their strong rival. These indicators cover both financial and non-financial aspects.

Financially, profitability ratios measure the total effectiveness of management in generating profits on sales, equity, assets, and owner's investment and considered as the most challenging and critical especially for SMEs [8 9,10]. Profitability is the monetary reflection on variables such as risk, expected demand, industry profit, technology cycles and competition density.

Growth ratio and profitability have a strong relationship due to the mechanism of higher level of profitability lead to increased growth rate [11, 12]. As growth covers market-share growth, assets growth, net revenue growth, net income growth and number of employee growth [4], it has great influence on profitability ratios.

Another financial indicator is market value that identified through terms of earning per share, stock price improvement, dividend yield, stock price volatility, and market value added [4]. Market value indicators reflect the capabilities of business and its competency in creating a unique way to utilize resources.

The other performance indicator is termed as competitive advantage indicators which measures the ability of organizations to gauge customers' and employees' satisfaction, managing environmental issues and social performance. These indicators are non-financial indicators that aims to see how internal resources especially technology are able to be exploited in achieving high customers' and employees' satisfaction, reduce and react to environmental pollution plus its capability to have positive image in the society.

2.2. *Technology Assessment (TA) Scale*

Based on various yet limited literatures available on how technology should be assessed, it is found that early studies circulate on the appropriate approach to value technology. Schot and Rip[13] for example, reviewed past studies and identified several approaches of TA, namely awareness TA, strategic TA, constructive TA and interactive/participatory TA. All these are based on the philosophy aiming to reduce (human) error and costs.

Later, Jolly [14,15,16] is able to build a strong foundation in the aspect of providing an appropriate scale in assessing technology. His researches construct the scales required in TA. The scale is divided into two parts which addressed technology attractiveness and technology competitiveness. As to date, his works are cited by numerous researchers who have the same interest and philosophy as his. Some of them are Padovani & Carvalho (2016), Nagano, Stefanovits & Guimaraes (2014, 2016), Dutra, Ribeiro & Cavalho (2016) and Shin, Coh & Lee (2013).

Jolly[16] emphasized that technology attractiveness does not depend on the firm's action due to the reason that it is beyond control (external variable) while, technology competitiveness depends on the firm's decision and behaviour and it is within firm's control[16]. Based on the author's early publication[14], 16 indicators are used for depicting technological competitiveness and another 16 are being used for describing technological attractiveness.

2.2.1. Technology Attractiveness

Technology attractiveness depends more on the firms external features [16]. The 16 indicators are being divided into four elements which are market potential, competitive situation, and technical potential and socio-political situation [14].

Market potential stressed that market, demand and customers are very important in decision making regarding the technological aspect. However, in the commercialization of new technology is quite challenging and face a lot of uncertainties [17]. Thus, choosing an appropriate of technology is the most important part that a manager needs to pay attention. This is because using an appropriate technology will directly link it to gain technology competence yet will sustain its position in the market [18]. Three criteria developed under market potential which are market volume, the span of application and market sensitivity to the technical factors.

Next, competition situation allows firms to understand the concept of competition intensity [16] in order to invent an initiative to enhance value creation [19]. There are six criteria need to be adopted for assessing competitiveness: number of competition, competitors' level of involvement, intensity of competitiveness, impact of technology on competitive issues, barrier to copy, and potential to develop the dominant design.

In addition, technical potential assessed five (5) criteria covering the element of the technology position in own lifecycle (the S-curve), the potential for progress, the gap with competitive technologies, the threat of substitute technologies, and the potential for unit-to-unit transfer.

Finally, the scale provided by Jolly[16] assessed how social-political affects technology implementation which consider societal stake and public support for technology development. This is important as technology will become more attractive if it able to gain public support in term of financial [20]. Based on these, the research will adopt attractiveness scale of technology by Jolly [16] as summarized in the Table 1 below.

Table 1: The four elements of technology attractiveness [16]

Environmental factors over which the company has a weak control	
Market potential	Market Volume opened by technology
	Span of applications opened by technology
	Market sensitivity to technical factors
Competitive situation	Number of competitors
	Competitors' level of involvement
	Competitive intensity
	Impact of technology on competitive issues
	Barrier to copy or imitation
	Dominant design
Technical potential	Position of technology in its own life-cycle
	Potential for progress
	Performance gap vis-à-vis alternative technologies
	Threat of substitution technologies
	Potential for unit-to-unit transfers
Socio-political situation	Societal stakes
	Public support for development

2.2.2. Technology Competitiveness

Technology competitiveness evaluate internal features of a firm through 16 indicators which are grouped into *value of technological resources* and *value of complementary resources* [14,16]

Competitiveness of technological resources covers nine (9) aspects including tangible assets, intangible asset, and human resources. The first criterion is the origin of the assets, whether it is dependent or independent in its development process. Dependent means that firm totally depends on the external third party asset while independent means that firm has their own asset to develop its technology and both of this method has its own pros and cons [16].

The second criterion includes relatedness to core business which is defined by the relationship between the contributions of technologies implementation toward the company's core business. According to Zhang and Liu [21], applying appropriate technology to the core business is really important since it will directly give high impact on the production efficiency, improve the stability of production, raise the equipment operation rate, reduce consumption of products, and improve the utilization of waste materials.

Third criterion under technological resources also includes the experience accumulated by a firm in the certain technological field. When a firm have an experience and broad knowledge in the specific field is considered as important aspect in order to handle the technology because it will directly reflect the level efficiency of using the technological resources.

The fourth criterion is about the pattern owned by the firm. A firm that patented their new technology is considered as a stronger firm as they can put obstacles toward others firm from access the resource for producing the technology [22]. Firms are considered to gain competitive advantage since their invention on new technology has been patented.

The fifth criterion of technological resource is the value of laboratories and equipment owned by a firm that emphasize about the expertise of R&D staff and its effectiveness in doing research. Mulero Mendigorri, García Valderrama and Rodríguez Cornejo [23] stress that having and R&D staff that has large expertise and know-how on certain aspect would lead to having a valuable outcome.

The sixth and seventh criteria highlight about the selection either applied research or fundamental research that will be more expensive to develop a technology [24,16] while the eighth criterion which is the development of team competencies. This is considered as most crucial part since it will show the successful of a technology [16].

Lastly is diffusion of technological knowledge in the firm. A firm must have a knowledge and capacity on handling technology. Lin and Tang [25] supported that knowledge is very important to improve intellectual capital to gain effectiveness in organizational performance.

Meanwhile, *complementary resources* are similarly important as technology resources. There are seven criteria embedded under this factor. The first criterion is the capability to keep up with fundamental science and technique (S&T). Firm will be able to be successful if they could keep up-to-date with the latest fundamental knowledge in S&T. The knowledge gap between science and technology in a firm will appear due to lack of application of scientific knowledge [26].

The second criterion is capacity of a firm to finance technology development. In this notion, R&D manager should be able to convince the financial manager that certain technology is able to attract and develop financing capability [16].

The third criteria stressed that the technology can be developed through the strong relationship between R&D department and marketing department yet it requires smooth communication between them in order to produce a better technology that could fit customer's need. Pérez-Luño & Cambra [27] found the company that constantly adopt an incremental innovation would be able to connect with the requirement of customer yet it will create competitive advantages which lead to the fourth criterion.

Technology also can be developed through the quality relationship between R&D with production department. Manager should control the transfer of knowledge in order to smooth the production [16]. The fifth criteria of complementary resources is capacity of the firm to protect their technology from being imitated by other by acquiring intellectual property protection of its invention via internal development effort or through external licensing to ensure the possibility of downstream technology

[28] as R&D function cannot disconnect its relationship with the market in producing their own design. Sixth criteria suggest that a firm is stronger when they are able to reach the stage where they could produce a product that become most preferable in design and dominant to the customers [16].

The final criteria are more prone to timing factor. According to Khalil and Shankar [1], action to develop, to industrialize and commercialize must be taken at the right time if a firm wants to succeed in a competitive marketplace. The summary of these two aspects of technology attractiveness is presented in the following table.

Table 2: Two elements of technology competitiveness (Adapted from Jolly[16])

Internal factors over which the company can exert a strong control	
Technological resources	Origin of assets
	Relatedness to the core business
	Experience accumulated in the field
	Registered patents
	Value of laboratories and equipment
	Fundamental research term competencies
	Applied research term competencies
	Development team competencies
	Diffusion in the enterprise
Complementary resources	Capability to keep up with fundamental S&T knowledge
	Financing capacity
	Quality of relationship between R&D & Production
	Quality of relationship between R&D & Marketing
	Capacity to protect against imitation
	Market reaction to the company's design
	Timetable relative to competition

3. Conclusion

The review above highlights the importance of understanding how certain criteria are really important in TA. The scale developed by Jolly [14,15,16] is seen as the most appropriate scale that relates TA towards evaluating firms' performance towards competitive advantage. Taking consideration of technology life cycle (S-curve), this scale is seen capable in addressing continuous TA needs at any stage of the life with specific proposed relationship of both elements of scale towards financial and non-financial performance.

References

[1] Khalil T and Shankar R 2000 *The key to competitiveness and wealth creation*. Mac Graw-Hill International Editions.

[2] Stewart H 2002 *Technology assessment: Making sure we get it right*. [Electronic Version]. EDUCAUSE Center for Applied Research. Research Bulletin, **21**: 1-13

[3] Phaal R, Farrukh C J and Probert D R 2001 Technology management process assessment: A case study *International journal of operations & production management* **21**(8):1116-32

[4] Santos J and Brito L 2012 *Toward a subjective measurement model for firm performance* BAR, Braz. Adm. Rev **9**: 95-117

[5] Teeratansirikool L, Siengthai S, Badir Y and Charoenngam C 2013 Competitive strategies and firm performance: the mediating role of performance measurement. *International Journal of Productivity and Performance Management* **62**(2): 168-84

[6] Martín-Rojas R, García-Morales V and Mihi-Ramírez A 2011 How can we increase Spanish technology firms' performance? *Journal of Knowledge Management* **15**(5): 759-78
[7] Vilanova M R N, Magalhaes Filho P and Balestieri, JA P 2015 Performance measurement and indicators for water supply management: Review and international cases *Renewable and Sustainable Energy Reviews* **43**: 1-12
[8] Moyer R C, McGuigan J and Kretlow W 2005 *Contemporary Financial Management* 10th Edition (New York: South-Western College Publication)
[9] Delen D, Kuzey C and Uyar A 2013 Measuring firm performance using financial ratios: A decision tree approach *Expert Systems with Applications* **40**(10): 3970-83
[10] Madison M 2016 Retrieved 30 August 2016, from http://smallbusiness.chron.com/evaluate-companys-profitability-10401.html
[11] Bottazzi G, Secchi A and Tamagni F 2008 Productivity, profitability and financial performance *Industrial and Corporate Change* **17**(4): 711-51
[12] Hu X and Liu C 2016 Profitability performance assessment in the Australian construction industry: a global relational two-stage DEA method *Construction Management and Economics*, 1-13
[13] Schot J and Rip A 1997 The Past and Future of Constructiove Technology Assessment. *Technological Forecasting and Social Change* **54**: 251-68
[14] Jolly D 2003 The issue of weightings in technology portfolio management *Technovation* **23**(5): 383-91
[15] Jolly D R 2008 Chinese vs. European views regarding technology assessment: Convergent or divergent? *Technovation* **28**(12): 818-30.
[16] Jolly D R 2012 Development of a two-dimensional scale for evaluating technologies in high-tech companies: An empirical examination *Journal of Engineering and Technology Management*, **29**(2):307-29.
[17] Hellman H and Vandenhoed R 2007 Characterising fuel cell technology: Challenges of the commercialisation process. International *Journal of Hydrogen Energy* **32**(3): 305-15
[18] Acur N, Kandemir D, De Weerd-Nederhof P and Song M 2010 Exploring the Impact of Technological Competence Development on Speed and NPD Program Performance *Journal of Product Innovation Management* **27**(6): 915-29
[19] Badawy M 2011 Is open innovation a field of study or a communication barrier to theory development?: A perspective *Technovation* **31**(1): 65-67.
[20] Dai X and Cheng L 2015 The effect of public subsidies on corporate R&D investment: An application of the generalized propensity score *Technological Forecasting and Social Change* **90**:410-19
[21] Zhang R and Liu J 2015 on the Core Competence of Chinese Manufacturing Industry under the New Formats of "Internet Plus"—A Case Study on Furniture Industry *Journal of Service Science and Management* **8**(6): 886- 93
[22] Zane L and DeCarolis D 2016 Social networks and the acquisition of resources by technology-based new ventures *Journal of Small Business & Entrepreneurship* **28**(3):203-21
[23] Mulero Mendigorri, E, García Valderrama T and Rodríguez Cornejo V 2016 Measuring the effectiveness of R & D activities. *Management Decision* **54**(2):321-62
[24] Organization for Economic Co-operation and Development 2005 *OECD in Figures: Statistics on the Members Countries* OECD Publishing
[25] Lin X and Tang H 2016 A Framework for Human Resource Configurations in Knowledge-intensive Organizations *IJBA* **7**(1)
[26] Subramanian A and Soh P 2010 An empirical examination of the science–technology relationship in the biotechnology industry *Journal of Engineering And Technology Management* **27**(3): 160-71
[27] Pérez-Luño A and Cambra J 2013 Listen to the market: Do its complexity and signals make companies more innovative? *Technovation* **33**(6-7):180-92

[28] Allarakhia M and Walsh S 2011 Managing knowledge assets under conditions of radical change: The case of the pharmaceutical industry *Technovation* **31**(2-3): 105-17

IOP Conf. Series: Materials Science and Engineering **215** (2017) 012018 doi:10.1088/1757-899X/215/1/012018

A STUDY OF USING HYDROGEN GAS FOR STEAM BOILER IN CHOLOR– ALKALI MANUFACTURING

Sasitorn Peantong[1] and Somkiat Tangjitsitcharoen[2]
Department Of Industrial Engineering,
Chulalongkorn University, Bangkok 10330, Thailand

E-mail address : sasitorn.peatong@gmail.com[1],somkiat.t@eng.chula.ac.th[2]

Abstract. Main products of manufacturing of Cholor – Alkali, which commonly known as industrial chemical, are chlorine gas (Cl_2), Sodium Hydroxide (NaOH) and hydrogen gas (H_2). Chorine gas and sodium hydroxide are two main products for commercial profit; where hydrogen gas is by product. Most industries release hydrogen gas to atmosphere as it is non-profitable and less commercial scale. This study aims to make the most use of hydrogen as a substitute energy of natural gas for steam boiler to save energy cost. The second target of this study is to reduce level of CO_2 release to air as a consequence of boiler combustion. This study suggests to install boiler that bases on hydrogen as main power with a high turndown ratio of at least 1:6. However, this case study uses boiler with two mode such as natural gas (NG) mode and mixed mode as they need to be flexible for production. Never the less, the best boiler selection is to use single mode energy of hydrogen. The most concerned issue about hydrogen gas is explosion during combustion stage. Stabilization measures at emergency stop is introduced to control H_2 pressure to protect the explosion.
This study varies ratio of natural gas to hydrogen gas to find the optimal level of two energy sources for boiler and measure total consumption through costing model; where CO_2 level is measured at the boiler stack. The result of this study shows that hydrogen gas can be a substitute energy with natural gas and can reduce cost. Natural gas cost saving is 248,846 baht per month and reduce level of NOx is 80 ppm 7% O_2 and 2 % of CO_2 release to air as a consequence of boiler combustion.

1. Introduction

Energy could be divided into two types, which are nonrenewable and renewable. Most energy consumption nowadays comes from nonrenewable energy such as fossil, crude oil, coal and natural gas. The nonrenewable energy is accounted as one of the factor that contributes to the Green House effect and global warming. On the other hand, renewable energy is a source of power that not only saves cost but also does not cause such global warming effect. Those energy include solar energy, wind energy, geothermal energy and hydrogen energy. Hydrogen is an ideal renewable energy carrier due to its high combustion heat and the fact that its only combustion product is water [1]. Natural gas contains 70% of methane. After separation process, it produces clean energy with complete combustion that causes less effect to environment, comparing to other types of petroleum energy. [2]The hydrogen gas is clean energy. The different property between hydrogen and natural gas is level of concentration of flammable content. Hydrogen must contain 4-75% of flammable content while natural gas has 5.3-15%. Therefore, natural gas is more dangerous as it has the shorter range of flammable combustion time frame. [3] Sources of hydrogen can come from many processes.
One of those processes is the exchange of proton by using electrical power and running water through membrane. The equation could be shown as [4]

$$\frac{3}{2}H_2+\frac{3}{4}O_2 \rightarrow \frac{3}{2}\ H_2O)\ \text{fresh water}\ (+214kJ\ )\text{electric energy}(\ +\ 214kJ\ )\text{heat energy} \quad (1)$$

And the chemical reaction in the fuel cells is:

$$H_2+\frac{1}{2}O_2 \rightarrow H_2O + \text{electric energy} + \text{heat energy} \quad (2)$$

However, this case study uses concentrated salt solution. When concentrated salt solution runs through membrane, the process separates anode and cathode from each other. The outcome of the process is soda, choline and hydrogen. Hydrogen gas, as by product, will be released to the air to balance the process. The equation is shown as

$$2Nacl + 2H_2O \rightarrow 2NaOH + H_2 + Cl_2 \quad (3)$$

According to many studies about hydrogen gas, it is considered clean energy, the combustion equation follow as below and applicable for combustion as show in figure1 and 2, respectively
Natural gas combustion as

$$CH_4 + 2O_2 \rightarrow CO_2 + 2H_2O \quad (4)$$

Hydrogen gas combustion as

$$H_2 + 1/2O_2 \rightarrow H_2O \quad (5)$$

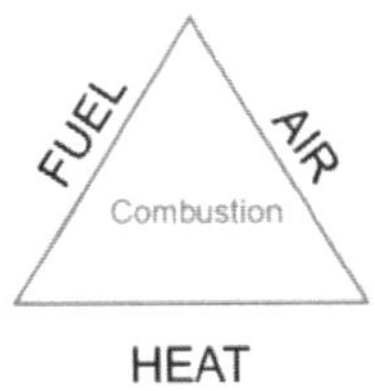

Figure1. Principle of combustion

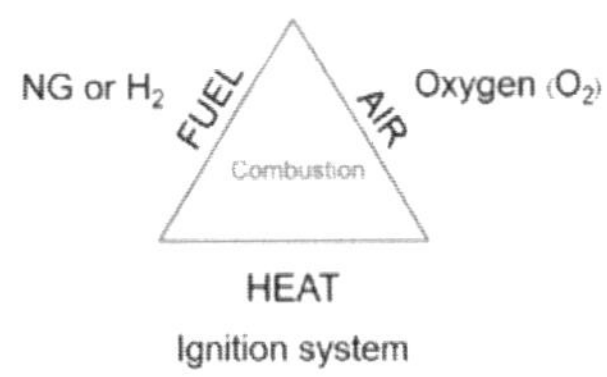

Figure2. Principle of combustion for hydrogen

Hydrogen gas is also known as high efficiency power that could generate outstanding yield. By using 2% of hydrogen gas, it could generate 100% of energy. In short, hydrogen gas is a low cost source of energy that encourages business for high return. [5]

In case study of chemical manufacturing of NaOH and Cl2 for domestic customers, that shows that process of 50 % Caustic soda solution and Caustic Soda Flake 98% uses steam energy from boiler and natural gas as much as 50.85 million baht per year. This study aims to reduce consumption of natural gas for the production by using hydrogen gas as substitute energy source for boiler's steam production. Expected result from using hydrogen s substitute is to reduce energy cost by 80% comparing to the existing method.

2. Analysis data and Experimental

2.1. Understanding of Current Situation

Experimental boiler in this case study has 6mt steam capacity per hour and powered by natural gas) NG(. Estimated energy consumption of boiler is 13,000 MMBtu, given a 22 years life span. The quantity of natural gas consumption from steam boiler is increasing every year. Figure 3 and 4 show the consumption data from the last 2 years. And the process flow of target facility and the structure of recycling hydrogen for boiler, respectively

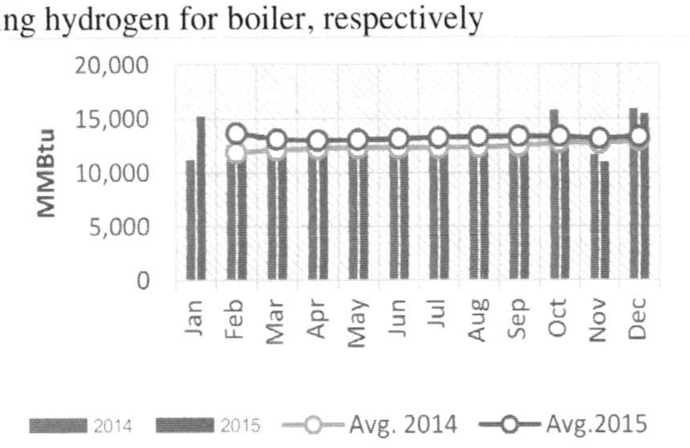

Figure 3. Quantity of natural gas consumption hydrogen for boiler.

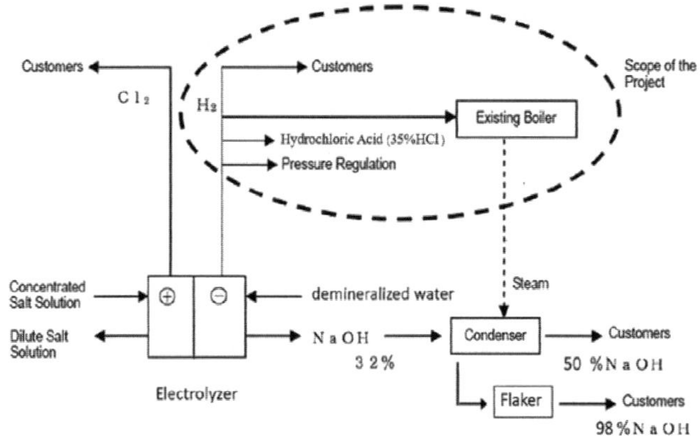

$$2NaCl + 2H_2O \Rightarrow 2NaOH + H_2 + Cl_2$$

Figure 4. Shows structure of recycling from steam production in 2014-2015

Hydrogen is produced after concentrated salt solution pass through electrolysis process. It is contained in hydrogen station with header and will be pressurized by root blower to transport to customer gas station. Later, hydrogen will be sent to synthetic tower for 35% hydrochloric acid production and the excess hydrogen from the system will be released to atmosphere to stabilize as show in figure 5

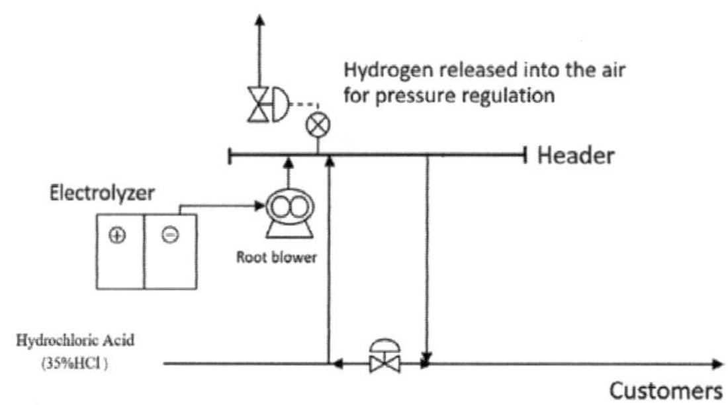

Figure 5. Hydrogen flow before implementation

2.2. Analysis of Current Situation

Production can generate hydrogen 6.94Mts per day per plant. The remaining from production is 1.32 Mts per day per plant. Hydrogen will be released to air to stabilize as show in figure 6

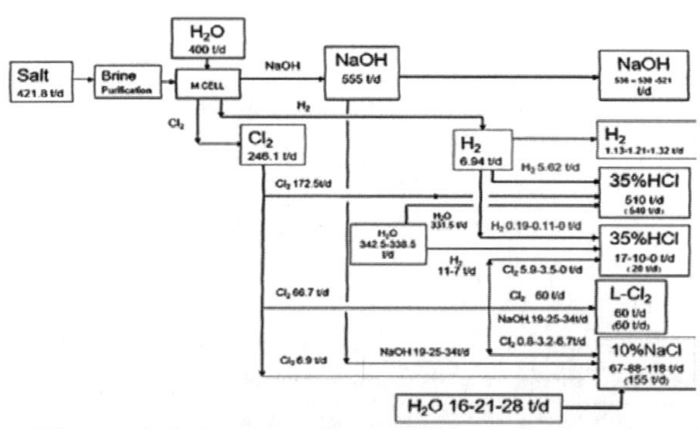

Figure 6. Balancing of material in case of study

The production capacity is relevant with electrical cost calculation from Provincial Electricity Authority (electricity power plant of Thailand) It accounts for peak consumption of electrical power. Monday to Friday between 9am to 10pm is considered peak hour and the production would run with 42% capacity while 10pm to 9am Monday to Friday, Saturday, Sunday and public holiday; which considered off peak, the production is on 100% capacity. As a result, quantity of hydrogen on each period of time are different as shown in figure 7.

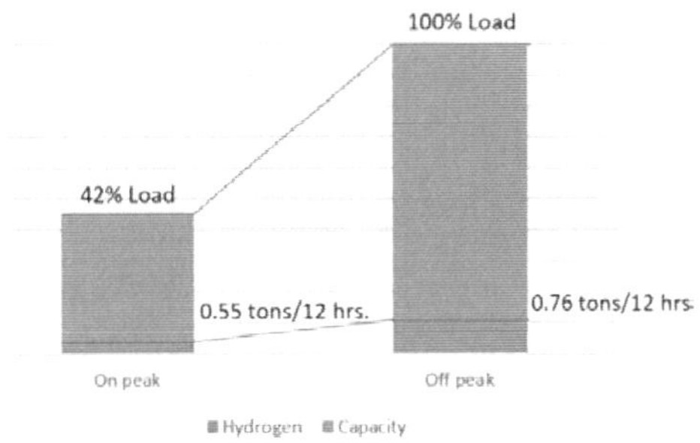

Figure 7. H_2 capacity base on Time of Use Tariff.

2.3 Experimental and Procedure

The experimental setup and Procedure are summarizing in Table 1

Target Settings	Procedure	Requirement
To reduce natural gas consumption and energy cost for boiler by using hydrogen as substitute energy.	Use boiler that base on hydrogen as main power.	* Selection of boiler with a high turndown ratio * Measures to reduce NO_x generation * Stabilization measures at emergency stop * Selection of boiler with three mode such as NG mode, H_2 mode and mixed mode * Determination of ratio for balance hydrogen gas and natural gas in case mixed mode
To reduce level of H_2 and CO_2 release to air as a consequence of boiler combustion	By using a new hydrogen boiler and adjust the fluctuated pressure.	* Determination of holder capacity and pressure regulation for balancing load

Table 1 The Target Settings, Procedure and requirement

2.4 Boiler selection

The most efficient way is to apply the concept with hydrogen base boiler. Never the less, this study introduces both hydrogen base energy boiler and mixed mode energy boiler.

Specification of boiler

Fire Tube, Cylindrical Horizontal, 3 pass, and wet back design

Fuel System:
Gaseous fuel: Natural Gas

Burner capacity :4.5 MW
Data for gaseous fuel
Heat value :31.4 MJ/m3
Standard volume flow max. :516 m3/hr
Standard volume flow min. :125 m3/hr

Gas temperature :15 ºC
Gas pressure max. : 3.8 bar
Gas pressure min. : 2 bar
Protection pressure :4 bar (by others)

Second fuel: Hydrogen (H_2)

Burner capacity : 3.5 MW
Heat value :10.76MJ/m3
Standard volume flow max. :1167 m3/hr
Standard volume flow min. : 300 m3/hr

Gas temperature :15 ºC
Gas pressure max. : 0.08 bar (controlled gas pressure)
Gas pressure min. : 0.07 bar (controller gas pressure)
Protection pressure :0.5 bar (by others)

Stabilization measures at emergency stop

As double electromagnetic valve, when the new boiler stops for an emergency, a maximum pressure of hydrogen equal 500 mbar with initial shut-off and minimum pressure monitor with initial shut-off and venting.

3. Combustion process analysis

the following average results were obtained from the operations in this study

Case 1. The operations for natural gas (NG)

Table 2. NG mode

Item	Steam Ton	Flow NG M-M_3	% Load (%)	NOx ppm 7% O_2	CO_2 (%)	Excess O_2 (%)
NG Fire mode	1,836	66,428	72	80	2	-

Case 2. The operations for mixed mode (H_2+NG)

Mixed mode H_2+NG details for mixed mode as following:

Mixed mode operation for 2 main fuels with priority logic. If the burner is supplied with more than one main fuel, in dual-fuel mode with two main fuels, one of them will be fired by priority and the other main fuel is fired along with it at minimum firing rate. When not enough of the priority fuel is available, the rest of the firing plant's power requirement is made up by the other main fuel. If the supply of the priority fuel drops below a set minimum, the fuel valves for it will be closed and will not be opened again until it is available again.

Table 3. H2 and NG mixed mode (80:20)

Item	Steam	Flow NG	Flow H2	NOx	CO2	Excess O2
	Ton	M3	M3	ppm 7%O2	(%)	(%)
H2 and NG mixed mode	840	19,478	119,278	-	-	-

4. Conclusion

This case study investigated the hydrogen energy performance of boiler and the aim is to use hydrogen efficiently.The experimental results shows that hydrogen could be used as substitute energy and reduce energy cost. Natural gas cost saving is 248,846 bath per month and reduce level of NOx and CO_2 release to air as a consequence of boiler combustion are 80 ppm 7% O_2 and 2 %.Hydrogen gas is clean energy. Hydrogen can implement substitute energy concept to other kind of industry.

5.Reference

[1] Huai-Jun Lin,Towards easily tunable hydrogen storage via a hydrogen-induced glass-to-glass transition in Mg-based metallic glasses, Acta Materialia 120 (2016) 68-74
[2] Cagdas Aksu. Extension of PREMIER combustion operation range using split micro pilot fuel injection in a dual fuel natural gas compression ignition engine: A performance-based and visual investigation, Fuel 185 (2016) 243–253
[3] M. FISCHER.SAFETY ASPECTS OF HYDROGEN COMBUSTION IN HYDROGEN ENERGY SYSTEMS. Int. J, Hydrogen Energy, Vol. 11. No. 9)1986(, page 593-601.
[4] Shani Elitzur. Combined energy production and waste management in manned spacecraft utilizing on-demand hydrogen production and fuel cells, Acta Astronautica 128 (2016) 580–583
[5] David Lewis. Hydrogen and its relationship with nuclear energy. Progress in Nuclear Energy 50 (2008) page 394-401

Effects of Perceived Risks, Reputation and Electronic Word of Mouth (E-WOM) on Collaborative Consumption of Uber Car Sharing Service

Mega Wati Hawapi[1, a], **Zuraidah Sulaiman**[2, b*], **Umar Haiyat Abdul Kohar**[3, b], **Noraini Abu Talib**[4, d]

[1, 2, 3, 4, 5]Department of Business Administration, Faculty of Management, Universiti Teknologi Malaysia (UTM), 81310 Skudai, Johor, Malaysia

[a]megawati.uts@gmail.com, [b]zuraidahs@utm.my, [c]umarhaiyat@utm.my, [d]m-aini@utm.my

Abstract. Current transition from traditional economic model of selling and buying to sharing economic business creates a huge impact on consumers' preferences to participate in collaborative consumption. The market entrance of sharing economic business is relatively new, thus it builds scepticism among consumers. Consumers' trust becomes the most crucial aspect in determining their willingness to participate in collaborative consumption. This study will reveal the effects of perceived risks (performance and social), reputation and Electronic Word-of Mouth (E-WOM) on Malaysian consumers' intention of collaborative consumption, especially for Uber car sharing service. This study inspires to enrich the literature for collaborative consumption and perceived risk theory. From the practical perspective, this study may provide insights in assisting the collaborative consumption service providers especially Uber car users on factors influencing the intention to engage in such service.

1. Introduction

The presence of Web 2.0 has enabled countless opportunities for new market entrance over the boundaries. One of the most widely discussed subjects regarding new market entrance as a consequence from the presence of Web 2.0 is "sharing economy business". This innovative economic model is describes as the ability of individuals or groups to generate income from underutilized assets. The demand of this innovative economic model has been derived from the growth of world population which resulting in a lack of resources. Thus, consumers are seeking for the best solutions to fulfil their needs.

Sharing economy business model also promise more sustainable way to utilize the resources. In the recent years, collaborative consumption which falls into the category of sharing economy business model has benefited from the existence of Web 2.0. This peer-to-peer based activity is conducted through online platform as a medium of interaction. Collaborative consumption is classified into three types of systems including product service systems, redistribution markets and collaborative lifestyles[1]. By focusing on product service system that emphasizes on product usage rather than ownership, this research stressed on car sharing service. The statistics of car sharing users recorded by Statisa in 2016 shows rapid increase of consumer acceptance towards collaborative consumption.

this study aims to measure the effects of perceived risk, reputation and Electronic Word of Mouth (E-WOM) on Malaysian's collaborative consumption of car sharing service and utilizes measurement items from the previous research.

1.1. Problem Statements and Objectives

The most dominant car sharing providers in Malaysia are Uber and GrabCar. Uber car sharing service covers several areas including Kuala Lumpur, Johor Bahru, Ipoh, Melaka, Penang and Kota Kinabalu. The growing market of collaborative consumption mostly in car sharing service has increased the competition of traditional business model that focused on product and service ownership such as car dealers, rental car companies and taxi service. Current market situation in Malaysia creates a lot of opposition from traditional business model. Some criminal issues arise from the conflict of taxi drivers who oppose the emergence of car sharing service in Malaysia. This situation also affects the trustworthiness of collaborative consumption among consumers. In order to provide better exposure and increase consumer participation in car sharing service, the risk associated with this new market should be identified.

The research trend regarding collaborative consumption has increased over the past several years. Previous researchers believe that consumer participation in collaborative consumption is highly related with the burden of ownership. The risk associated with product ownership is the most important factor that makes consumer switch into collaborative consumption[2]. Other researchers focusing on trust model in identifying the risk associated with collaborative consumption[3]. However, the component of perceived risk is not widely used in determining the factors that affecting consumer intention to participate in collaborative consumption.

Based on the arguments above, this study is specifically aims to fulfill the following objectives:

1. To identify the relationship between the independent variables (performance risk, social risk, electronic word-of-mouth (E-WOM) and reputation) on consumer intention toward collaborative consumption.
2. To recognize the most important factor that influence consumer intention toward collaborative consumption.
3. To investigate the influence of demographic profile groups (age, gender, race, income, education background, household size, and car ownership) on consumer intention toward collaborative consumption.

2. Literature Review

2.1. Sharing Economy Business

This new economic model is no longer focusing on product ownership but rather emphasize on the product usage. Sharing economy is often associated with peer-to-peer model where information technology becomes the platform for people to offer and share underutilized resourceswhich are financial, social and environmental[4]. Although sharing economy has resulted limited access to the product, but this innovative concept also reduced the transaction cost because the cost of ownership is not incurred[5]. Temporary access right of the product in sharing economy business is classified as accessibility-based system which is contrary to the concept of traditional market[6].

Sharing sites act as a platform for sharing economy business provider to connect directly with consumer and create disintermediation across the supply chain[7]. Since digital platform has become the enablers of sharing economy, technology acceptance will also affect the adoption level of sharing economy business. 'Technophilia' or dependency on internet to do things more efficiently and easy is one of the factors that affect their adoption on sharing economy[5].

Sharing economy business is often associated with sustainability which defined as the ability of resources to meet current needs without compromising the ability of future generations to meet their needs. Through sharing economy, everyone is able to fully utilize their resources by mitigating the cost of ownership through the act of sharing, renting, lending, subscribing, reselling, swapping and/or donating[8]. Environmental consciousness is also one of the factors that derive the demand on this innovative economic model[4].

2.2. Performance Risk

Performance risk refers to the chances of item failing to meet the performance requirements originally intended of the purchase[9].Performance risk is highly affected customer satisfaction level when the service purchased is not completed in the manner as they expected. Performance risk is similar with functional risk which involves the consumer's belief that a purchased product or service will not offer preferred benefits to a consumer[10]. This risk is perceived more prominently when the consumers are unable to try the product or service before purchasing.

Consumers are concern of their privacy in collaborative consumption because they are utilizes items which are shared by others[11]. However, other perspectives believe performance risk of product ownership will resulted in consumer to shift from traditional economic model to access based consumption.

Most of car sharing drivers are non-professional drivers and they are likely to be unfamiliar with certain road or places. Statistics from Benenson Strategy Group shows the percentage of Uber drivers without previous professional driving experience is made up of 64%. Consumers also concern about the availability of the car at their desired time. The risk of product availability is classified as perceived product scarcity risk[12]. Since the activity of collaborative consumption is done through the use of online platform, the complexity an online platform application is also positively influenced a consumer decision[13].

2.3. Social Risk

Social risk is the external factor of psychological risk which described as the likelihood of the purchase resulting in others thinking of the consumer less favourably[9]. Social risk is also defined as the potential loss of status in social group as a result of adopting a product or service. Social risk is concerned with the possibility of attracting unfavourable attention and response from purchasing a particular product. Social risk also presence as the results from the social embarrassment and loss of social esteem when the user's friends or family comparing the brand image with the image they have on them, as well as the internal psychological disappointment at oneself to make a purchase which is not consistent with their image[14].

Car sharing service is often associated as the biggest rival for taxi and other licensed transportation service. The rise of the sharing economy is predicted to have a major societal impact, and thus holds relevance to both practitioners and policy makers[6]. One of the reasons in refusal of social acceptance towards collaborative consumption is because of the unfair tax imposition and licensing.

As of mid-2015, protests had been staged in Germany, Spain, France, and the United Kingdom and dangerous incidents involving passengers have been documented regarding car sharing service. Similar to other countries, car sharing service such as Uber and GrabCar is a new evolution in Malaysia that gain different acceptance among citizen. Therefore the decision to participate in car sharing service is often depends on the perceived social risk that clarify the acceptace of a consumer social circle towards their decision[12].

2.4. Reputation

Reputation is described as consumer's perceived character of certain brand or overall perception of the substantial characteristics about the companies or brands[15]. Reputation is also explained as the value judgment from a valuer based on accumulated past behaviour of the valuee. There is a time lag effect in reputation where it is accurate after the latest transaction, dependable on previous transaction and become less accurate over the time. Overall assessment based on long term behaviourhas distinguished reputation from risk, which is associated with temporary behaviour assessment[3].

Reputation of collaborative consumption increased in accordance with a sharp rise in consumer participation over the years. Based on the statistics from The Economist in 2013,sharing economy business specifically in peer-to-peer rental market is valued as 26 billion dollars U.S. MIT Sloan Experts in 2011 estimated peer-to-peer exchange goods and services is to dominate the market with the total value of 110 billion dollars U.S. Since collaborative consumption has become one of the biggest threats for traditional economic model, it is important to sustain positive reputation. Due to the product quality that cannot be measured easily, the feeling of uncertainty in consumer mind will increase during the purchase decision process[16]. Thus, reputation plays a vital role in creating first impression that will affect consumer decision.

There are several ways that have been implemented by car sharing service companies in order to improve their reputation including online rating system; whereas adding photos and videos is another tool that has been use to build reputation in collaborative consumption[17]. By using the medium of online platform, these testimonials are able to reduce the risk of uncertainty and make collaborative consumption as a safer choice. Some collaborative consumption companies use the medium of third party in order to enhance their reputational ratings. As a conclusion, reputation in collaborative consumption is a collection of recommendation which prevent some malicious recommendations from affecting overall reputation of platform provider[18].

2.5. Electronic Word of Mouth (E-WOM)

Word-of-mouth (WOM) is initially defined as physical communication between non-commercial entities about a product of the company. The term of Word-of-Mouth evolved in line with adoption of technology where internet become enablers for consumer to channel their thoughts regarding consumption experience in collective and unbiased manner[19].By referring to the Dell Hell scandal in the early stage of Web 2.0 development, it is undeniable that Electronic Word-of-Mouth creates powerful impact on brand image.

Electronic Word-of-Mouth plays an important role mostly for intangible product or service such as collaborative consumption. By focusing on hospitality and tourism industry to illustrate intangibility of product, previous researcher described WOM as the most influential information source that affect consumer decision[20]. This is because the level of difficulty in evaluating intangible product is very high prior to consumption. In order to provide better platform for consumer to channel their opinion or thoughts regarding collaborative consumption, platform providers usually develop an online rating system where consumer are able to give the reviews regarding their consumption. Electronic Word-of-Mouth about collaborative consumption may also occur in several form including photos and video sharing[17]. In the other hand, E-WOM alsohas positive influence on perceived value and perceived risk of collaborative consumption[13]. Since car sharing service is relatively new in Malaysian market, this research believes that car sharing service is highly determined by Electronic word-of-mouth.

2.6. Collaborative Consumption

Collaborative involve two or more parties in order to achieve mutual objective at the same time. Whereas consumption can be define as the usage of particular resources. Collaborative consumption is one of the branches in sharing economy business. Among other activities that involved in sharing economy business are cooperatives, co-housing, co-creation, and Wikinomics[21]. Collaborative consumption is defined as "peer-to-peer based activity of services, coordinated through community-based online services"[22].

There are several factors that motivates consumer to participate in collaborative consumption including cost saving, sustainability, self-belonging towards the society and to avoid the burdens of ownership. In term of sustainability, the resource efficiency in production and technological innovations are inadequate to reduce the current use of natural resources. Thus, collaborative consumption will increase to potential to conserve resources[23].

The growing use of information technologies in the Web 2.0 also generates tremendous impact on collaborative consumption. The Web 2.0 Era has increased the amount of user-generated content, as well as amount of information that is created and consumed online[24]. In fact, sharing platform is describes as a system where content generation is highly distributed and scattered due to strong user's personalized control[25].

3. Research Hypotheses

This research will examine the effect of independent variables which are perceived risk, reputation and Electronic Word of Mouth (E-WOM) on intention of collaborative consumption. However, only two relevant components of perceived risk will be used as independent variables, which are performance risk and social risk. The following statements are hypothesis that will be used in this study:

H1 *There is a significant and positive relationship between performance risk and consumer intention of collaborative consumption.*

H2 *There is a significant and positive relationship between social risk and consumer intention of collaborative consumption.*

H3 *There is a significant and positive relationship between reputation and consumer intention of collaborative consumption.*

H4 *There is a significant and positive relationship between electronic word of mouth (E-WOM) and consumer intention of collaborative consumption.*

The following Figure 1 is a conceptual framework depicting all the hypotheses outlined above:

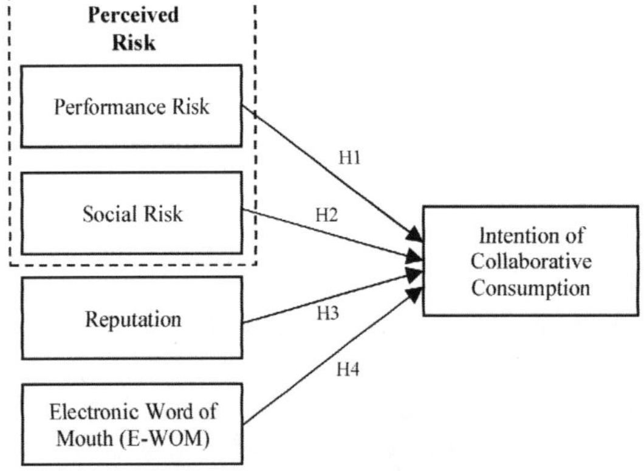

Figure 1: Conceptual Framework

4. Research Instrument

The questionnaire will be distributed to the respondents through physical and online platform. The questionnaire consists of four sections. Section A will include screening question, while Section B contains four factors of independent variables (performance risk, social risk, reputation and electronic word-of-mouth (eWOM). Section C will cover several questions regarding the dependent variable (intention towards collaborative consumption). Finally, the respondents will answer several demographic questions including age, gender, race, income, education background, household size, and car ownership. All scale items are adopted and adapted from previous research.

4.1. Performance risk

1. There is high likelihood that the car I want will not be available when I want it.
2. There is high likelihood that the car I want will not arrive on time.
3. I cannot examine the quality of Uber car.
4. I may have problems when riding in a stranger's car.
5. It is too complicated to use Uber application.

4.2. Social Risk

1. Sharing a car allows me to be a part of a group of like-minded people.
2. Sharing a car makes me feel unfair to taxi drivers.
3. By sharing a car, my peers will regard me as a stingy person.
4. My friends would approve of the sharing option.
5. My family would approve of the sharing option.

4.3. Reputation

1. Uber service is as good as expected.
2. Uber service is accurate, but with some degradation, e.g. delay for service.
3. I have experience where Uber rejected my request.
4. I have experience where Uber gave the wrong or malicious response for my request.
5. I have a better option than using Uber.

4.4. Electronic Word of Mouth (E-WOM)

1. I constantly follow specific online media that provide information about Uber service.
2. I often read other rider's online reviews to know whether Uber makes a good impression on others.
3. I often consult other user's review to use Uber service.
4. I frequently gather online information before I choose to request an Uber service.
5. If I do not read others' online reviews, I worry about my decision to use Uber service.

4.5. Intention of Collaborative Consumption

1. Collaborative consumption through car sharing service is a better mode of consumption rather than buying a car.
2. All things I considered, I expect to continue collaborative consumption often in the future.
3. My participation in collaborative consumption benefits me financially.
4. My participation in collaborative consumption saves my time.
5. Collaborative consumption helps save natural resources.

5. Conclusions

This study aims to identify the relationships between performance risk, social risk, reputation, electronic word-of-mouth, and other demographic factors toward consumer intention of collaborative consumption. This article outlines the problem statement and objectives that become the cause of the research being conducted and reviews the relevant marketing literature on collaborative consumption, risk, reputation, and E-WOM. It then highlights the hypotheses and scale items to measure the proposed relationships. It is envisaged that future articles to be generated from this study will discuss in greater details on the descriptive and inferential results, as well as theoretical and managerial implications of this study. Findings from this study will assist the collaborative consumption service providers especially Uber car users in improving their services in order to compete in the transportation market strategically.

References

[1] Botsman, R., & Rogers, R. (2010). What's mine is yours. The Rise of Collaborative Consumption.
[2] Schaefers, T., Lawson, S. J., & Kukar-Kinney, M. (2015). How the burdens of ownership promote consumer usage of access-based services. Marketing Letters, 1-9.
[3] Liang, Z., & Shi, W. (2005, January). PET: A personalized trust model with reputation and risk evaluation for P2P resource sharing. In System Sciences, 2005. HICSS'05. Proceedings of the 38th Annual Hawaii International Conference on (pp. 201b-201b). IEEE.
[4] Cohen, B., & Kietzmann, J. (2014). Ride on! Mobility business models for the sharing economy. Organization & Environment, 27(3), 279-296.
[5] Schor, J. B., & Fitzmaurice, C. J. (2015). 26. Collaborating and connecting: the emergence of the sharing economy. Handbook of Research on Sustainable Consumption, 410.
[6] Dervojeda, K., Verzijl, D., Nagtegaal, F., Lengton, M., Rouwmaat, E., Monfardini, E., & Frideres, L. (2013). The sharing economy: accessibility based business models for peer-to-peer markets. European Commission Business Innovation Observatory," September.
[7] Demary, V. (2015). Competition in the sharing economy (No. 19/2015). IW policy paper.
[8] PricewaterhouseCoopers, L. L. P. (2015). The sharing economy. Report, Consumer Intelligence Serice.
[9] Pires, G., Stanton, J., & Eckford, A. (2004). Influences on the perceived risk of purchasing online. Journal of Consumer Behaviour, 4(2), 118-131.
[10] Kim, L. H., Kim, D. J., & Leong, J. K. (2005). The effect of perceived risk on purchase intention in purchasing airline tickets online. Journal of Hospitality & Leisure Marketing, 13(2), 33-53.
[11] Bardhi, F., & Eckhardt, G. M. (2012). Access-based consumption: The case of car sharing. Journal of Consumer Research, 39(4), 881-898.
[12] Lamberton, C. P., & Rose, R. L. (2012). When is ours better than mine? A framework for understanding and altering participation in commercial sharing systems. Journal of Marketing, 76(4), 109-125.
[13] Liang, L. J. (2015). Understanding Repurchase Intention of Airbnb Consumers: Perceived Authenticity, EWoM and Price Sensitivity (Doctoral dissertation).
[14] Mitchell, V. W., & Harris, G. (2005). The importance of consumers' perceived risk in retail strategy. European Journal of marketing, 39(7/8), 821-837.
[15] Fombrun, C. J., & Rindova, V. P. (2000). The road to transparency: Reputation management at Royal Dutch/Shell. The expressive organization, 7, 7-96.
[16] Erdem, T., & Swait, J. (1998). Brand equity as a signaling phenomenon. Journal of consumer Psychology, 7(2), 131-157.
[17] Tovey, M., & Masum, H. (2011). The reputation society.
[18] Wang, Y., & Vassileva, J. (2003, September). Trust and reputation model in peer-to-peer networks. In Peer-to-Peer Computing, 2003.(P2P 2003). Proceedings. Third International Conference on (pp. 150-157). IEEE.
[19] Hennig-Thurau, T., Gwinner, K., Walsh, G., & Gremler, D. (2004). Electronic word of mouth via consumer- opinion platforms: what motivates consumers to articulate themselves on the Internet? Journal of Interactive Marketing, 18(1),38–52
[20] Litvin, S. W., Goldsmith, R. E., & Pan, B. (2008). Electronic word-of-mouth in hospitality and tourism management. Tourism management, 29(3), 458-468.
[21] Buczynski, B. (2013). Sharing is good: How to save money, time and resources through

collaborative consumption. New Society Publishers.
[22] Hamari, J., Sjöklint, M., & Ukkonen, A. (2015). The sharing economy: Why people participate in collaborative consumption. Journal of the Association for Information Science and Technology.
[23] Leismann, K., Schmitt, M., Rohn, H., & Baedeker, C. (2013). Collaborative consumption: towards a resource-saving consumption culture. Resources, 2(3), 184-203.
[24] Kaplan, A. M., & Haenlein, M. (2010). Users of the world, unite! The challenges and opportunities of Social Media. Business horizons, 53(1), 59-68.
[25] Rodrigues, R., & Druschel, P. (2010). Peer-to-peer systems. Communications of the ACM, 53(10), 72-82.

IOP Conf. Series: Materials Science and Engineering **215** (2017) 012020 doi:10.1088/1757-899X/215/1/012020

E-WOM Review Adoption: Consumers' Demographic Profile Influence on Green Purchase Intention

Roslin Abdul Rahim[1,a], Zuraidah Sulaiman*[2,b], Thoo Ai Chin[3,c], Mohd Shoki Mohd Arif[4,d] and Mohd Hakim Abdul Hamid[5,e]

[1,2,3,4] Department of Business Administration (Marketing), Faculty of Management, Universiti Teknologi Malaysia (UTM), 81310 Skudai, Johor, Malaysia.
[5] Department of Computer System and Communication, Faculty of Information and Communication Technology, University Teknikal Malaysia Melaka (UTEM),76100 Melaka, Malaysia.

[a]roslinrahim89@gmail.com, [b]zuraidahs@utm.my, [c]acthoo@utm.my, [d]m-shoki@utm.my, [e]mohdhakim@utem.edu.my

Abstract. Nowadays, green products are getting popular in their acceptance by the Malaysian consumers. Due to the advancement of the Internet technologies and the wide spread of electronic word of mouth (E-WOM), consumers seem to be more influenced in purchasing the green products. In this study, consumers' demographic profiles, such as age, gender, income, education background, and occupation are being explored to investigate their influences on consumers' green product purchase intention. The purpose of this paper is to showcase the results of the differences between several demographic profile groups on green product purchase intention using descriptive analysis, ANOVA and independent sample T-Test. T-test results showed that there is a statistically significant difference between gender on consumers' green product purchase intention. Meanwhile, the results generated by ANOVA indicated that there are no significant differences between age, income, education background and occupation on consumers' green product purchase intention. These results shed light on the potential market segment that should be targeted by marketers and producers of green products in Malaysia.

1. Introduction

Electronic Word of Mouth (E-WOM) has been regarded as an important resource for consumers to seek information about products or services in the marketplace. Personal blogs, discussion forums, online communities, personal e-mail, chat rooms, instant messaging, social media sites (e.g. Twitter, Facebook, Instagram) and online customer review are used to communicate and disseminate E-WOM. By using E-WOM, a consumer can share his or her opinions and experiences related to a product with complete strangers who are socially and geographically dispersed. E-WOM has become an important factor for consumers in purchasing intention.

Currently, environmental issues receive a high attention not only from the governments, but also from firms and companies, whereby they begin to incorporate green values into their marketing strategies. Green products are known to have less environmental impact or are less detrimental to human health than traditional products. "Green" can be defined as low polluting, recyclable and resource saving [1]. The purchase intention of the green products and attitudes toward environmental friendly products are also affected by the socio economic characteristics such as gender, age, education, income and occupation. For instance, seniors or the older generation tend to have higher buying decision for green products because they tend to spend more time on reading and engage in recycling than young people [2]. This is an example how a person's environmental belief and behaviour can be explained by his age [3]. This study aims to investigate the influences of consumers' demographics profile, such as gender, age, education, income and occupation on the consumers' green purchase intention.

2. Demographic Profile and Green Purchase Intention

It is known that demographics are based on factors such as age, race, gender, education, income, education and occupation of a population. Demographic variables such as age, gender, race, income, education and occupation have a strong relationship towards a consumers' purchase intention [4].

2.1. Gender and green purchase intention

In purchasing a product, females and males have their own tastes and preferences. Females are more likely to intend to purchase green products because they are more concerned about environmental attitudes and environmental concerns compared to male consumers [5]. A research by [4] indicated that female consumers have a higher purchase intention for an environmentally friendly food product than male consumers. It is shown in their research that gender has a strong relationship with purchase intention.

2.2 Age and green purchase intention

Based on the previous research in [6], age has an impact on buying products which show that younger people make simpler product purchasing decisions compared to older people whose purchasing decisions are more complex because they are more cautious in choosing a product. Young consumers are willing to accept new and innovative ideas, which most of the supporters of green products are [5].

2.3 Education level and green purchase intention

In deciding on the green purchasing, education level signifies an important decision to the consumer. High level of education is an important background factor that facilitates green purchase intentions [2]. Previous study shows that consumers with higher education are more knowledgeable about green products and know the benefits of green products. Education level is one of the important factors for consumers to be more concerned with environmental issues [7]. Educational are important for building attitudes, activity levels, and knowledge about nature and the environment to consumers. With this knowledge, consumers intend to purchase green products are high [8].

2.4 Income level and green purchase intention

Income level has a positive relationship with environment sensitivity in that consumers with high incomes are more willing to buy green products compared to consumers with low income [9]. Green products are known as being a bit expensive compare to non-green products. Most low income consumers do not intend to buy green products and will normally buy the cheapest product. Hence, it is predicted that the low income consumers will not prefer to purchase the green products.

3. Methodology

This research adopted a conclusive research design. Specifically, this research used the descriptive research method. The major objective of descriptive research is to describe the data, usually to describe market characteristics or functions [10]. In this research, demographic profile (gender, age, occupation, education and income) were investigated whether there are differences between the groups on consumers' green purchase intention. Cross-sectional design was used as this research involved one-time collection of information from the Malaysian population. Non-probability sampling technique which is convenient sampling was used in this research. To get the right sample size from the Malaysian population, the calculation is made according to [11]. Based on the calculation, the minimum number of respondents needed was 175. However, a sample size of 200 respondents was recommended to be sufficient for data analysis [10]. Data was collected online using Google forms and the link to the survey was shared in social media such as Facebook and Twitter. Data was then analyzed by using the Statistical Package for Social Science software version 22.0.

4. Research result

A total of 209 respondents were received and used in subsequent analysis. In this study, the age range was from 21 to 60 years old above. The research model was tested using the SPSS 22.0 for descriptive, reliability, ANOVA and independent t-test analysis.

4.1 Instrument reliability

Cronbach's Alpha Coefficient was used to test the reliability of the variables. Table 1 shows the summary of item reliability.

Table 1 Summary of items reliability

Variables	Final Items in Questionnaire	Cronbach's Alpha Coefficient Value
Independent Variables		
Sender Expertise	8	0.944
Receiver Search Extent	5	0.917
Message Trustworthiness	7	0.923
Dependent Variable		
Green Purchase Intention	5	0.897
Mediator Variable		
E-WOM Review Adoption	4	0.909

Based on Table 1, Cronbach's Alpha for the variables ranged from the lowest of 0.897 (Green Purchase Intention) to 0.944 (Sender Expertise). In general, if the value of Cronbach's Alpha is less than 0.60, the scale is considered to be poor while scores that exceed 0.60 to 0.70 are acceptable [11]. In conclusion, the result showed that the scores of the Cronbach's Alpha for all the variables used in this research exceeded the preferable scores of 0.70 and they indicated that the measurements scales of the variables were reliable for this research.

4.2 Descriptive result

Descriptive result shows the characteristic of the respondents for this research. The respondents' profile is presented in Table 2. Gender ratio of the respondents was 68.4% female and 31.6% male. The results showed that 77.1% of respondents attained high education level. 74.2% of the respondents were 21-30 years old, and 20.1% of the respondents were between 31-40 years old. Around 38.8% of the respondents had an average monthly income of below RM 2,000 and most of the respondents are employed by 64.6%. Figure 1 until Figure 5 show the graph of each demographic of respondents.

Table 2 Demographic profile of respondents

Demography	Number of Respondents	Percentages (%)
Gender		
Male	66	31.6
Female	143	68.4
Age		
21-30 years old	155	74.2
31-40 years old	42	20.1
41-50 years old	9	4.3
51-60 years old	3	1.4

60 years old above	0	0
Education		
Primary School	0	0
Secondary School	8	3.8
Certificate	7	3.3
Diploma	33	15.8
Degree	117	56.0
Master/PHD	44	21.1
Income		
RM 2,000 above	81	38.8
RM 2,000-RM 2,999	67	32.1
RM 3,000-RM 3,999	32	15.3
RM 4,000-RM 4,999	23	11.0
RM 5,000 above	6	2.9
Occupation		
Employed	135	64.6
Self-employed	8	3.8
Non-employed	7	3.3
Student	59	28.2
Marital status		
Single	129	61.7
Married	78	37.3
Divorce	2	1.0

Based on Figure 1 below, the male respondents represented 31.6% (66 male respondents) of the total respondents while the female respondents represented 68.4% (143 female respondents) of the total respondents.

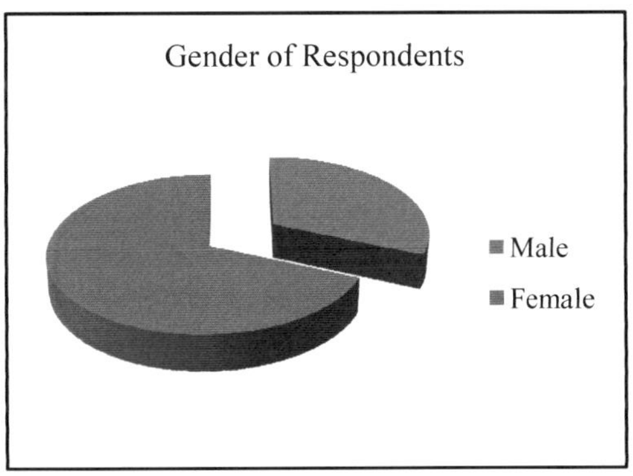

Figure 1 Gender of Respondents

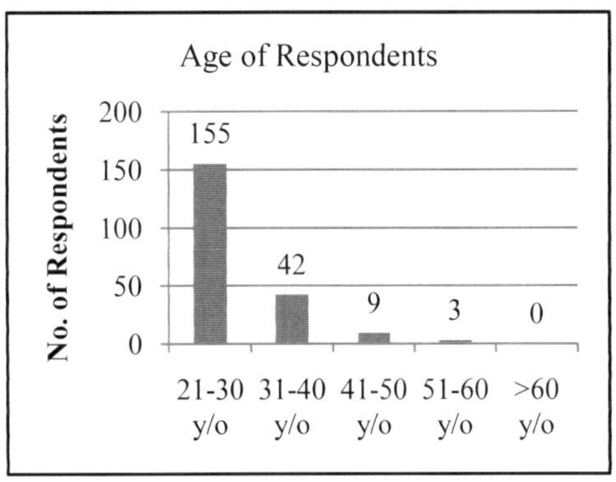

Figure 2 Age of Respondents

Figure 2 shows the age of the respondents whereby 21-30 years old (74.2%) was the highest respondent's age group, followed by 31-40 years old (20.1%); 41-50 years old (4.3%) and over 51 years old (1.4%).

Figure 3 illustrates the graph of the education level of the respondents. The highest amount of respondents were degree holders with 117 respondents (56%), followed by Master/PHD holders with 44 respondents (21.1%), Diploma holder with 33 respondents (15.8%), secondary holders with 8 respondents (3.8%) and the least respondents are certificate holders with only 7 respondents (3.3%).

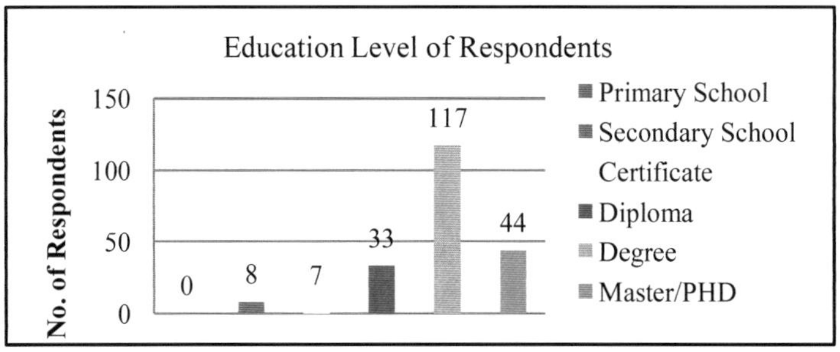

Figure 3 Education Level of Respondents

Lastly, Figure 4 and Figure 5 show the monthly income and occupation of respondents that participated in this research. 81 respondents (38.8%) earn less than RM2,000 per month. 67 respondents (32.1%) between RM2,000 until RM 2,999 per month, 32 respondents (15.3&) earn between RM3,000 until RM3,999 per month, 23 respondents (11%) earn between RM 4,000 until RM4,999 per month and more than RM5,000 per month with 6 respondents. In this study, mostly of the respondents are employed with 64.6% and follow by student by 28.2%. Then, 3.8% respondents work as self-employed and 3.3% are non-employed.

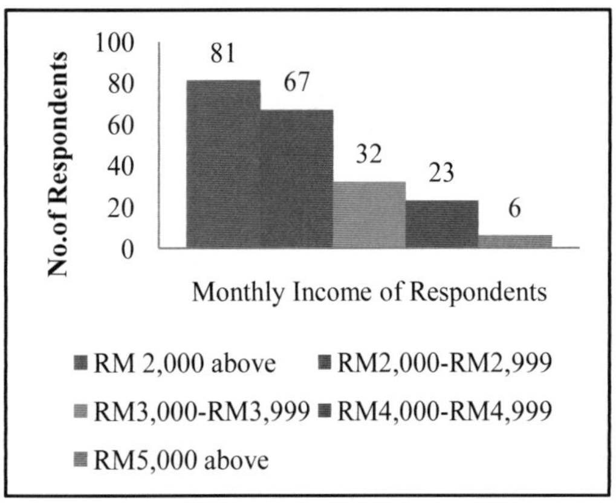

Figure 4 Monthly Income of Respondents

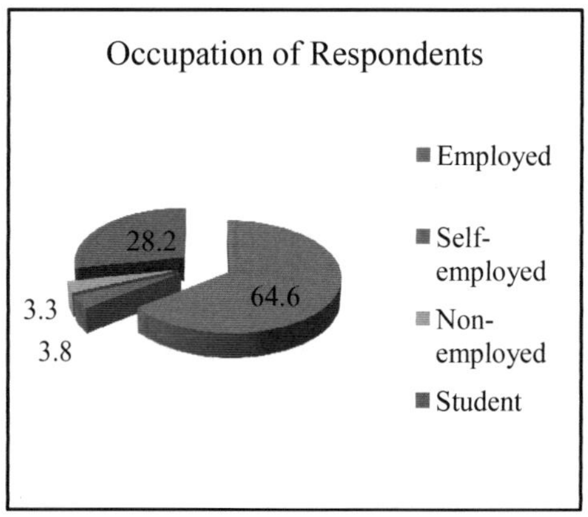

Figure 5 Occupation of Respondents

4.3 Inferential result

Independent t-test and ANOVA were used to analyze the differences between demographic profiles such gender, age, education, income and occupation. The results were concluded as follows:

Table 3 Summary of the Hypotheses Result

Hypotheses	Statistical Test	Significant Value	Result
H1: There is a difference between gender groups on green purchase intention	Independent sample t-test	0.014	Significant
H2: There is a difference between age groups on green purchase intention	ANOVA	0.557	Not significant
H3: There is a difference between education groups on green purchase intention	ANOVA	0.088	Not significant
H4: There is a difference between income groups on green purchase intention	ANOVA	0.565	Not significant

H5: There is a difference between occupation groups on green purchase intention	ANOVA	0.563	Not significant

Based on Table 3, an independent sample t-test was conducted to compare the independent variable gender and dependent variable green purchase intention. As the value of p is less than 0.05, there is statistically significant difference between gender groups on the green purchase intention. Meanwhile, the result test by ANOVA shows that there is no significant difference between age, education, income and occupation and green purchase intention because the level of significance is above than 0.05 as stated in Table 3 above.

5. Discussion and Conclusion

In this study, SPSS version 22.0 was utilised to run the descriptive, reliability, ANOVA and independent t-test analysis. The differences between the demographic profiles (gender, age, education, income level and occupation) and green purchase intention have been investigated. In conclusion, this study concludes that there is a different between male and female consumers on their purchase intention of green products. This is probably due to the fact that female consumers possess higher environmental concerns compared to male consumers as in [5]. Meanwhile, consumers with different age, education, income and occupation groups do not have any significant difference with regards to green purchase intentions. The results are contradicting to the literature, hence it can be concluded that green products are being accepted by Malaysian consumers in these recent years regardless of their personal backgrounds due to the popularity and increase in marketing campaigns for such products. As green products are discussed in abundant online and in social media, E-WOM plays a role in influencing consumers from all walk of life to purchase green products.

References

[1] Tsay Y Y 2009 The Impacts of Economic Crisis on Green Consumption in Taiwan. *Management of Engineering and Technology*. p. 2367- 2374.

[2] Promotosh B and Sajedul I M 2011 *Young Consumers' Purchase Intentions of Buying Green Products A study based on the Theory of Planned Behavior*. Master Thesis. UMEA University More references.

[3] Johnson R B, & Onwuegbuzie A J 2004 Mixed methods research: A research paradigm whose time has come. *Educational researcher,* **33**(7), p 14-26.

[4] Teng P K and Ow M W 2014 Consumers Awareness and Purchase Intention Towards environmental Friendly Food Products. *5th Int. Conf. on Business and Economic Research Proceeding*. P.p 904-912.

[5] Lee S H 2009 How do online reviews affect purchasing intention?. *African Journal of Business Management*. Vol 3(10). p 576-581.

[6] Mariëlle E H C 2010 The importance of product aspects in choice: the influence of demographic characteristics. *Journal of Consumer Marketing*. **27**(1). p. 26 – 34. Emerald Group Publishing Limited.

[7] Nguyen T H G 2014 Understanding Vietnamese Consumers' Purchase Intentions toward Green Electronic Products in Ho Chi Minh City. *Proc. of the First Asia-Pacific Conf. on Global Business, Economics, Finance and Social Sciences*. www.globalbizresearch.org.

[8] Kaufmann H R, Panni M F and Orphanidou Y 2011 Factors Affecting Consumers□ Green Purchasing Behavior: An Integrated Conceptual Framework. Vol. **XIV**.No. 31.

[9] Junaedi S 2012 The Role of Income Level in Green Consumer Behavior: Multigroup Structural Equation Model Analysis. *Int. Conf. on Business and Management, Phuket*. p 372-382.

[10] Malhotra N K 2010 *Marketing research: an applied orientation* 6th ed. Upper Sadle River, N. J.: Pearson Education Inc.

[11] Tabachnick B. G. and Fidell L S 2007 *Using multivariate statistics* 5th ed. Boston: Pearson Education.

Materialism, Altruism, Environmental Values, Learning Strategies and Sustainable Claim on Purchase Intention of Energy Efficient Vehicle (EEV) – A Literature Review

Muhamad Syakir Shukor[1,a], **Zuraidah Sulaiman***[2,b], **Thoo Ai Chin**[3,c], **Norhayati Zakuan**[4,d], **Farrah Merlinda Muharam**[5,e]
[1, 2, 3, 4, 5] Department of Business Administration, Faculty of Management, Universiti Teknologi Malaysia (UTM), 81310 Skudai, Johor, Malaysia.

[a]syakirshukor92@gmail.com, [b]zuraidahs@utm.my, [c]acthoo@utm.my, [d]norhayatimz@utm.my, [e]merlinda@utm.my

Abstract. One of the toughest challenges in social marketing is behaviour intervention. Previous research have developed various models and theories to simultaneously examine behaviour changes and their effects. Due to resources scarcity and global warming, automakers have come out with an innovative idea of Energy Efficient Vehicle (EEV) which has been a great improvement in the automotive industry. This invention targets for behavioral change or behavioral adoption for consumers to adjust their preferences from conventional vehicle to EEV. High market growth in automotive industry have encouraged social marketers, policymakers, governments and academics to propose suitable intervention approach in motivating preferences toward EEV. This study will explore the causal model of Environmental Responsible Behaviour (ERB) in measuring the purchase intention of EEV in Malaysia. In specific, this study focuses on two types of EEV – hybrid car and fuel efficient car. This study will hopefully add onto the body of knowledge for value orientation that influences green behaviour. From the practical perspective, this study may provide insights in assisting the stakeholders and automotive industry players on promoting the pro-behaviour toward EEV.

1. INTRODUCTION

Malaysia has been declared as having one of the highest carbon emissions per capita in the world, with 12.3 tons per capita compared to world average of 7.9 tons for per capita in 2011 (Malaysia's CO2 Emission Data, 2016). This scenario is seen globally as petroleum consumption has become third main source of global carbon emission (Fifth Assessment Report of the Intergovernmental Panel on Climate Change, 2014). As an innovative resolution, automaker came out with the idea of an EEV range in which the engine technology used is moving toward alternative energy that simultaneously consume less petroleum fuel and produce less carbon emission. EEV includes fuel-efficient internal combustion engine (ICE) vehicles, hybrid, electric vehicles (EV) and alternative fuelled vehicles powered by Compressed Natural Gas (CNG), Liquefied Petroleum Gas (LPG), Biodiesel, Ethanol, Hydrogen and Fuel Cell.

However, the issue of EEV market penetration will depend heavily on customer perception. Consumers often tend to be sceptical when new technologies been introduced such as EEV. In addition, this technology could also be seen as novel as the mass-markets consumers had little exposure [5]. With all the environmental issue that have been highlighted, it has often been perceived as premium and luxury buying [6] [7] [8] [9].

Building upon Value Belief Norm (VBN) model [21] [22] and New Environmental Paradigm (NEP) Model [21], this research will look into the extended ERB model by redefining the gap between value (materialism and altruism) with environmental belief. In defining materialism value effect on environmental belief there is still a gap in explaining the reason how two different lifestyles can be related to one another. De Groot and Steg [42] discussed the role of altruism values in supporting environmental belief. In addition, this research will look into a gap in defining learning strategies that occur due to increasing environmental concern. Suki [6] discussed the ability of sustainable claims in enhancing the role of learning. It had been supported by Ajzen [11] in that stated claim would strengthen learning strategies in term of comparing options available

2. Literature Review

2.1. Materialism

Materialism had been reviewed and developed over a few decades. Earliest notion about materialism is by Lange [46] in Richins & Dawson [47] that defined it as the philosophical notion that nothing exists except matter and its movements. Belk [48] defined materialism as the importance a consumer attaches to worldly possessions. Some researchers defined it as a cultural system in which material interests are not made subservient to other social goals [49] [50] [51].

Materialism that occur in market societies lead to environmental decline [54]. In western industrial societies, materialism been viewed as one-dimensional model and its achievement is a primary societal objective [55]. However, Jones [56] argues that, because materialism is deeply embedded in the institutional structures of industrial societies, the institutions themselves must be examined. As a result, admonishing consumers to consume less for personal, social, or environmental reasons is likely to be ineffective in changing behaviour. Culturally embedded institutions continuously reinforce and reward materialism as a mode of consumption. Consequently, one who is materialistic would find little reason for altering consumption behaviours in order to be more accommodating to the environment.

Stiglitz [57] and Kilbourne [58] argue that the globalization process is accelerating the spread of free market liberalism that enhances materialism as the solution to the environmental problem, and, as a result, environmental degradation is continuously accelerating through the consequent increase in trade related economic growth [57]. Thus, a more thorough examination of materialism and its role in environmental decline is necessary. Kilbourne et al. [59] add a necessary dimension to the values problem, however. In a multi-national study of university students, they demonstrate that general values orientations are related to materialism. Specifically, they demonstrate an inverse relationship between transcendent values and materialism and a direct relationship between enhancement values and materialism using Richins and Dawson's [47] conceptualization of materialism. Kilbourne et al. [59] did not examine the relationship between materialism and environmental beliefs. Thus, while Fisk [60] and Porritt [61] both suggest that consumption is an important factor in environmental degradation, little empirical evidence relating the various types of consumption to environmental behaviours exists. Based on the above arguments, the hypothesis below is advanced:
H1: There is a negative relationship between materialism and environmental belief

2.2. Altruism

Most of altruism studies were based on Schwartz's [32] [33] universal value system or on social value orientations as proposed in the social dilemma literature [34] [35]. Schwartz's [32] [33] universal value system proposed ten motivational values related to human being that can been divided into two dimensions. First dimension is openness to change versus conservatism, distinguishes values that stress independence, such as self-direction and stimulation, from values that emphasize tradition and conformity. The second dimension distinguishes social or self-transcendent values, such as universalism and benevolence, from those that pursue personal interests or self-enhancement, such as power and achievement.

Other scholars have proposed a similar distinction into three value orientations [26] [41]. For example, Stern [21] [22] [23] [26] argued that three different value orientations may affect beliefs: an egoistic, a social-altruistic, and a biospheric value orientation. Intensive study had been done by Stern and et al [21] [22] [25] [42] [43] regarding the distinction between these three values. However empirically and theoretically this not been proven yet. According to De Groot and Steg[42], there is still further validation needed in revealing whether a biospheric value orientation is emerging independently from an altruistic value orientation and whether the three value orientations provide a distinct basis for environmental beliefs, intentions, and behaviour.

As individual with a social-altruistic value orientation can be defined as individual that orients their decision to behave pro-environmentally or opposite due to perceived costs and benefits for other people [21] [22] [62]. Schwartz [32] defines altruism as feelings and behaviour that show a desire to help other people and lack of selfishness. Schwatz's value theory [63] describes the dimension of altruism in four dimension. The first is equality which mean equal opportunity for all. Second, world at peace without war and conflict. Third, social justice which means correcting injustice and caring for the weak. Lastly, being helpful by working for the welfare of others.

Previous study had test altruism with two other value orientation which are biospheric and egoistic. However, it been proven that altruism and environmental belief are correlated and seem to be clearly differently related to environmental belief when three values were tested together, in which value transcended into belief. [21][42][30][31].

De Groot and Steg [42] explain the value tested as altruism has been slightly underrepresented in that the altruism value is based on benevolence motivational type and universalism motivational type. This will limit the understanding of its relationship toward environmental belief. However universalism values have been proven to have relationship toward environmental and social behaviour than benevolence values. Therefore, altruism has a unique relationship toward environmental belief, hence the hypothesis below is advanced:

H2: Altruism moderates the relationship between materialism and environmental belief

2.3. Environmental Belief

Previous study had shown that environmental beliefs are subsequent to values [26][43]. General beliefs are about the human–environmental relationship and refer to folk wisdom about the environment [21]. Most research uses Dunlap and Van Liere's [64] new environmental paradigm (NEP) to measure general beliefs. Specific beliefs are beliefs about the existence of environmental problems such as water shortages, ozone depletion, and global warming.

The focus in this study is on specific beliefs. Logically, concern would not arise unless preceded by the belief that environmental problems exist. However, an individual may believe in a human–environmental relationship characterized by ecological folk wisdom without ever being concerned that problems actually do exist. In both Stern et al. [26] and Dietz et al [43] models, specific beliefs and attitudes precede intentions and behaviours. Stern [21] further argues that the link from values to environmentalism mediates beliefs because of perceived consequences to whatever the individual values activate norms. Thus, measures of beliefs in this study refer to specific adverse consequences for valued environmental objects. This suggests that individuals might be in a state of cognitive dissonance [65] if their desired consumption behaviour is perceived to cause negative environmental consequences. As evidence of negative environmental consequences emanating from increasing consumption accumulates, cognitive dissonance should increase in materialistic individuals. How this dissonance is resolved depends upon the circumstances relating to the relative attractiveness of the alternatives [66]. If the individual perceives that the environment as a valued object is threatened, environmental concern will increase. This will then increase the likelihood of more environmentally friendly consumption behaviour. According to the explanations above, this hypothesis is proposed:

H3: There is positive relationship between environmental belief and environmental concern

2.4. Environmental Concern

Alwitt and Pitts [67] state that a gap between environmental attitudes and behaviours exists within the US, and that this is relevant for policy makers and marketers alike. This is because social marketers and public policy makers seek to reduce environmentally negative behaviour, and if their intentions are not successful, then more market regulation may be necessary. Thus, a more thorough understanding of the relationship between concern and behaviour is necessary.

Studies of environmental concern in marketing began in the early 1970s with a series of studies relating to various dimensions of the problem. Anderson and Cunningham [68]; Kinnear et al. [69], and many others sought to characterize environmentally concerned consumers. Since that time, environmental concern has never declined as an environmentally related construct within the marketing literature. Fujii [70] describes that environmental concern can be associated with the awareness of the consequences of a given behavior, such as knowing the consequence of producing carbon dioxide emission. Mostafa [71] describes that environmental concern appeared in late 1960 and early 1970 in west communities. Some of it is associated with society, industry and modern technology, partly as an impact of the first world's oil crisis [72]. Based on these arguments, this hypothesis is developed:

H4: There is positive relationship between environmental concern and purchase intention of EEV

2.5. Learning Strategies

In purchase intention process, customers will always be provided with sufficient information to evaluate their alternative solution toward their need. Previous research has concluded this can be started by consumer research in which two learning strategies can occur, intentional and incidental learning strategies [83][84][85]. Petty and Cacioppo, [86] elaborate intentional learning as active and deliberate acquisition of knowledge and this form of learning facilitates the central route of information processing Petty and Cacioppo [86], while incidental learning occurs when knowledge acquisition occurs as an unintended consequence of engaging in another task. These learning strategies happen in first stage of purchase intention process which is seeking for information.

2.5.1 Intentional Learning

Green-house warming has a tremendously effect on goods selection in the market. Thus, consumer's environmental concern had increased in which goods and services consumption becomes more towards environmentally sensitive goods and services, and consumers change their purchasing preferences toward greener options [44]. As a result, customers are looking to make well-informed decisions about their purchasing choices that proclaim environmental attributes. The choices of goods may refer to review of credible, independent third parties [88][89]or comparing their available choices by their eco labelling schemes [90]. In the market, the product may be same within the same category. However, consumers with high environmental concern will differentiate them by their environmental impact [87]. These activities or processes are assumed to be an intentional learning strategies in which consumers are actively involved in an attempt to identify sufficient or additional information about environmental attributes of the available purchase choices. Thus, the hypothesis below is advanced:

H4: Intentional learning mediates the relationship between environmental concern and environmental purchase intentions.

2.5.2. Incidental Learning

Previous study had shown that Malaysians had started to acknowledge the environmental problem. Hence, product consumption had become environmental concern oriented. On a daily basis, consumers are exposed to various promotions by governments and automakers on environment related issues. In such instance, consumers will been attracted to the information to which they have been exposed, such as environmental visuals used by firms as their promotional tool [91][97]. This reflects that consumers obtain their information passively as a sequence from another task that can be defined as incidental learning. Moreover, any information that is exposed to consumers via incidental learning is important in subsequent environmental related purchase decisions [94][96].

Many firms make statements regarding their product or environmental policies in order to cultivate their image as an environmental company. In addition, a company's environmental reputation does effect consumer incidental learning. Consumers will assess the product or service provided depending on environmental reputation that are related to consumption consequences toward the [92][95]In particular, consumers with high environmental concern will be highly alert to a firm's reputation and environmental statement [96]. Based on the justifications above, the following hypothesis is proposed:

H5: Incidental learning mediates the relationship between environmental concern and environmental purchase intentions.

2.6. Sustainable Claim
Ministry of International Trade and Industry (MITI) Malaysia had listed 56 international automakers brands in Malaysia in the final quarter 2015. Thus, consumers have been provided with a huge range of automotive products compared to 35 automotive brands in 2013[2]. Simultaneously, this scenario occurs in high environmental concerned societies. This will encourage consumers to differentiate their choice by their effect toward environment. Some EEV have been classified as premium products, with prices thirty percent higher than conventional vehicle [97] such as hybrid and semi hybrid vehicle. On the other hand, EEV such as FEV have been priced as affordable vehicle [6]. Thus sustainable claim will factored in when purchasing an EEV.

Consumers with intentional learning will be alert to environmental impact caused by their consumption. They were highly involved in information seeking process and in evaluating alternatives processes. This group of consumers will react differently toward any information especially if their attitude toward sustainability can be classified as environmental concern [42][89]. Sustainable claims made by firms will strengthen their belief and help in making purchase choice. This will help them in differentiating the products that are available in their market compared to the products that not have any sustainable claim. Thus:

H6a: Sustainable claim moderates the relationship between intentional learning and purchase intention of EEV

Consumers that acquire information via incidental learning are highly concerned about the environment. Environmentally concerned consumers always keep their preferences in accuracy and credibility of a firm's sustainable claim [92][100][101][102]. With climate change becoming worst day by day, consumers will be highly alert to any claim made in the market.As shown by Stern [21], social cognitive consistency theory explained that the greater extent that a person has high environmental concern, the easier it will be to change their motivation and hence their behaviour. As the concern increases, they will change their preferences. Therefore:

H6b: Sustainable claim moderates the relationship between incidental learning and purchase intention of EEV

2.7. Purchase Intention
According to Peter and Olsen [73], the key process in consumers' decision making is the integration process by which knowledge is combined to evaluate two or more alternatives and eventually to select one. Belch and Belch [74] describe purchase intention as a matching of purchase motives with brand or product characteristics under consideration. Purchase intention is not yet an actual purchase. Once a customer has decided to buy something, he or she needs to implement the decision and make the actual purchase. Kotler et al. [75] stated that automotive

products in the industry involve complex buying behaviour, as it involves high involvement of the consumers.

3. Methodology

3.1. Research Design and Procedure

This research will employ a sequential mixed method via the application of interview technique (qualitative) followed by survey experiment (quantitative). This study will use qualitative data first in order to obtain an in-depth understanding on altruism value in the context of Environmental Responsible Behaviour (ERB) of Energy Efficient Vehicle (EEV). An exploratory interview will be conducted to several groups of respondents such as academia and experts (i.e. automaker, environmentalist and advertiser). Next, a survey experiment will be conducted. Malhotra [76] stated that experiment is the process of manipulating one or more independent variables and measuring their effect on one or more dependent variables. This research design was chosen to suit the aims of the research and to get the respondents' reaction when different advertising is used.

3.2. Population and Sampling Procedure

According to Hair et al [79] and Malhotra (2012), the suitable range for problem solving research is between 280 to 500 respondents. For this research, 300 survey questionnaires will be distributed via a convenience sampling to a target population of respondents aged between 25 to 45 years. As this research is focusing on high involvement product, the target population is suitable because they have higher need for transportation and higher disposable income.

Referring to current study in Malaysia on EEV, this research will choose three states with the highest GDP. High disposable income is one of factor that influencing purchase decision of EEV such as hybrid car[80] [81] [82]. Referring to Malaysia Statistical Department (2014), the three states with the highest GDP by the end of 2014 are Johor (99,887 million), WP Kuala Lumpur (164,663 million) and Selangor (245,080 million).

4. Conclusion

The Malaysia Automotive Association stated that the sales number of passengers car in Malaysia had increased tremendously from 416,692 units in 2005 to 576,657 units in 2013 [1]. This shows that the Malaysian car industry had grown well in the past 8 years, as the involvement of multinational companies had led to competitions among the car manufacturer companies. This research will benefit the marketers in the automotive industry by highlighting to them the importance of social intervention on EEV.

References

[1] Malaysia Automotive Association Vehicle Sales, 2016, available from <http.paultan.org/16/04/05/Malaysia-vehicle-sales-data-by-brand> [26 April 2016]
[2] Ministry of International Trade and Industry (2014) National Automotive Policy, 5 pg 1-29
[3] Malaysia's CO2 Emission Data, 2016, Available from <http://data.worldbank.org/indicator/EN.ATM.CO2E.PC> [2014]
[4] Fifth Assessment Report of the Intergovernmental Panel on Climate Change, 2014, Available from<https://www.ipcc.ch/pdf/assessmentreport/ar5/wg3/ipcc_wg3_ar5_https://www.ipcc.ch/pdf/assessment-report/ar5/wg3/ipcc_wg3_ar5_> [2014]
[5] Schuitema, G., Anable, J., Skippon, S., & Kinnear, N. (2012). The adoption of electric vehicles. In 5th International Conference on Traffic & Transport Psychology.
[6] Suki, N M. (2015). Consumer environmental concern and green product purchase in Malaysia: structural effects of consumption values. Journal of Cleaner Production.

[7] Ramayah, T., Lee, J. W. C., & Mohamad, O. (2010). Green product purchase intention: Some insights from a developing country. Resources, Conservation and Recycling, 54(12), 1419-1427. doi: http://dx.doi.org/10.1016/j.resconrec.2010.06.007
[8] Suki, N. M. (2013). Green Awareness Effects on Consumers 'purchasing Decision: Some Insights from Malaysia. IJAPS, 9(2), 49-63.
[9] Yusof, J. M., Musa, R., & Rahman, S. A. (2012). The Effects of Green Image of Retailers on Shopping Value and Store Loyalty. Procedia - Social and Behavioral Sciences, 50, 710-721. doi: http://dx.doi.org/10.1016/j.sbspro.2012.08.074
[10] Abdul Ghafar M N (2016). Research approach toward behavioural research, Penerbit UTM
[11] Ajzen I, Fishbein MA. (1980) Understanding Attitudes and Predicting Social Behavior. Englewood Cliffs, NJ: Prentice-Hall;
[12] Klockner, C.A., (2013). How powerful are moral motives in environmental protection? An integrated model framework. In: Heinrichs, K., Oser, F., Lovat, T. (Eds.), Handbook of Moral Motivation.Theories, Models and Applications.Sense Publishers, Rotterdam, NL.
[13] Klockner, C.A., Blo¨baum, A., (2010). A comprehensive action determination model—towards a broader understanding of conservationist behaviour. Journal of Environmental Psychology 30, 574–586.
[14] Han, H., Hsu, L., Sheu, C., (2010). Application of the theory of planned behavior to green hotel choice: Testing the effect of environmental friendly activities.Tourism Management 31, 325–334
[15] L Chan & B Bishop (2013), A moral basis for recycling: Extending the theory of planned behaviour. Journal of Environmental Psychological 36 (2013) 96-102
[16] Sheeran, P., and S. Orbell. (2000). Using implementation intentions to increase attendance for cervical cancer screening. Health Psychology 19 (3): 283–89.
[17] Smith, B. N., and M. F. Stasson. (2000). A comparison of health behavior constructs: Social psychology predictors of AIDS-preventive behavioral intentions. Journal of Applied Social Psychology 30 (3): 443–62.
[18] Klockner, C.A., Friedrichsmeier, T., (2011). A multilevel approach to travel mode choice—how person characteristics and situation specific aspects determine car use in a student sample. Transportation Research Part F: Traffic Psychology and Behaviour 14, 261–277.
[19] Klockner, C.A., Nayum, A., Mehmetoglu, M., (2013). Positive and negative spillover effects from electric car purchase to car use. Transportation Research Part D 21, 32–38.
[20] Klockner, C.A., Oppedal, I.O., (2011). General versus domain specific recycling behaviour—applying a multilevel comprehensive action determination model to recycling in Norwegian student
[21] Stern PC. (2000), Toward a coherent theory of environmentally significant behavior. Journal Social Issues;56(3):407–24
[22] Stern, P. C., Dietz, T., & Guagnano, G. A. (1998). A brief inventory of values. Educational and Psychological Measurement, 58(6), 984-1001.
[23] Stern, P. C., T. Dietz, and J. S. Black. (1985–86). Support for environmental protection—the role of moral norms. Population and Environment 8 (3–4): 204–22.
[24] Stern, P. C., & Dietz, T. (1994). The value basis of environmental concern. Journal of Social Issues, 50, 65-84.
[25] Stern, P. C., Dietz, T., & Kalof, L. (1993). Value orientations, gender, and environmental concern.Environment and Behavior, 25, 322-348.
[26] Stern PC, Dietz T, Kaloff L, Guagnano GA.(1995), Values, beliefs, and proenvironmental action: attitude formation toward emergent attitude objects. Journal of Applied Social Psychology;
[27] Rokeach M. (1973)The nature of human values. New York: Free Press;
[28] Fransson N, Garling T. (1999) Environmental Concern: Conceptual Definitions, Measurement Methods, and Research Findings. J Environment Psychology 19(4):369–82.

[29] Joireman, J. A., Lasane, T. P., Bennet, J., Richards, D., & Solaimani, S. (2001). Integrating social value orientation and the consideration of future consequences within the extended norm activation model of proenvironmental behaviour. British Journal of Social Psychology, 40, 133-155.
[30] Nordlund, A. M., & Garvill, J. (2003). Effects of values, problem awareness and personal norm on willingness to reduce personal car use. Journal of Environmental Psychology, 23, 339-347.
[31] Schultz, P. W., & Zelezny, L. C. (1998). Values and pro-environmental behavior. A five country survey. Journal of Cross-Cultural Psychology, 29, 540-558.
[32] Schwartz, S. H. (1992). Universals in the content and structure of values: Theoretical advances and empirical tests in 20 countries. In M. Zanna (Ed.), Advances in experimental social psychology (pp. 1-65). Orlando, FL: Academic Press.
[33] Schwartz, S. H. (1994). Are there universal aspects in the structure and contents of human values?Journal of Social Issues, 50(4), 19-45.
[34] McClintock, C. G. (1972). Social motivation—A set of propositions. Behavioral Science, 17(5), 438-454.
[35] Messick, D. M., & McClintock, C. G. (1968). Motivational basis of choice in experimental games. Journal of Experimental Social Psychology, 4, 1-25.
[36] Leopold, A. (1949). A Sand County almanac. New York: Oxford University Press.
[37] Naess, A. (1989). Ecology, community, and lifestyle: An outline of an ecosophy. Cambridge, UK: Cambridge University Press.
[38] Reid, L. (1962). The sociology of nature. In N. Nelissen, J. Van der Straaten, & L. Klinkers (Eds.),Classics in environmental studies. An overview of classic texts in environmental studies(pp. 57-67). Utrecht, The Netherlands: International Books.
[39] Singer, P. (1975). Animal liberation: A new ethics for our treatment of animals. New York: New York Review.
[40] Merchant, C. (1992). Radical ecology. The search for a livable world. New York: Routledge.
[41] Axelrod, L. J. (1994). Balancing personal needs with environmental preservation: Identifying the values that guide decisions in ecological dilemmas. Journal of Social Issues, 50(3), 85-104.
[42] De Groot, J. I. M., & Steg, L. (2008). Value orientations to explain beliefs related to environmental significant behavior how to measure egoistic, altruistic, and biospheric value orientations. Environment and Behavior, 40(3), 330-354.
[43] Dietz T, Stern PC, Guagnano GA.(1998) Social structural and social psychological bases of environmental concern. Environmental Behaviour; 30(4):450–71.
[44] Kilbourne, W., & Pickett, G. (2008). How materialism affects environmental beliefs, concern, and environmentally responsible behavior. Journal of Business Research, 61, 885e893.
[45] Vaske, J. J., & Donnelly, M. P. (1999). A value-attitude-behavior model predicting wildland preservation voting intentions. Society & Natural Resources, 12(6), 523-537.
[46] Lange, Frederick Albert (1865;1925), The History of Materialism,London: Routledge & Kegan Paul.
[47] Richins ML, Dawson S. A consumer values orientation for materialism and its measurement: Scale development and validation. Journal of Consumer Research 1992;19:303–16(December).
[48] Belk RW. (1985) Materialism: trait aspects of living in the material world. Journal of Consumer Research; 12:265–80 (December).
[49] Polanyi, Karl (1944), The Great Transformation, New York:Rinehart.
[50] Sahlins, Marshall (1976), Culture and Practical Reason, Chicago: University of Chicago Press.
[51] Mukerji, Chandra (1983), From Graven Images: Patterns of Modern Materialism, New York: Columbia University Press.
[52] Ahuvia AC, Wong NY. (2002)Personality and values based materialism: their relationship and origins. Journal of Consumer Psychology; 12(4):389–402.

[53] Ger, G., & Belk, R. W. (1996). Cross-cultural differences in materialism.Journal of economic psychology, 17(1), 55-77.
[54] Inglehart R. (1981). Post-materialism in an environment of insecurity. America Politic Science Review; 75:880–900.
[55] Schmookler AB. (1991), The insatiable society: materialistic values and human needs. Futurist;25(4):17–20.
[56] Jones A. (1987), The violence of materialism in advanced industrial society: an ecosociological approach. Sociology Review 1987;35(1):19–47.
[57] Stiglitz JE. (2002),Globalization and its discontents. New York, NY: W. W. Norton & Company;.
[58] Kilbourne W. (2004), Globalization and development: an expanded macromarketing view. J Macromark;24:122–35 (December).
[59] Kilbourne WE, Grünhagen M, Foley JA. (2005), Cross-cultural examination of the relationship between materialism and individual values. Journal of Economic Psychology;26(5):624–41
[60] Fisk G. (1973)Criteria for a theory of responsible consumption. Journal of Marketing; 37(1):24–31.
[61] Porritt J. (1984), Seeing green: the politics of ecology explained. Oxford, UK: Basil Blackwell;
[62] Van Vugt, M., Meertens, R. M., & Van Lange, P. A. M. (1995). Car versus public transportation? The role of social value orientations in a real-life social dilemma. Journal of Applied Social Psychology, 25, 258-278.19,255-265.
[63] Schwartz, S. H. (1994). Are there universal aspects in the structure and contents of human values?Journal of Social Issues, 50(4), 19-45.
[64] Dunlap RE, Van Liere K.(1978) The "New Environmental Paradigm": a proposed instrument and preliminary results. Journal of Environmental Education; 9(1):10–9.
[65] Festinger L. (1957.)A theory of cognitive dissonance. Stanford, CA: Stanford University Press;
[66] Festinger L, Carlsmith JM. (1959) Cognitive consequences of forced compliance. Journal of Abnormal Social Psychology; 58:203–10.
[67] Alwitt LF, Pitts RE. (1996) Predicting Purchase Intentions for an Environmentally Sensitive product. J Consumer Psychology; 5(1):49–64.
[68] Anderson Jr, W. T., & Cunningham, W. H. (1972). The socially conscious consumer. The Journal of Marketing, 23-31.
[69] Kinnear, T. C., Taylor, J. R., & Ahmed, S. A. (1974). Ecologically concerned consumers: who are they?. The Journal of Marketing, 20-24.
[70] S. Fujii (2007), Environmental Concern, Attitude Toward Frugality And Ease Of Behaviour As Determinants Of Pro-Environmental Behaviour Intention, Journal Of Environmental Psychology, 26, 262-268.
[71] Mostafa, M. M. (2007). A hierarchical analysis of the green consciousness of the Egyptian consumer. Psychology & Marketing, 24(5), 445-473.
[72] Grunert, S. C., & Juhl, H. J. (1995). Values, environmental attitudes, and buying of organic foods. Journal of economic psychology, 16(1), 39-62.
[73] J. Peter, Olson (2002), Consumer Behaviour and Marketing Strategy, 6th Edition, Mcgraw-Hill.
[74] G.E. Belch, M.A. Belch (2012), Advertising and Promotion, An Integrated Marketing Communication Perspective, 9th Ed, Mcgraw-Hill. Singapore, 290-294
[75] Kotler, P., Bowen, J. T., Makens, J. C., Xie, Y., & Liang, C. (2006).Marketing for hospitality and tourism (Vol. 893). New Jersey: Prentice hall.
[76] N.K. Malhotra (2012), Basic Marketing Research, 4th Edition, Pearson International Edition, USA

[77] Creswell, J. W., Klassen, A. C., Plano Clark, V. L., & Smith, K. C. (2011). Best practices for mixed methods research in the health sciences.Bethesda, MD: National Institutes of Health, 10.
[78] Daniel M. Oppenheimer, Tom Meyvis, Nicolas Davidenko (2009)Instrumental Manipulation Check; Detecting Satisficing To Increase Statistical Power, Journal Of Experimental Social Psychology,467-872
[79] Hair JF, Anderson RE, Tatham RL, Black WC. (2010.) Multivariate data analysis.(7 ed.). Upper Saddle River, NJ: Prentice Hall;
[80] Hong, Y., Khan, N., & Abdullah, M. (2013). The Determinants of Hybrid Vehicle Adoption: Malaysia Perspective. Aust J Basic Appl Sci, 7(8), 447-454.
[81] Soon, W. L., Seng, W. T., Luen, W. K., & Siang, J. M. L. D. (2013). Hybrid vehicle adoption? a conceptual study. Journal of Education and Vocational Research (JEVR), 4, 165-168.
[82] Yee, C. J., & San, N. C. (2011). Consumers' perceived quality, perceived value and perceived risk towards purchase decision on automobile. American Journal of Economics and Business Administration, 3(1), 47.
[83] Hutchinson, J.W., & Alba, J.W. (1991). Ignoring irrelevant information: Situational determinants of consumer learning. Journal of Consumer Research, 18, 325–345.
[84] Hutchinson, J.W., & Eisenstein, E.M. (2008). Consumer learning and expertise. In C.P. Haugtvedt, P.M. Herr, & F.R. Kardes (Eds.), Handbook of consumer psychology (pp. 103–131). New York: Lawrence Erlbaum Associates.
[85] Gursoy, D., & McCleary, K.W. (2004). An integrative model of tourists' information search behavior. Annals of Tourism Research, 31, 353–373.
[86] Petty, R. E., & Caccioppo, J. T. (1984). The effects of involvement on responses to argument quantity and quality: central and peripheral routes to persuasion. Journal of Personality and Social Psychology, 46, 69e81
[87] Stoll, M.L. (2011). Green chemistry meets green business: A match long overdue. Journal of Business Ethics, 99(S1), 23–28.
[88] Parguel, B., Benôit-Moreau, F., & Larceneux, F. (2011). How sustainability ratings might deter 'greenwashing': A closer look at ethical corporate communication. Journal of Business Ethics, 102, 15–28.
[89] Mohr, L. A., Eroglu, D., & Ellen, P. S. (1998). The development and testing of a measure of scepticism toward environmental claims in marketers' communications. Journal of Consumer Affairs, 32, 30-55.
[90] Leire, C., & Thidell, Å. (2005). Product-related environmental information to guide consumer purchases – a review and analysis of research on perceptions, understanding and use among Nordic consumers. Journal of Cleaner Production, 13, 1061–1070.
[91] Kong, Y., & Zhang, L. (2014).When does green advertising work? The moderating role of product type. Journal of Marketing Communications, 20, 197–213
[92] Newton, J. D., Tsarenko, Y., Ferraro, C., & Sands, S. (2015). Environmental concern and environmental purchase intentions: The mediating role of learning strategy. Journal of Business Research, 68(9), 1974-1981.
[93] Pickett-Baker, J., & Ozaki, R. (2008). Pro-environmental products: Marketing influence on consumer purchase decision. Journal of Consumer Marketing, 25, 281–293.
[94] Prakash, A. (2002). Green marketing, public policy and managerial strategies. Business Strategy and the Environment, 11, 285–297.
[95] Magnier, L., & Schoormans, J. (2015). Consumer reactions to sustainable packaging: The interplay of visual appearance, verbal claim and environmental concern. Journal of Environmental Psychology, 44, 53-62
[96] Toms, J.S. (2002). Firm resources, quality signals and the determinants of corporate environmental reputation: Some UK evidence. The British Accounting Review, 34, 257–282.

[97] Xie, G. -X., & Kronrod, A. (2012). Is the devil in the details? The signaling effect of numerical precision in environmental advertising claims. Journal of Advertising, 41,103–117
[98] Yong Hoe Hong,Nasveen Khan,Muhamad Madi Abdullah (2013), The Determinants Of Hybrid Vehicle Adoption;Malaysian Perspective, Australian Journal Of Basic And Applied Sciences,347-454
[99] Bickart, B.A., & Ruth, J.A. (2012). Green eco-seals and advertising persuasion. Journal of Advertising, 41, 51–67.
[100] Chen, Y., & Chang, C. (2013). Greenwash and green trust: The mediation effects of green consumer confusion and green perceived risk. Journal of Business Ethics, 114, 489–500.
[101] Crane, A. (2000). Facing the backlash: Green marketing and strategic reorientation in the 1990s. Journal of Strategic Marketing, 8, 277–296.
[102] Shrum, L.J., McCarty, J.A., & Lowrey, T.M. (1995). Buyer characteristics of the green consumer and their implications for advertising strategy. Journal of Advertising, 24, 71–82.

Printable candlelight-style organic light-emitting diode

J H Jou*, M Singh, W C Song and S H Liu

Department of Materials Science and Engineering, National Tsing Hua University, Hsinchu 30013, Taiwan
E-mail: jjou@mx.nthu.edu.tw

Abstract

Candles or oil lamps are currently the most friendly lighting source to human eyes, physiology, ecosystems, artifacts, environment, and night skies due to their blue light-less emission. Candle light also exhibits high light-quality that provides visual comfort. However, they are relatively low in power efficacy (0.3 lm/W), making them energy-wasting, besides having problems like scorching hot, burning, catching fire, flickering, carbon blacking, oxygen consuming, and release of green house gas etc. In contrast, candlelight organic light-emitting diode (OLED) can be made blue-hazard free and energy-efficient. The remaining challenges are to maximize its light-quality and enable printing feasibility, the latter of which would pave a way to cost-effective manufacturing. We hence demonstrate herein the design and fabrication of a candlelight OLED via wet-process. From retina protection perspective, its emission is 13, 12 and 8 times better than those of the blue-enriched white CFL, LED and OLED. If used at night, it is 9, 6 and 4 times better from melatonin generation perspective.

Keywords: Candlelight, High Light-Quality, Melatonin Suppression, Retina Exposure Limit, Printable OLED.

1. Introduction

Intense visible light that consists of high energy blue radiation can penetrate cornea and lens of human eyes and damage the retinal cells. It can also cause an age-related macular degeneration (AMD), an irreversible blindness.[1-2] More exposure to modern lighting is increasing the percentage of vision loss due to AMD. Several medical studies also showed concern about intense white light, which can cause circadian disruption and sleep disorder by suppressing the melatonin secretion.[3-5] Kloog et al. reported that women exposed to bright light at night were found to have a higher risk of breast cancer occurrence than those to the least light exposure.[6] Brainard et al. reported that over exposure of light at night would disrupt the circadian clock by suppressing the melatonin-secretion, and cause breast cancer among women.[7-8] Moreover, some studies have been done on artifacts in Van Gogh museum and found that intense white light had deteriorated the actual colors of the oil paintings through bleaching.[9-10] The International Dark-Sky Association reported that high color temperature modern light disrupts the living-behavior of diverse nocturnal creatures and destructs the ecosystem.[11] The light pollution is a serious threat to nocturnal species, especially for fireflies, tree-frogs, monarch butterflies, sea turtles, and Atlantic salmon.

Steps toward modernization and development are also leading to the more exposure to the blue emission. People are frequently using portable display devices, TVs, computer screens, and artificial lighting. Blue emission hazards is increasing day by day. Therefore, the development of a good light that is at least safe to retina, physiology and psychology, should become a priority of R&D in solid state lightings. In addition to energy-saving issues, US Department of Energy had also addressed the need for a human-friendly light source.[12]

It should be noted that candles have been used for more than 5000 years. Its continuous emission is free from ultraviolet and deep blue radiation. Candle exhibits low color temperature (1,850 K) chromaticity[13] and create a pleasant sensation, which may originate from the naturally occurring melatonin secretion and helps relax people. Therefore, a solid state lighting source that mimics the positive characteristics of candlelight, such as little deep-blue or violet emission, low color temperature, and continuous and diffused emission, and avoids the unpleasant smoke and burning hazards would be an ideal direction. Organic light emitting diode (OLED) is a preferred option due to its inherent design flexibility and easily tunable chromaticity.

OLEDs can be fabricated either through dry-process or wet-process. Dry-process involved thermal evaporation of organic materials under high vacuum. This approach is mostly preferred for small molecules to achieve high device performance, including high efficiency and long lifetime. However, there is some issues like confined area, high vacuum requirement, more material consumption, low throughput, and difficulties in multiple dopants co-evaporation etc. In contrast, wet-processed enables flexible, large area-size OLED devices. Wet-processed OLED is capable of realizing printable feasibility. By employing printing technique such as ink-jet, roll-to-roll, and letterpress application, OLED device fabrication can achieve cost-effective mass production.

In 2013, Jou's group reported an innovative organic light-emitting diode (OLED) "candle light-style OLED", which can minimize the blue-emission hazards.[14] The device showed a color temperature of 1,970 K with an efficacy of 24.2 lm/W. They also reported a blue-hazard free general lighting based on candle light OLED with a color temperature of 2,279 K, a power efficiency of 85.4 lm/W.[15] In 2014, Zhang's group report low color temperature OLED 1,883 K with an efficacy of 8.3 lm/W[16] and Ma's group reported 54.6 lm/W a 1,910 K OLED.[17] All these efforts were to fabricate low color temperature OLEDs using hybrid process, multiple emissive layer, and interlayer.[17-19] A few efforts have been attempted to realize a low color temperature illumination via wet-process. In 2012, Jou's group reported a physiologically-friendly low color-temperature (1,773 K) OLED with an efficacy of 11.9 lm/W.[20] The device exhibited a much lower color temperature than incandescent bulb (2000-2500 K).

In this study, we have fabricated very-high light-quality candle light-style OLEDs via wet-processable emissive layer. An emissive layer was composed by four blackbody-radiation complementary emitters, namely sky blue, green, deep red, and yellow, doped in a host 4, 4, N N-dicarbazolebiphenyl (CBP). The fabricated devices showed a low melatonin suppression sensitivity (MSS) and high maximum retina admissible exposure limit "t" than those of blue enriched white CFL, LED and OLED. These high light-quality candle light-style OLEDs enabled a cost-effective method for healthy illumination.

2. Theoretical

2.1 Maximum retina admissible exposure limit "t"

The maximum retina admissible exposure limit "t" (sec) was formerly reported by International Electrotechnical Commission (IEC).[21] The value of "t" can be determined according to the IEC 62471 standard,[21] as shown below.

$$t = \frac{100}{E_B} \qquad \text{..........(1)}$$

Where, E_B is defined as the blue-light weighted irradiance, and is calculated from the spectral irradiance and blue-light hazard function.

The permissible retinal exposure is calculated from the radiance of the given light source that is directed to the human eye. For eye protection, the blue-light effective radiation must be limited for a duration. The maximum exposure duration for blue light could be equal to or lower than 100 s if the human eye is directed to a light source of radiation $E_B = 1$ W/m^2.[22] If the radiance is less than 1 W/m^2, the exposure limit will exceed 100 s. The calculated exposure limit "t" can be classified into four risk groups namely, Risk Group 0, Risk Group 1, Risk Group 2, and Risk Group 3, if "t" is greater than 10,000 s, between 10,000 and 100 s, between 100 and 0.25 s, or less than 0.25 s, respectively. Risk group 0 is assigned for "no risk" to retina.

2.2 Melatonin Suppression Sensitivity determination

The action spectrum of MLT suppression per photon quanta, S_{PQ}, was first presented by Jou in the US patent.[23] The action spectrum of melatonin suppression power per photon quanta, S_{PQ} for a given monochromatic light, λ, relative to that of the reference light, λr, can be expressed as follows:

$$S_{PQ}(\lambda) = 10^{(\lambda r-\lambda)/C} \qquad \text{............. (2)}$$

Where, C is a fitting constant.

The photopic luminosity function V (λ) convert S_{PQ} (λ) into the melatonin suppression power per lux, S_{LC} (λ), in order to give it a practical meaning for general illumination purpose. The action spectrum of melatonin suppression per lux, $S_{LC}(\lambda)$

$$S_{LC}(\lambda) = \lambda S_{PQ}(\lambda)/V(\lambda) \qquad \text{..............(3)}$$

The action spectrum of MLT suppression per lux for a given polychromatic light and can be expressed as following:

$$S_{LC}(\lambda) = \frac{\int \lambda S_{PQ}(\lambda) S_I(\lambda) d\lambda}{\int V(\lambda) S_I(\lambda) d\lambda} \qquad \text{......... (4)}$$

Where, $S_I(\lambda)$ is the power spectrum of the studied light.

Melatonin suppression sensitivity of a given light relative to the reference blue light of 480 nm i.e.

$$\text{Relative melatonin suppression sensitivity} = \frac{S_{LC}(\lambda)}{S_{LC}(480\text{ nm})} \times 100\ \% \qquad \text{....... (5)}$$

2.3 Light-quality, SRI determination

The light quality of a given light can be quantified by the natural light spectrum resemblance index, SRI, according to Jou et al.[24] For human eye's perspective, the power spectrum of a given light source is converted to the luminance spectrum. The SRI, which is based on a direct comparison of the luminance spectrum of a given light source with the reference blackbody-radiation at the same color temperature, can be defined as following:

$$SRI = \frac{\int L(\lambda,T)\, d\lambda}{\int LBR(\lambda,T)\, d\lambda} \times 100\% \qquad \ldots\ldots (6)$$

Where, $L_{BR}(\lambda,T)$ is the luminance spectrum of the blackbody-radiation, and $L(\lambda,T)$ is the overlapping area between the luminance spectra of the studied light source and its corresponding blackbody-radiation.

3. Experimental

3.1 Device structure

Figure 1 showed the corresponding energy level diagram of all the studied candle light-style OLED devices. Wet-processed OLEDs consisted with a cross-linkable hole-transporting material as well as electron-confining layer of 3,6-bis(4-vinylphenyl)-9-ethylcarbazole (VPEC). The devices were composed by four candlelight complementary dyes, namely sky-blue, green, yellow, and deep-red. The emissive layer consisted of a host 4,4,NN-dicarbazolebiphenyl (CBP) doped with green dye tris(2phenyl-pyridine) iridium (Ir(ppy)3), sky-blue dye bis[3,5-difluoro-2-(2-pyridyl)phenyl]-(2carboxypyridyl) iridium(III) (Firpic), deep-red dye bis(1-phenylisoquinolinolato-C2,N) iridium (acetylacetonate) (Ir(piq)2(acac)), and yellow dye Iridium(III) bis(4-phenylthieno[3,2-c]pyridinatoN,C2) acetylacetonate (PO-01).

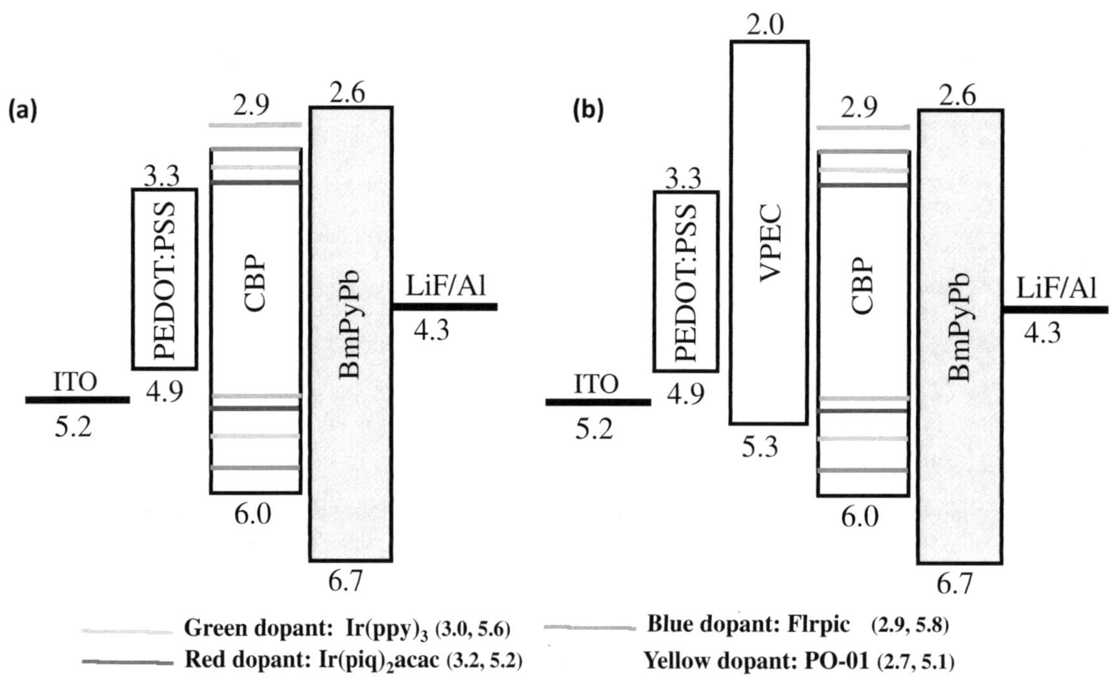

Figure 1. Schematic diagrams of the energy levels of the candle light-style OLEDs with a wet-processed emissive layer (a) without and (b) with a cross-linkable electron confining layer, VPEC.

3.2 Device fabrication

Each device consisted of a pre-cleaned 125 nm indium tin oxide (ITO) anode layer. A hole-injection layer (HIL) of an aqueous solution of poly(3,4-ethylenedioxythiophene)-poly(styrenesulfonate) (PEDOT:PSS) was spin-coated at 4000 rpm for 20s. The resulting hole-injection layer was baked at 160 °C for 40 mins. VPEC polymer dissolved in the solvent chlorobenzene and resulting solution was spin-coated at 3,000 rpm for 20s onto the per dried PEDOT:PSS layer to form a 10 nm hole transporting layer (HTL). The HTL was baked at 120 °C for 20 mins to remove residual solvent, then heated at 230 °C for 40 mins for crosslinking reaction.[25] Thereafter, an emissive layer solution was spin-coated at 2500 rpm for 20 s. A 32 nm 1,3-bis(3,5-dipyrid-3-yl-phenyl) benzene (BmPyPb) electron transporting layer, a 1 nm lithium fluoride (LiF) electron injection layer and a 100 nm aluminum (Al) cathode layer were deposited via a thermal evaporation process at the pressure of 10^{-4} Torr.

3.3 Characterization

All devices were measured under atmospheric condition. The current density-voltage and luminance (J-V-L) characteristics of the resultant devices were measured through a Keithley 2400 electrometer with Minolta CS-100A luminance-meter, while the spectrum and CIE color chromatic coordinates were measured by using PR-655 spectrascan spectroradiometer. Further, quality of emission spectrum were defined in terms of color rendering index, CRI and natural light-style spectrum resemblance index, SRI. Emission spectrum of the devices was also characterized for melatonin suppression sensitivity (MSS) and maximum permissible retina exposure (MPE) limit "t".

4. Results and discussion

Table 1. Effect of doping concentrations on color temperature (CT), light quality, maximum permissible exposure limit (MPE), and melatonin suppression sensitivity (MSS %) of the studied wet-processed candlelight-style OLEDs. Devices 2 and 3 represent VPEC based OLEDs.

Device	Dopant wt%				CIE(x,y)	CT(K)	SRI	CRI	MLT suppression Sensitivity (%)	Max. Exposure limit "t" @ 100 lx
	Blue (Flrpic)	Green ($Ir(ppy)_3$)	Red ($Ir(piq)_2acac$)	Yellow (PO-01)	@ 1,000 cd/m²					
1	14	0.3	0.9	0.5	(0.51, 0.43)	2300	90	90	9.8	2535
2	14	0.3	0.9	0.5	(0.56, 0.41)	1807	88	91	3.0	4248
3	12	0.3	1.1	0.5	(0.55, 0.41)	1862	89	92	3.2	4328

The OLED devices were fabricated by using various doping concentrations of deep-red dye, Ir(piq)2acac, and sky-blue dye, Flrpic, with a fixed composition of two other complementary dyes: green and yellow as shown in Table 1. With 14 wt.% of sky-blue, 0.3 wt.% of green, 0.9 wt.% of deep-red, and 0.5 wt.% of yellow dye, Device 1 exhibited a color temperature of 2,300 K with an SRI and CRI of 90. The employment of multiple emitters in a single emissive layer extended the emission from 380 to 780 nm (Figure 2a) and realized a very-high CRI and SRI. Emission in the short wavelength region from 380 to 450 nm contributed to a higher color temperature 2,300 K than that of the candlelight (1,850 K) counterpart.

The incorporation of a cross-linkable hole-transport layer of VPEC reduced the color temperature by suppressing the emission in deep blue region (up to 450 nm) while maintained the desirable high light-quality. As can be seen in Figure 2 (a), the wide electroluminescent spectra exhibited four emission bands with a plateau in the long wavelength side. The resulting Device 2 showed a color temperature of 1,807 K with an 88 SRI and a 91 CRI. The lower color temperature OLED showed a decrement in melatonin suppression sensitivity from 9.8 to 3% and an increment of 67% (from 2535 to 4248s) in maximum retina exposure limit. Further, with 12 wt% sky-blue dye and 1.1 wt.% of deep-red dye, Device 3 showed a color temperature of 1,862 K, a very-high CRI of 92, an 89 SRI, and chromaticity (CIEx,y: 0.55, 0.41) analogous to the black-body radiation, as shown in figure 2(b). The device showed a 3.2% melatonin suppression sensitivity and 4328 s (72 minutes) exposure time to the retina.

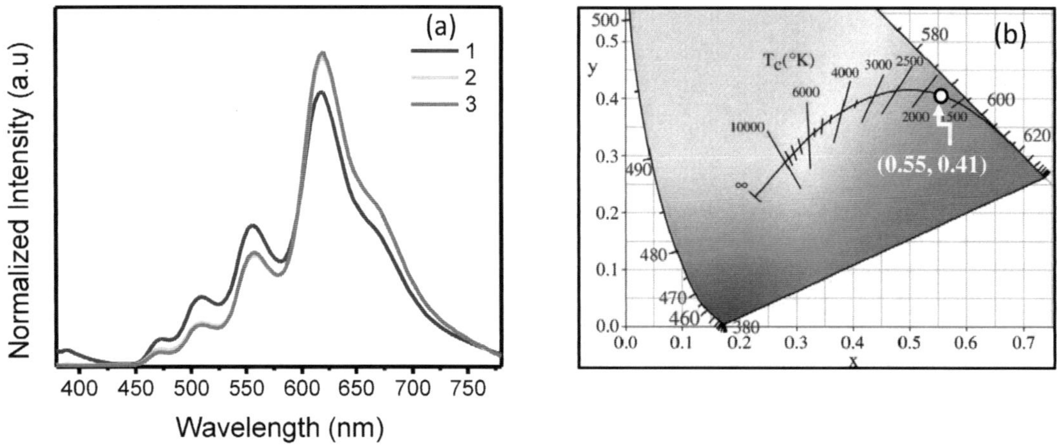

Figure 2.(a) Electroluminescent spectra of wet-processed candlelight-style OLEDs and (b) CIE chromaticity coordinates (0.55, 0.41) of Device 3.

Table 2: Comparison of a printable candlelight-style OLED (1,862 K) with current lighting measures in terms of maximum permissible exposure limit (s) and melatonin suppression sensitivity (%). The presented OLED is a promising candidate to safeguard human eye and physiology due to its blue-hazard free or low color temperature emission.

Light Source	Emission Spectrum	Maximum permissible exposure limit (s)		Melatonin Suppression Sensitivity (%)
		@100 lx	@500 lx	
Candlelight OLED (1,862 K)	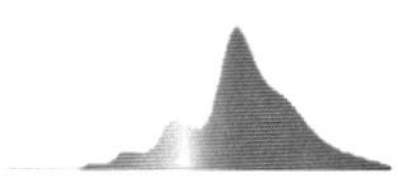	4328	847	3.2
Candle (1,850 K)	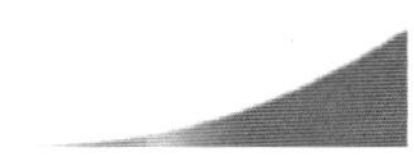	2616	523	4
Cold-white OLED (5,000 K)	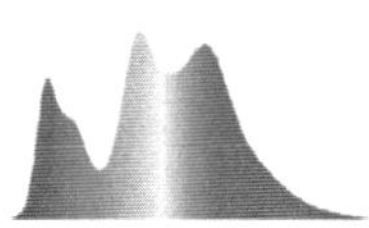	546	109	12
Cold-white LED (5,501 K)		343	67	20
Cold-white CFL (5,921 K)		316	63	29
	400 500 600 700			

The fabricated candlelight OLED (1,862 K) is compared with current lighting measures such as cold-white CFL, LED and OLED in terms of melatonin suppression sensitivity (%) and maximum permissible exposure limit (s) at 100 lx and 500 lx. As seen in Table 2, melatonin suppression sensitivity of the candlelight OLED is 3.2% to that of the 480 nm blue light. Whilst, it is 4% for candles. The candlelight OLED is 9, 6 and 4 times safer than those of the cold-white CFL, LED and OLED, respectively. At 100 lx, the OLED showed 13, 12 and 8 times more retina exposure time than those of the blue-enriched white CFL, LED and OLED. By increasing the applied illuminance from 100 to 500 lx, exposure limit would be decreased by 5 times. It should be noted that both color temperature of the emission spectrum and applied illuminance affect the maximum permissible exposure limit to retina.

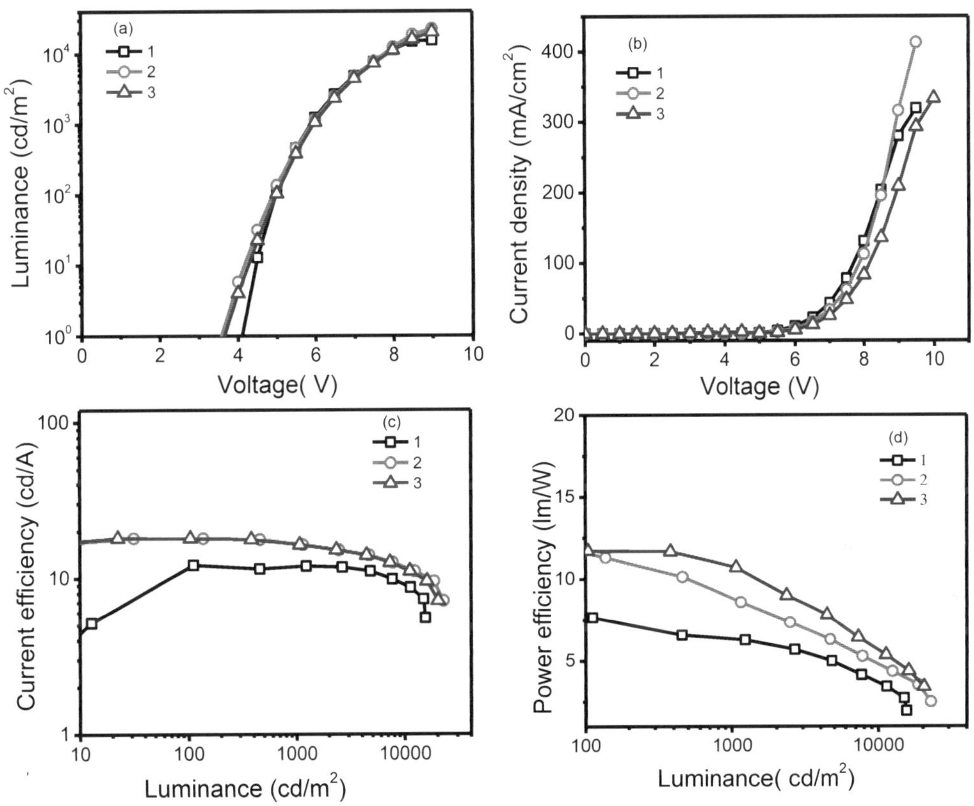

Figure 3. Comparison of (a) luminance, (b) current density, (c) current efficiency, and (d) power efficiency of the fabricated wet-processed candlelight-style OLED 1, 2 and 3.

The measured current density–voltage–luminance characteristics for Devices 1, 2 and 3 are shown in figure 3. Initially, Device 1 exhibited a maximum luminance of 15,520 cd/m^2, a power efficiency of 7.2 lm/W and a current efficiency of 11.3 cd/A at 100 cd/m^2. The comparatively lower device efficiency may be due to concentration quenching effect.[26] However, there existed triplet exciton energy transfer routes, including energy transfer from the sky-blue emitter to the host, CBP, due to its high triplet energy (2.65 eV), as comparing 2.55 eV for CBP, and then from the host to the other three emitters. Additionally, the electron-transport layer, BmPyPb, functioned in two ways; i) it provided a balanced charge injection because of its high electron mobility (1x10^{-4} cm^{2}V^{-1}s^{-1})[27] and ii) its deep HOMO level (-6.7 eV) and high triplet energy (2.65 eV) blocked the holes and confined the excitons in the emissive layer, respectively. Exciton confinement by the electron-transport layer could lead to a high luminous OLED device.

By incorporating a hole-transport layer, VPEC, the device showed a maximum luminance of 22,610 cd/m^2, and a power efficiency of 11.8 lm/W and a current efficiency of 18 cd/A. As seen in the current density-voltage result, Figure 3b, a 60% improvement in current density was observed as a VPEC layer was incorporated in OLED device. That is because the hole-transporting VPEC layer provided a relatively lower energy barrier (0.4 eV) as comparing with the large energy barrier (1.1 eV) between the interface of PEDOT:PSS and CBP host (Device 1), and hence greatly improved the injection of hole, leading to the observed higher current density. The effect of having a high current density can be seen in the luminance-voltage plot shown in Figure 3 a, wherein the VPEC based devices (2 and 3) showed slightly lower driving voltages and improved maximum luminescence. The incorporation of the cross-linkable hole-transport layer of VPEC enhanced the device performance by facilitating a balanced carrier injection in the emissive layer. A relatively lower (0.7 eV) HOMO level of VPEC than that of CBP host allows a smoother hole injection from the anode to emissive layer and a high (-2.0 eV) LUMO level can effectively prevent overflow of the electrons from the emissive layer. The OLEDs with VPEC also showed better performance either from the viewpoint of current efficiency or power efficiency (Figures 3c and d). Device 3 showed a slightly higher efficacy as comparing with Device 2 at high brightness. It is because of the high triplet energy (2.88 eV)[25] of VPEC effectively can prevent the out-diffusion of exciton formed in the sky-blue emitter and enhance efficacy. In summary, the performance enhancements may be attributed to the hole transport material, VPEC, that enables a favorable hole-injection as well as effective electron- and exciton-confinement.

5. Conclusion

In this study, we have fabricated wet-processed candlelight-style OLED devices. Multiple complementary emitters in a single emissive layer provided a wide-wavelength spectrum ranging from 450 to 780 nm. The fabricated candlelight-style OLED exhibited a blue-hazard free, human eye- and physiological-friendly low color temperature emission. Its emission spectrum should presumably permit 13 times more exposure duration to retina before causing any damage to retina cells and shows 6 times less suppression sensitivity to melatonin-secretion as comparing with the high color temperature cold-white CFL. The presented candlelight-style OLEDs showed a very-high light-quality emission that provides visual-comfort without hydrocarbon burning issues. It can be used as a human- and environment-friendly lighting source.

Acknowledgement

The authors are grateful for the financial support in part by Ministry of Economic Affairs through grant MEA 104-EC-17-A-07-S3-012 and Ministry of Science and Technology through grant MOST 104-2119-M-007-012, 103-2923-E-007-003-MY3, and 105-2119-M-007-012.

References

[1] Singerman L J and Miller D G, *Rev Ophthalmol.* **10,** 88 (2003).
[2] Melton R, *Rev opt.* **2**, 151 (2014).
[3] Pauley S M, *Med. Hypotheses.* **63**, 588 (2004).
[4] Mills P R, Tomkins S C and Schlangen L J M, *J. Circadian Rhythm.* **5**, 1 (2007).

[5] Sato M, Sakaguchi T and Morita T, *Biol. Rhythm. Res.* **36**, 287 (2005).
[6] Kloog I, Haim A, Stevens R G, Barchanade M and Portnov B A, *Chronobiology Intl.* **25**, 65 (2008).
[7] Stevens R G, Brainard G C, Blask D E, Lockley S W and Motta M E, *CA Cancer J. Clin.* **64**, 207 (2014).
[8] Davis S, Mirick D K and Stevens R G, *J. Natl. Cancer Inst.* **93**, 1557 (2001).
[9] http://www.vangogh.ua.ac.be
[10] Monico L et al. *S. Anal. Chem.* **85**, 851 (2013).
[11] http://www.darksky.org
[12] Solid-State Lighting Research and Development: Multi-Year Program Plan, Office of Energy Efficiency and Renewable Energy, U.S. Department of Energy, Washington, DC p. 52 (April 2013).
[13] Jou J H, Su Y T, Liu S H, He Z K, Sahoo S, Yu H H, Chen S Z, Wang C W and Lee J R, , *J. Mater. Chem.* C **4**, 6070 (2016).
[14] Jou J H, Hsieh C Y, Tseng J R, Peng S H, Jou, Y C, Hong J H, Shen S M, Tang M C, Chen P C, and Lin C H, *Adv. Funct. Mater.* **23**, 2750 (2013).
[15] Jou J H, Kumar S. An C C, Singh M, Yu H H, Hsieh C Y, Lin Y X, Sung C F and Wang C W, *Optics Express.* **23**, 576 (2015).
[16] Y Gong Y, Liu J, Zhang Y, He G, Lu Y, Fan W B, Yuan W Z, Sun J Z and Zhang Y, *J. Mater. Chem.* C **2**, 7552 (2014).
[17] Hu Y, Zhang T, Chen J, Ma D and Cheng C H, *Isr. J. Chem.* **54**, 979 (2014).
[18] Jou J H, Wang H C, Shen S M, Peng S H, Wu M H, Chen S H and Wu P H, *J. Mater. Chem.* **22**, 8117 (2012).
[19] Jou J H, et al. *J. Mater. Chem.* **21**, 17850 (2011).
[20] Jou J H, *Org. Electron.* **13**, 1349 (2012).
[21] International Electrotechnical Commission. "IEC 62471: 2006." Photobiological safety of lamps and lamp systems. Geneva: IEC (2006).
[22] ICNIRP, "ICNIRP guidelines on limits of exposure to incoherent visible and infrared radiation", Health Physics 105 (1), (2013).
[23] Jou J H, Melatonin suppression extent measuring device. Patent S20120303282 A1 (2012).
[24] Jou J H, et al. A universal, easy-to-apply light-quality index based on natural light spectrum resemblance, Appl. Phys. Lett. 104, 203304-203309, DOI: 10.1063/1.4879635 (2014).
[25] Jou J H, Hwang P Y, Wang W B, Lin C W, Jou Y C, Chen Y L, Shyue J J, Shen S M and Chen S Z, *Org. Electron*, **24**, 254 (2015).
[26] Jou J H et al., *Appl. Phys. Lett.* **96**, 143306 (2010).
[27] Su S J, Gonmori E, Sasabe H. and Kido J, *Adv, Mater.* **20**, 4189 (2008).

IOP Conf. Series: Materials Science and Engineering **215** (2017) 012023 doi:10.1088/1757-899X/215/1/012023

Super-Hydrophobic Green Corrosion Inhibitor On Carbon Steel

H. Hassan[*a], A. Ismail[a], S. Ahmad[a] and C. F. Soon[b]
[a]Faculty of Mechanical and Manufacturing Engineering (FKMP)
[b]Faculty of Electric and Electronic Engineering (FKEE)
Universiti Tun Hussein Onn Malaysia (UTHM), 86400 Parit Raja, Johor Darul Takzim, Malaysia.

Email: azzura@uthm.edu.my

Abstract. There are many examples of organic coatings used for corrosion protection. In particular, hydrophobic and super-hydrophobic coatings are shown to give good protection because of their enhanced ability to slow down transport of water and ions through the coating. The purpose of this research is to develop water repellent coating to avoid direct contact between metal and environment corrosive and mitigate corrosion attack at pipeline system. This water repellent characteristic on super-hydrophobic coating was coated by electrodeposition method. Wettability of carbon steel with super-hydrophobic coating (cerium chloride and myristic acid) and oxidized surface was investigated through contact angle and inhibitor performance test. The inhibitor performance was studied in 25% tannin acid corrosion test at 30°C and 3.5% sodium chloride (NaCl). The water contact angle test was determined by placing a 4-μL water droplet of distilled water. It shows that the wettability of contact angle super-hydrophobic with an angle of 151.60° at zero minute can be classified as super-hydrophobic characteristic. By added tannin acid as inhibitor the corrosion protection on carbon steel becomes more consistent. This reveals that the ability of the coating to withstand with the corrosion attack in the seawater at different period of immersions. The results elucidate that the weight loss increased as the time of exposure increased. However, the corrosion rates for uncoated carbon steel is high compared to coated carbon steel. As a conclusion, from both samples it can be seen that the coated carbon steel has less corrosion rated compared to uncoated carbon steel and addition of inhibitor to the seawater provides more protection to resist corrosion attack on carbon steel.

Keywords: Carbon Steel, Super-hydrophobic, Electrodeposition, Contact Angle, Linear Polarization Resistance

1. Introduction
Carbon steel or plain-carbon steel is the largest amount of metal alloy which is used in oil and gas industrial. Carbon steel piping is damaged due to corrosion mechanism. In US industries, corrosion cost estimated to reach 170 billion a year [1]. The deterioration of the corrosion attacks every component at every stage in the life of every oil and gas field [2]. In this research, the super-hydrophobic film is functionally as a barrier to the corrosion attack. Super-hydrophobic coating can be seen through two aspects which are under atmospheric environment and as a self-cleaning property. Under atmospheric environment, air surrounding environment of the substrate following by some water droplets contact that are appear on the surfaces can be prevent by super-hydrophobic film. Hence, super-hydrophobic film and green inhibitor acts as a barrier and prevents water droplet from contacting surface substrate [3].

2. Methodology
In this research, the experimental material was a carbon steel which is commonly used as a construction material especially as a structure and pipelines. The chemicals has been used to fabricate super-hydrophobic film are 0.1M myristic acid and 0.038M cerium chloride diluted in ethanol. Super-hydrophobic on the surface of carbon steel was fabricated by electrodeposition method. Substrate was abraded by using emery paper until the surface was shining mirror. In the electrodeposition process there are two electrode cells which is consist of carbon steel as anode and platinum wire as a cathode. Then, super-hydrophobic film was fabricated by applying a cell voltage of 30V between anode and cathode for 1 hour at room temperature. In this research, there are three experimental which are contact angle test, electrochemical test and immersion test. For contact angle test, the measurement on coated carbon steel that are modified with super-hydrophobic surfaces was determined by drop of 4-μL water droplet of distilled water. It can be seen the different degrees of wettability super-hydrophobic on carbon steel surface by distilled water droplet on surface inclined at 20°C. For electrochemical experimental, a copper wire was soldering into the non-working surface. Then, the sample was coated with epoxy. The artificial seawater

was synthesized with composed of 3.5% sodium chloride (NaCl) are mixed with 25% tannin acid that react as a green inhibitor to reduce corrosion rate. The experimental temperature is taken about 30°C. Meanwhile for immersion test, a small section of the carbon steel sample that are coat with super-hydrophobic has been exposed to the artificial seawater solution. The samples has been immersed for duration of 10 days, 20 days and 30 days. There are 3 samples for each period of days. Then, the loss of weight of the material was measured for a period of time.

3. Results and Discussion

Contact Angle Test Measurement. The contact angle test is a test that measure the contact angle of a test liquid in contact with super-hydrophobic film of carbon steel substrate. Contact angle meter test was observed by drop of distilled water droplet on surface inclined at 20°C. Water contact angle test was measured with 3 samples for uncoated carbon steel and 3 samples for coated carbon steel. The water contact angle test was determined in room temperature by placing a 4-μL water droplet of distilled water. Based on the Figure 1, it can be simplified that the coated carbon steel shows wettability contact angle super-hydrophobic with an angle of 151.60° at zero minute.

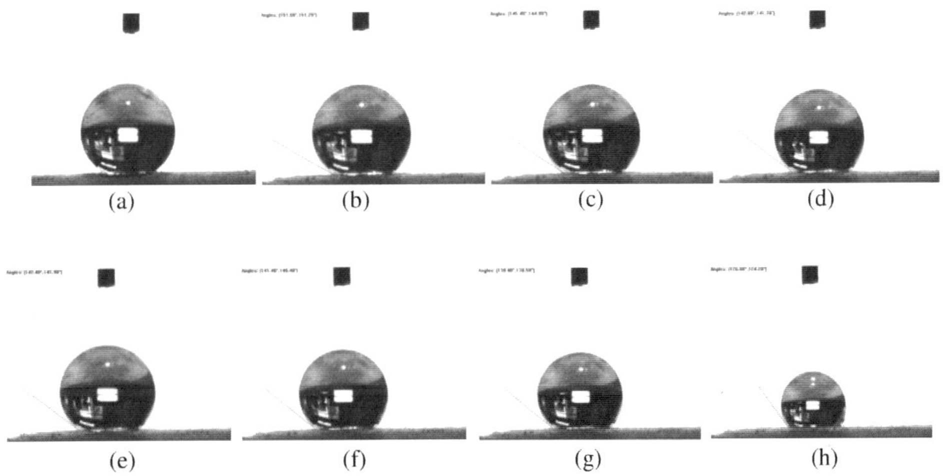

Figure 1: Water droplet on a super-hydrophobic carbon steel samples inclined at 20°C in an room temperature after (a) 0 minute, (b) 2 minutes, (c) 4 minutes, (d) 8 minutes, (e) 12 minutes, (f) 16 minutes, (g) 20 minutes and (h) 40 minutes.

The surface on carbon steel sample proved that the sample is a super-hydrophobic by present super-hydrophobic property with an evidenced that the water droplet was easily rolls off from the surface substrate. However, the contact angle of super-hydrophobic sample was not last long to remained as a super-hydrophobic. It can be seen that from the first zero minute it is super-hydrophobic with angle of 151.60°. Unfortunately, the contact angle from 151.60° at the minute zero has turn to 126.00° in 40 minutes. It results that the contact angle were decreased when a time increased at per minute. Table 4.1 shows the reduction of the contact angle in every minute.

Table 1: A reduction of contact angle at every minute.

Time duration per (minute)	Contact angle of the super-hydrophobic surfaces
0	151.60°
2	151.60°
4	145.40°
8	142.80°
12	142.40°
16	141.40°
20	139.90°
40	126.00°

Theoretically, super-hydrophobic coatings have the advantage of maintaining a contact angle of 150° even the duration of water droplets on the surface took a long time. Meanwhile, the contact angle of super-hydrophobic surface was decreased due to the surface has a cracked structure. A drop of distilled water has been absorbed and causes a reduction in the contact angle. It can be simplified that the attraction between a water droplet and a super-hydrophobic surface is affected by micro-morphology which determined the contact mode of the water droplet on super-hydrophobic surface.

Electrochemical Corrosion Test. For this electrochemical test, Tafel extrapolation method has been used to obtain the data because the substrate that are used is an active material. The solution that has been used for this electrochemical test is artificial seawater which is synthesized 3.5% NaCl mixed with 25% tannin acid. The electrochemical test was done with temperature of 30°C. In this electrochemical test, the polarization graph has determined at a sweep rate of 1 Mv s^{-1}. Figure 2 shows a comparison Tafel plot for uncoated carbon steel and coated carbon steel.

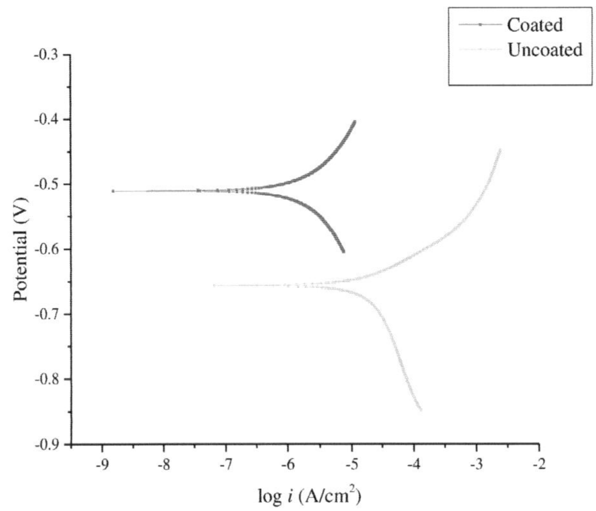

Figure 2: Comparison Tafel plot for uncoated and coated carbon steel.

Figure 2 shows the corrosion current densities were obtained by extrapolation of the anodic and cathodic polarization curves to corrosion potential. From the observation, the data of the samples are consistent. There are no noised that can be seen from the Tafel plot. While for Table 2 below shows a data of corrosion rates for uncoated and coated carbon steel that has been obtained from AutoLab potentiostat/galvanostat device.

Table 2: Results for open circuit potential and corrosion rates of carbon steel.

Samples	OCP (V)	Corrosion rates (mm/year)
Uncoated	-0.647	0.023
Coated	-0.505	0.003

It can be noted that the polarization current density of super-hydrophobic sample remains lower than uncoated carbon steel. Super-hydrophobic film shown that it can reduce corrosion rate from 0.023 mm/year of uncoated carbon steel to 0.003 mm/year of coated carbon steel. On the other hand, the lower corrosion rate of the super-hydrophobic also is caused of the used of 25% tannin acid in an artificial seawater. It reacted as a green inhibitor to slow down corrosion potential. Hence, it is effective in reducing corrosion rate of carbon steel. It simplified that the super-hydrophobic film presents a good reaction performance against corrosion of the underlying carbon steel substrate.

Immersion Corrosion Test. Immersion corrosion test has been conduct to determine the weight loss of the uncoated carbon steel and coated carbon steel. Through this test the weight of the uncoated carbon steel and coated carbon steel before and after was taken to identify the weight reduction in order to calculate the corrosion rate of the sample for a period time. Table 3 shows a weight loss and corrosion rates data that has been obtained from immersion test.

Table 3: Data obtained from immersion test for weight loss and corrosion rate.

Samples	Days	Weight loss (g)	Corrosion rates (mm/year)
Uncoated carbon steel	10	0.00249	0.24
	20	0.01529	0.15
	30	0.04536	0.43
Coated carbon steel	10	0.00303	0.02
	20	0.01340	0.13
	30	0.01508	0.14

Table 3 shows that there was a reduction of samples weight caused by corrosion attack to the uncoated carbon steel and coated carbon steel. For the first 10 days, the weight lost was low and not obviously can be seen but for the next 20 days the weight lost was decreased and for the last 30 days the weight loss became more noticeable and became higher. The time duration of carbon steel exposed to the corrosive environment affect the rate of corrosion to occur. The corrosion attacked only occurred on top of the exposed surface area but it can be seen that the super-hydrophobic coating layer slow down the corrosion rate of carbon steel during the immersion test even though the coated surface was not fully protect the top layer of the surface area. It simplified that, the corrosion rate of the immersion test sample was increased significantly as the period of immersion increased.

Metallographic Observation. In this research, the metallographic observation has been observed by using optical microscope (OM) and scanning electron microscope (SEM) device. Surface morphology for electrochemical test shows that the corrosion products occur on top of carbon steel. This corrosion occurs when the sample is supplied by a current and reacted with 3.5% NaCl which are mixed with 25% tannin acid solution. It clearly shows that the corrosion product formed on the surface of both samples during electrochemical test. Figure 3 shows uncoated and coated carbon steel after electrochemical test.

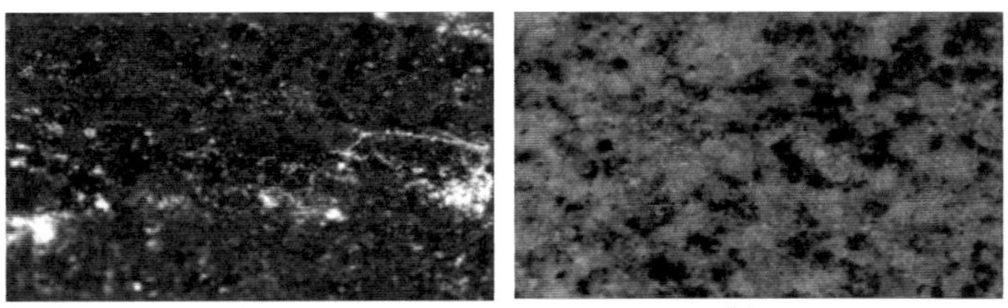

(a) (b)

Figure 3: (a) Uncoated carbon steel, (b) Coated carbon steel surface condition by using OM device after electrochemical test.

For immersion test, Figure 4 shows a surface morphology of uncoated carbon steel after immersion test on 10 days, 20 days and 30 days by using a SEM device. From observation on 20 days, the surface morphology shows that the surface start to stretching and turns to cracked on 30 days.

Figure 4: Surface morphology of cracks, porous, stretching and pitting corrosion on uncoated carbon steel surface (a) 1200x magnification on 10 days, (b) 5000x magnification on 20 days and (d) 2000x magnification on 30 days.

For coated carbon steel, the coating surface started to stretch and turns to propagate. The propagated factor caused a coating surface start to become a cracked surface and has a high tendency to corrosion attack. Porous and crack of corrosion are formed on the surface of the samples. Figure 5 shows a surface morphology of uncoated and coated carbon steel in after immersion test on 10 days, 20 days and 30 days.

Figure 5: Surface morphology of cracks, porous and pitting corrosion on coated carbon steel surface (a) 300x magnification on 10 days, (b) 1000x magnification on 20 days and (d) 2000x magnification on 30 days.

4. Conclusion

The weight of the sample was reduced when the time of immersion is increased because of the corrosion still occurs on top of the exposed sample surface even the surface already covered by corrosion product. The passive layers that occurs on the exposed surface after the corrosion product occurs slow down the corrosion rate but not prevent the corrosion from keep expanding to the other area at the sample surface that is why the weight of the sample continuously reduced for several days. The corrosion rate for uncoated carbon steel is 0.27 mm/year and coated carbon steel is 0.09 mm/year. It simplified that super-hydrophobic coating is more corrosion resistance compared to uncoated carbon steel. For the electrochemical test, the data that has been obtained shows that the corrosion rate of super-hydrophobic can reduce corrosion rate from 0.023 mm/year of uncoated carbon steel to 0.0035 mm/year coated carbon steel. The corrosion rate of super-hydrophobic is lower that uncoated carbon steel. The super-hydrophobic film presents a good reaction performance against corrosion of the underlying carbon steel substrate. Table 4 shows a comparison of corrosion rates between immersion and electrochemical test. The corrosion rates between both tests are different caused of the solutions that have been used is different. For immersion test, the solution is without tannin acid while electrochemical test used a solution with 25% tannin acid. It proved that tannin acid was reacted to slow down corrosion potential. Hence, it is effective in reducing corrosion rate of carbon steel.

Table 4: Comparison of corrosion rate between immersion and electrochemical for carbon steel.

	Corrosion rates (mm/year)	
	Uncoated	**Coated**
Immersion	0.27	0.09
Electrochemical	0.023	0.0035

Acknowledgement

The authors would like to acknowledge the Office for Research, Innovation, Commercialization and Consultancy Management (ORICC) from Universiti Tun Hussein Onn Malaysia for funded this research under grant FRGS 1460.

References

[1] D. Matthew Stuart, F. Asce, and F. Sei. 2013. "Metal Deterioration". Online. Vol. 174, pp. 1-2. Retrieved on October 16 from http://www.PDHOnline.org.

[2] A. Ismail (2014). Effect of oxygen concentration on corrosion rate of carbon steel in seawater, Corrosion Science 90, pp. 64-67. Retrieved on October 23 from http://www.ajer.org.

[3] A. Ismail (2015). Banana trunk as green corrosion inhibitor for carbon steel in subsea applications. Advanced materials research Vol 1087 (2015), pp 227-231. Retrieved on November 2 from www.scientific.net/AMR.108.227

IOP Conf. Series: Materials Science and Engineering **215** (2017) 012024 doi:10.1088/1757-899X/215/1/012024

Production process stability – core assumption of INDUSTRY 4.0 concept

F Chromjakova, R Bobak, D Hrusecka[1]

[1] Department of Industrial Engineering and Information Systems, Faculty of Management and Economics, Tomas Bata University in Zlin, Zlin, Czech Republic

chromjakova@fame.utb.cz, bobak@fame.utb.cz, hrusecka@fame.utb.cz

Abstract. Today's industrial enterprises are confronted by implementation of INDUSTRY 4.0 concept with basic problem – stabilised manufacturing and supporting processes. Through this phenomenon of stabilisation, they will achieve positive digital management of both processes and continuously throughput. There is required structural stability of horizontal (business) and vertical (digitized) manufacturing processes, supported through digitalised technologies of INDUSTRY 4.0 concept. Results presented in this paper based on the research results and survey realised in more industrial companies. Following will described basic model for structural process stabilisation in manufacturing environment.

1. Introduction

In recent years, we are speaking a lot about Internet of Things and about Industry 4.0 concept. All initiatives in this fields are concerned on the right process parameters compatibility. By implementation of both concepts we are confronted with core requirement: to have stabile processes in manufacturing and by manufacturing supporting enterprise processes. Many scientific papers describe the future state, but we need to find right way, how to prepare our industrial enterprises for input (implementation stage) of Industry 4.0 – we are speaking at this place often about revolution in managing, thinking and living in the frameworks of our traditional business models. Daily in our practice we can monitor the important change: computers will no longer be a tool, but in each industrial enterprise they will be "a digital management". And if we will initialize this state in the framework of Industry 4.0, we need to have aligned designated internal and external enterprise processes. Future target is a vision of adequate digital functionality.

Companies can be successful in any industry, but only if they use their intellectual capital to find sources of competitive advantage. Process innovation in production and administration processes must be reflected in human performance systems, processes and completely production technology. Implementation of Industry 4.0 concept strong depends from the right understanding of people, integrated in this new change processes in 4th industrial revolution.

Main goal of presented paper is to present core parameters of process innovation methodology, which enables to achieve real results in relative short time. Core output of presented methodology is a complexly way of mutual connection of people knowledge and skills, effective connection of man-man production conception integrated with e-processes and administration systems in enterprise. Results demonstrate, that SMEs have limited awareness of the risks associated with underinvesting in process innovation system.

IOP Conf. Series: Materials Science and Engineering **215** (2017) 012024 doi:10.1088/1757-899X/215/1/012024

2. Theoretical background of process stabilisation model

Process is important (Grieves, 2011), but it really doesn't describe the really interesting thing that people do, engage in practices, that is the real driver in creating, building, supporting and disposing of products. A successful product (Brissaud, Tichkiewitch, Zwolinski, 2006) must be balanced: marketing, technology, and user experience all play critical roles, but one cannot dominate the others. There is a big difference between infrastructure products, which we call non-substitutable goods, and traditional products, substitutable goods. With traditional goods, a company can survive with a stable, but non-dominant market share. With infrastructure goods, there can be just one.

Different factors are important at different stages in the development of a technology. In the early days, technology dominates. Who cares if it is easy to use? All that matters are better, faster, cheaper, more powerful technology. In the middle stages, marketing dominates. And in the end, mature stages – where the technology is a commodity. User experience can dominate, user experience and marketing. The customers want convenience and value for their money. They want ease of use, emotional appeal. The customer is not well served. As companies (Ernst & Young, 2012) find themselves in the behavioral change business, they would do well to leverage behavioral economics, a discipline that is rich with actionable insights on the biases that guide our behavior and on the levers – from the immediate feedback of social networks to the rewards of games – that actually work. Process innovation can dramatically improve business performance. Its goals are ambitious – companies embarking on process innovation often seek tenfold improvement in cost, time, or quality.

Estimates show that firms enter the market experiencing high productivity growth, and that above-average growth rates tend to last, although progressively weakened, for many years. The estimates also point out that productivity growth of surviving firms converges to different values according to activities and, on average, to almost 1.5% annually. Process innovations at some point then lead to some extra productivity growth, which tends to persist, although somewhat attenuated, for a number of years. If the introduction of process innovations then stops, however, innovation appears to be associated with an end to all productivity growth in the following years (Huergo, Jaumandreu, 2004).

Industry 4.0 concept can be characterized through the following words: big data, networking, digitalization, automation, Wi-Fi connection, cloud computing. A lot of effort will be invested in the restoring of old data to the latest IT standards. Special layers are used to transfer the experience of all employees into comprehensive IT systems. A lot of money will be invested in development of exclusive software solutions that help us in industrial enterprises look forward to the digital future.
In recent times, traditional manufacturing is upgrading and adopting Industry 4.0, which supports computerization of manufacturing by round-the-clock connection and communication of engineering objects (Shafiq, Sanin, Szczerbicki, Toro, 2016). Decisional DNA-based knowledge representation of manufacturing objects, processes, and system is achieved by virtual engineering objects, virtual engineering processes, and virtual engineering factories. Decisional DNA features as tools for effective Industry 4.0 implementation can facilitate in real time critical, creative, and effective decision making.

Manufacturing IT and Industry 4.0 is the Fourth Industrial Revolution with a potential of 12 bn Euros in Germany's chemicals industry (Gentner, 2016). But Switzerland is currently the best prepared of all countries in Europe. Many of the ideas are still very vague. As projects in Manufacturing IT and Industry 4.0 are different from the classical technical projects other strategies, for example agile project management, are necessary to secure success.

The fourth Industrial standard is based on advanced automation and robotics, sensor based computer technology, interconnected by wireless communication, and supported by big data solutions (Sandengen, Estensen, Rodseth, Schjolberg, 2016). Effective management and human cooperation i.e. teamwork and all processes will be increasingly important in the future. There will be important new form of industrial standard in relation to overall equipment effectiveness (OEE), predictive maintenance and total performance related to all equipment in the industrial factory processes. Not a few highly automated machines, but the system as a whole. OEE is a well-known standardized tool for performance measurement throughout the industry. In order to utilize data systems as ERP (Enterprise Resource Planning) and PLM (Engineering Systems), they must be integrated with business systems. Management normally acts on the bases of facts and financial performance. In most companies increase profit is the overall goal. Future management systems aim for a fully integrated and automated data flow based on advanced sensor technology. Increased use of digital systems enables quicker and better decisions.

Industry 4.0 describes the organisation of production processes based on technology and devices autonomously communicating with each other along the value chain: a model of the 'smart' factory of the future where computer-driven systems monitor physical processes, create a virtual copy of the physical world and make decentralised decisions based on self-organisation mechanisms. The concept takes account of the increased digitalisation of manufacturing industries where physical objects are seamlessly integrated into the information network, allowing for decentralised production and real-time adaptation in the future (European Parliament: Directorate-General for Internal Policies. Policy Department Economic and Scientific Policy).

Inter-connectivity and adaptivity in production are forming a strong foundation for Industry 4.0. The term adaptivity is synonymous with a new form of flexibility and adaptability of production processes and process chains which are self-adapting and optimizing. In its research and development projects, the Fraunhofer IPT addresses the challenges associated with planning both individual steps and the entire production process, in a virtual environment and in a simulation-assisted process then implements these in the corresponding machines, equipment and software systems (Fraunhofer IPA Stuttgart).

With regard to indi- vidual Industry 4.0 processes and technologies, however, it appears that across the board, regardless of company size and branch, little use is made of the evaluation of large data streams to optimise processes or for downstream services. Accordingly, little use is made also of higher level cloud services that are useful for that purpose, in contrast to SMEs elsewhere in Europe (Schroder, 2017).

To capture emerging opportunities and keep pace with the rapidly advancing technological frontier, industrial players need to act in three dimensions:

- Reach the next horizon of operational effectiveness
- Adapt business models to capture shifting value pools and
- Build foundations for the organization's digital transformation by developing digital capabilities, enabling collaboration in the ecosystem, managing data as a valuable asset, and coming to grips with cybersecurity.

In the realistic, hands-on production environments of our digital model factories, we will accompany on digital journey – across all stages, from interactive training in a digital diagnostic, through to an evaluation of your digital solutions, through to the process of building organization's digital capabilities (McKinsey Company).

3. Research study oriented on the process structure stabilisation in industrial enterprises

In according to the realization of research study we identify following research goals, concentrated on the fact, that production system oriented on the implementation on Industry 4.0 concept must fulfill following parameters:

- Information processes supporting manufacturing and production process planning – integrated information systems should create virtual reality managed by digital technologies in real time and place
- Communication processes in digitalized form – machines, technologies, processes, information and material flows should connect effective virtual information world with real technological world

- Operational management oriented on stabile flexible production – optimal performance of production system in according to the problem solving and decision making with right integration of human
- Production management based on relevant defined operational production processes – integrated value chain oriented on the standardized decision making and autonomous operability of production components
- Data security of used digitalized manufacturing technologies
- A high degree of reliability and stability of realized production processes without integration barriers
- Production systems IT maintenance

In according to the proposal of process stabilization model we identify following testing hypothesis – defined research questions:

a) **Industry 4.0 concept is based on stabile process managed structures and bring the expected cumulative benefits from digitization in the form of lower costs and increased revenues.**
Our survey was realized in 120 industrial companies in Czech and Slovak Republic (40% automotive, 30% mechanical, 20% external supply for automotive, 10% other industry). A lot of industrial enterprises is today confronted with directly requirement – to integrate three core areas:

- Enterprise organization structure
- Process / Product technological structures
- Digitized processes, supported all manufacturing integrated processes
-

Results achieved by individual interviews in selected industrial companies in Czech and Slovak Republic showed, that a lot of managers have not clear vision about the optimal combination of mentioned three core areas, the managers are oriented daily and strategically primarily on the achievement of defined company goals connected with effective productivity and performance preparedness for flexible production. They have small opinion, but this isn't enough for complexly problem solution in according to the implementation of INDUSTRY 4.0 concept. Medium companies made some steps according to digitization of machines and selected technologies, but it absent the right connection with enterprise information systems.

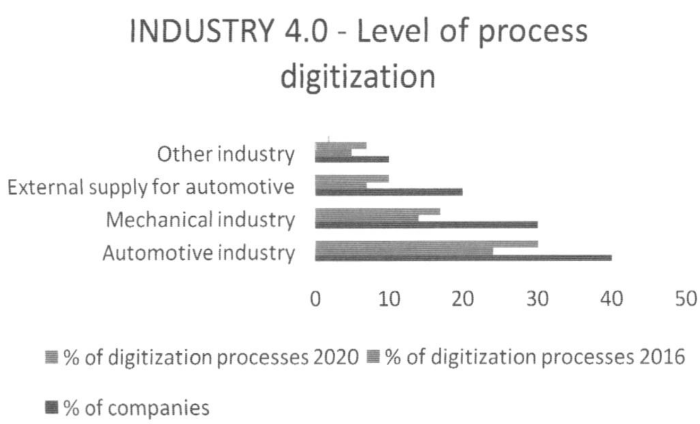

Fig. 1: Industry 4.0 – Level of process digitization – comparison 2016 and 2020 (source: Chromjaková, Bobák, Hrušecká)

Results verified mentioned hypothesis and showed, that positive cost regulation and revenues management lies in in optimal combination of advanced connectivity (14%), better "internet of things" services (26%), computer integrated processes (35%), cloud computing and advanced automation (15%) and user-friendly process standardization (15%). Right combination of all mentioned parameters secures and bring the potential of lower cost and increased revenues.

b) **Industrial company has clear identified the core processes supported the organization – process – digitized structures as a basis for implementation of Industry 4.0 concept.**
Verification of given hypothesis has brought a clear signal to the issue of identification of key business processes. The hypothesis was not confirmed, because all companies have identified key processes mostly in relation to the production process (horizontal core processes), but not in according to the implementation of Industry 4.0 concept as supporting managerial company process (vertical processes).
In according to the identification of core processes we mentioned by our survey the fact, that a lot of people in industrial companies have a problem with right understanding of the "process" definition. Traditional we are oriented on the production processes, supporting processes or we distinguish between managerial or operational processes. In the environment of Industry 4.0 concept we should make radical change in our thinking, because we are speaking about digital company – digital culture – digital processes. According to this fact we speak about new type of "process content" according to the digital enterprise environment. From our survey we achieved secondary the following knowledge: most industrial companies have respect before traditional enterprise culture, standards infrastructure, intellectual property protection by workplace, personnel leadership and coaching. This is in direct correlation with traditional model of personnel security or personnel integrated management and decision making processes. By this processes was the responsibility and delegated competencies by human, now in the Industry 4.0 concept there is necessary to transfer all important operational business competencies, responsibilities connected with production processes planning, scheduling and organization to the computer technologies and digital processes. Big accent will be given by process stabilization model proposal to the integration of professionals from IT departments. All these people should achieve clear opinion about future required process statement.

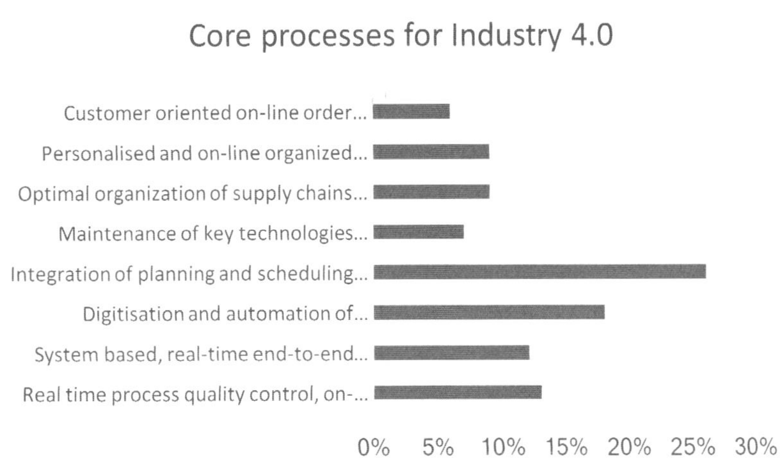

Fig. 2: Identification of core processes for Industry 4.0 implementation – process standardization and stabilization (source: Chromjaková, Bobák, Hrušecká)

IOP Conf. Series: Materials Science and Engineering **215** (2017) 012024 doi:10.1088/1757-899X/215/1/012024

c) **Industrial company has a clear vision about process steps, their mutual combination and optimal functionality, they know exactly the core process stabilization pillars before the implementation of Industry 4.0 concept.**

A lot of companies have clear vision about content of Industry 4.0 concept, they are waiting on positive experiences from other companies, that absolved first stages of implementation process. The realized questionnaire showed the accuracy of given hypothesis with an important signal to give to the companies more knowledge and skills from successful industrial companies having experiences with first implementation steps.

Where is in enterprise competent to identify the vision and define process steps of Industry 4.0 by concept implementation? This was our principally question during realization of our research in industrial enterprises. Basic impulse for this question was the fact, that we know responsible person as a director in each enterprise for the production department (evtl. production process), but we don't know the director for implementation of supporting business processes – for example – for implementation of new IT projects. In more companies we found during our survey product responsible person for implementation of IT project (78%), in better case for implementation of Industry 4.0 concept (14%). From point of process stabilization this is a crucial moment, because nobody from product managers can't be responsible in industrial enterprises for combination of horizontal and vertical integrated processes. Firstly, as we can see, we should identify right person, which will be responsible complexly for the process of Industry 4.0 concept preparation and implementation.

If we speak about process stabilization pillars, we are oriented on right mutual combination of three key pillars:

- Main production processes (stories, workflow, workplace organization, quality, TPM,etc.)
- Supporting administrative processes (production planning, logistics, economics, etc.)
- Production organization and management processes (human management, customers, etc.)

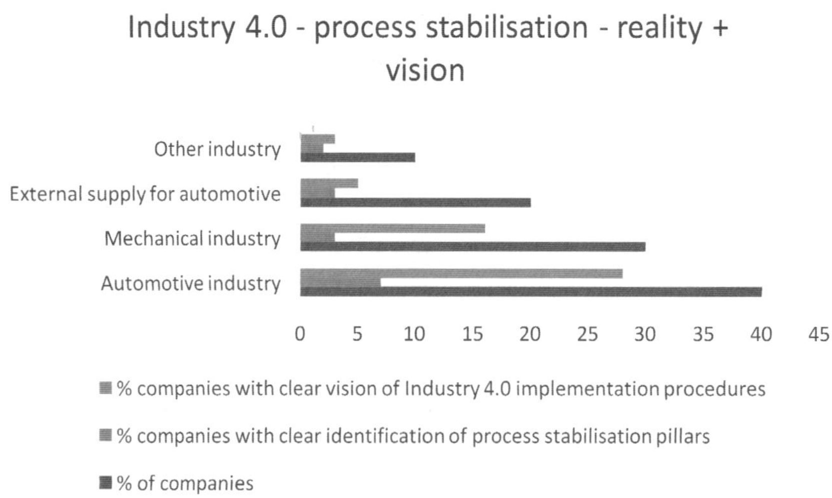

Fig. 3: Industry 4.0 – reality + vision – comparison between number of industrial companies (source: Chromjaková, Bobák, Hrušecká)

d) **Production process stability of INDUSTRY 4.0 concept enables flexible and optimal implementation of mentioned concept in manufacturing processes**

Research results identify by monitoring of production process stability as core value following metrics of process stability, used in analyzed companies:

- Production amount realized per one scheduled production cycle (74% companies)
- Productive throughput time per one scheduled production cycle without disruptions (65% companies)
- Stability of work/job standards per planned and realized totally customer order (54%)
- Number of flexible tasks identified through the just realized process (57%)

4. Proposal of process stabilization model

Definition: Process stabilization model build a background for preparation, implementation, testing and improvement of complexly integrated vertical and horizontal enterprises processes, connected with installation of digitized business processes.

The objective of the process stabilization model is to propose a mechanism by which a clear structure of processes realized by implementation of Industry 4.0 model is made, a strategic aims and objectives are identified, activities are sequenced and prioritized and the resources required are identified. It is designed in according to the right management and coordination of all realized processes from the planning procedures to the realization procedures.

By model development we take into account the following fundamental principles:

- Knowledge of principles of system oriented digitalized model of processes and process elements based on Industry 4.0
- Clear vision in according to the digitized operations, data delivery safety, data complexity, digital standards, norms and certificates
- Standardization of maintenance system for computerized and digitalized processes and operations through MES and MRP
- Skills with a human and his activity in the digitized processes by workplace

Important part of each model definition there is the first diagnosis of the production system, identification of core parameters connecting with stabilization, standardization of production processes, operations, workplaces, machine technologies, information and material flows, layouts, human. Right availability all process components build background of next process stabilization from the Industry 4.0 concept point of view.

From the Industrial Engineering point there is important to implement the same methods of management and process improvement: methods, which we used for management and process improvement in production processes should be the same for management and implementation of computered or digitized technologies. That means, firstly we should take into account parameters as system oriented thinking, simultaneous process management and organization, process owner, workplace identification and other important facts, that are crucial for the right functionality of proposed technologies or systems integrated into production processes with strong connection on digital technologies.

In according to the stability we identify important knowledge: each stabilized process or process pillar is confronted in real practice with the phenomena of allowable values for each defined index or relevant process parameter. Process stabilization model can be used than as a tool for process new definition or as a preventive tool for process management, in most cases we can use it for daily operative process management, when some unforeseen process conflicts occur.

IOP Conf. Series: Materials Science and Engineering **215** (2017) 012024 doi:10.1088/1757-899X/215/1/012024

Structure of process stabilization model was defined by research team as follows:

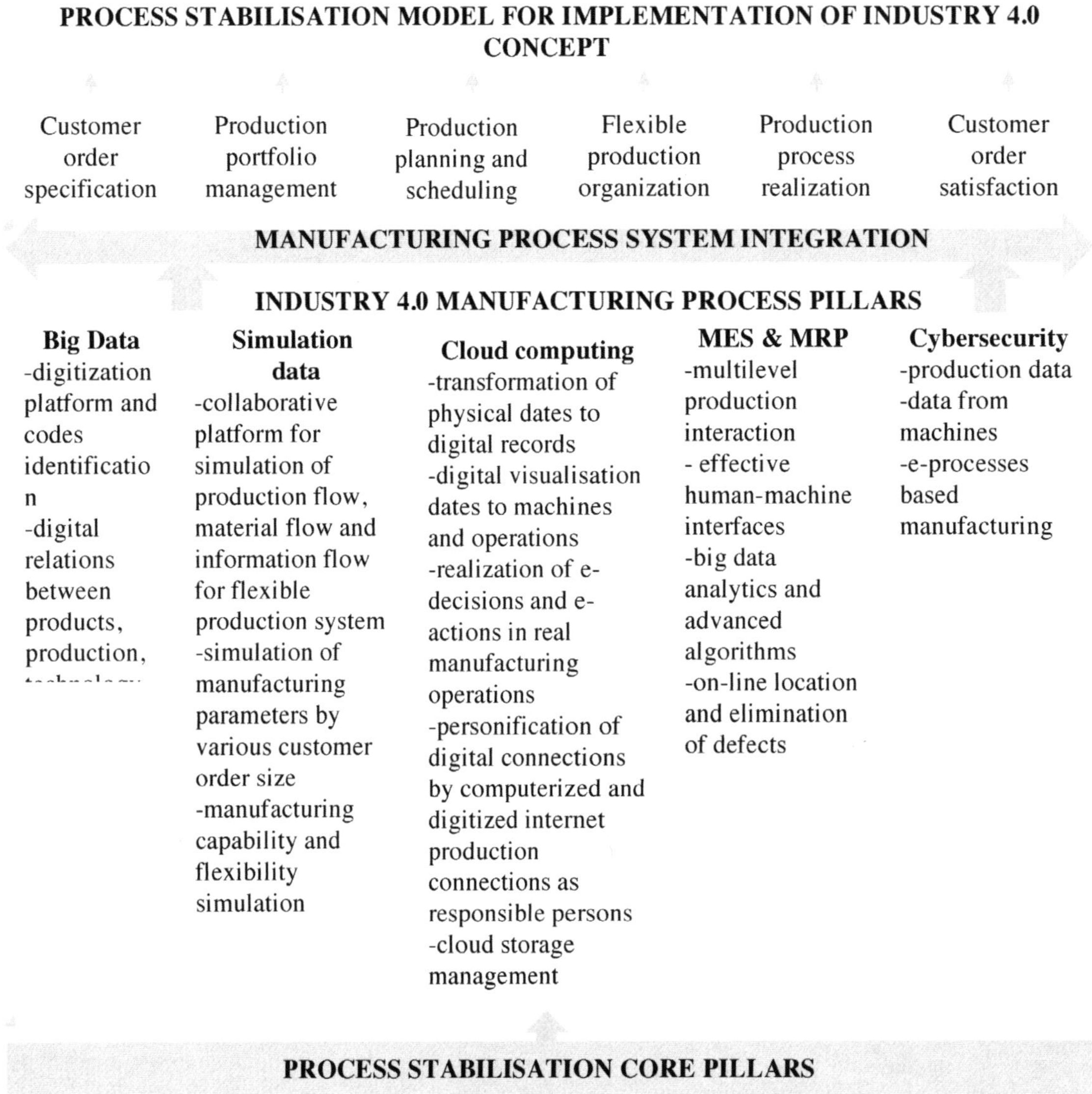

Fig. 4: Process stabilization model structure for Industry 4.0 concept implementation in industrial company (source: Chromjaková, Bobák, Hrušecká)

Process stabilization model proposal targets are concerned on following criterions:

- Coordination of production plan in according to the flexible production processes with available production capacities in production and supporting manufacturing processes
- On-line task prioritization and optimal sequencing between digitized processes and manufacturing workplaces ("just in sequence optimization")
- Physical objects are referenced, scheduled and managed by an ID
- Co-operative production planning in triangle "customer – producer – supplier" realized by web developed interactive digitized environment and data storehouse

- Optimization of production planning and scheduling through utilization of big data storehouses and cloud computing technologies in according to the customer order satisfaction in real time
- Optimization of supplier connections, transport and logistics costs
- Improvement of customer relationships and customer intelligence through the programmed on-line connection before production process
- Storage and inventories are regulated by different ID types for each component and by product component number through optimal virtual dynamic behavior knowledge
- Effective data maintenance, data security and responsibility for timeliness of data
- Assigned human responsibility for each component integrated in process model in according to the management and organization of each component

Process stabilization model should bring in the practice positive effects in the area of customer order specification, production planning and organization. We can develop our internal calculation model of benefits achieved after implementation of first stages of Industry 4.0 as follows:

Process stabilization model should bring in the practice positive effects in the area of customer order specification, production planning and organization. We can develop our internal calculation model of benefits achieved after implementation of first stages of Industry 4.0 as follows:

Industry 4.0 system real availability:
ISA = PD – ST - PSM [minute]
(PD - planned time of availability, ST – service time, PSM – preventive service maintenance)

Throughput time of e-process operation
TePO = IITS + DeOP + WTSR + IINO [minute]
(IIS – instruction input time in the system, DeOP – duration of e-operation, WTSR – waiting time on system reaction, IINO – instruction input time for next e-process operation)

Average reaction time on the system incident
ARTSI – TSU – ID - IE [minute]
(TUS – Time of system unavailability, ID – incident diagnosis, IE – incident elimination)

Index of data completeness availability for process realization in Industry 4.0 system
DCA = SRDO / RDI
(SRDO – Number of successful realized digitized operations given into systems as requirements, RDI – Returned no realized digitized operations)

Complex model of process stabilisation for implementation of Industry 4.0 concept:

$$\text{Max ISA} = 0{,}2\ COS + 0{,}1\ PPM + 0{,}2\ PPS + 0{,}3\ FPO + 0{,}1 PPR + 0{,}1\ COS$$

$$BD_{COS} + BD_{PPM} + BD_{PPS} + BD_{FPO} + BD_{PPR} + BD_{COS} = 1{,}0 \text{ (Big Data Condition)}$$

$$SD_{COS} + SD_{PPM} + SD_{PPS} + SD_{FPO} + SD_{PPR} + SD_{COS} = 1{,}0 \text{ (Simulation Data Condition)}$$

$$CC_{COS} + CC_{PPM} + CC_{PPS} + CC_{FPO} + CC_{PPR} + CC_{COS} = 1{,}0 \text{ (Cloud Computing Condition)}$$

$$MM_{COS} + MM_{PPM} + MM_{PPS} + MM_{FPO} + MM_{PPR} + MM_{COS} = 1{,}0 \text{ (MES\&MRP Condition)}$$

$$CS_{COS} + CS_{PPM} + CS_{PPS} + CS_{FPO} + CS_{PPR} + CS_{COS} = 1{,}0 \text{ (Cybersecurity Condition)}$$

(COS – Customer Order Specification, PPM – Production Portfolio Management, PPS – Production Planning and Scheduling, FPO – Flexible Production Organization, PPR – Production Process Realization, COS – Customer Order Satisfaction)

Index values given in maximisation function: 0,2 COS + 0,1 PPM + 0,2 PPS + 0,3 FPO + 0,1PPR + 0,1 COS build size of value added contribution to the whole complexly Industry 4.0 chain. These values were derived from the proposal, implementation, observation and testing of interval limit values by simulation project in the framework of RO Research Project, realized by Industrial Engineering and Information System Department by Tomas Bata University under title: "Parameters modelling for effective production and administrative processes in industrial companies based on Industry 4.0 concept". This project is realised from 2016 and till 2017 by Tomas Bata University in Zlin.

Identification of all core process components integrated in Industry 4.0 concept
Availability of all required functionalities of computerized and digitized technologies
Number of intelligent drive units integrated in the Industry 4.0 concept
Defined technical functionality, virtual functionality, communication capability and model process structure
Communication ability verification by each Industry 4.0 system component
Standardization of e-connections, standardization of input and output process parameters
Number of objects managed through digitized technology in one e-chain (ID-chain) and by one ID-administrator
Number of process conflicts in pre-implementation stage of Industry 4.0 concept (technical conflicts, technological conflicts, interface conflicts, data cybersecurity)
Number of total digitized processes in production
Number total digitized machines integrated in Industry 4.0 chain
Defined human responsibility for each process component integrated in the Industry 4.0 chain
Real time of system communication by realization of production process through Industry 4.0 system
React time on delivery of system components availability for realization of production process after input of customer order into system (specification and commitment of customer order)
Testing and validation of digitized processes in according to the flexible planning and organization of production process
Ability to re-plan the production process virtually by given instructions through ID competencies for flexible production organization
Availability of all relevant data and on-line data corrections availability in integrated process components
React time on process defect in system between process component owner (ID) and digitized workplace in production (identified by ID)
Grade of standardization of interfaces and abilities of units for and digitized regulation of flexible production system
Stability and security of defined standards, technological and technical rules, mutual process e-communication and e-management
Number of digital certificates for authentication of realized operations
Number of identities with login data for maintenance and management of operational Industry 4.0 system
Definition of system responsibilities for human – guaranty of system timeliness and usability
Number of virtual instances for recovery functions and security incidents elimination in Industry 4.0 system

5. Conclusion

Industry 4.0 concept enables radical improve the productivity and efficiency of complexly production value chains, enable to focus on the creative and strategic oriented business activities. Positive effects we can see by workers – this concept enable to organize flexible work in production processes and contributes to higher satisfaction of employees at all enterprises levels through the better work-life balance.

In our model proposal were integrated the verified research goals:

- Industry 4.0 concept is based on stabile process managed structures and bring the expected cumulative benefits from digitization in the form of lower costs and increased revenues.
- Industrial company has clear identified the core processes supported the organization – process – digitized structures as a basis for implementation of Industry 4.0 concept.
- Industrial company has a clear vision about process steps, their mutual combination and optimal functionality, they know exactly the core process stabilization pillars before the implementation of Industry 4.0 concept.
- Production process stability of INDUSTRY 4.0 concept enables flexible and optimal implementation of mentioned concept in manufacturing processes

Generated potential savings for companies are radical, special in according to the investment and return of investment in the IT technologies, customer relationships, production process planning and scheduling and flexible production outputs effects. An important role play here the phenomena of "collaborative factory" enables to realize the jobs and process operations in virtual reality with use of mobile workplaces. Each master, supervisor, team leader or shop floor employee can use in their work the assistance of multimodal, user-friendly interfaces through the complexly oriented computerized and digitized technologies, used in Industry 4.0. Industry 4.0 concept integrates the three key lines: internet of things, internet of services and internet of people – networks objects, people and systems. Optimal combination of strategic pillars will combine in the industrial company best principles from smart factory, business and social platforms. This combination guarantee right functionality of production process initialization, planning and scheduling and production process realization with according to the effective and real-time necessary data exchange, safety and reliability.

Our future work will be concentrated on the initialization of flexible digital process parameters, connected with the continuous implementation of INDUSTRY 4.0 concept.

References

[1] Rahmani Z, Mousavi S: Enhancing the innovation capability in the organization: A conceptual framework. 2nd International Conference on Education and Management Technology. IPEDR, vol. 13, IACSIT Press, Singapore, 2011
[2] Drucker P F: Innovation and Entrepreneurship: Practices and principles. Harper and Row New York, 1985
[3] Marshall A: Principles of economics: An introductory volume. Mc Milan, 1920
[4] Adler P, Shenhar A: Adapting your technological base: The organizational challenge. Sloan Management Journal, 32(1), pp. 25-37, 1990
[5] Gibson R: Rethinking the Future. Nicholas Brealey Publishing, London, 1998, ISBN 978-185788-108-0
[6] Zelený M.: The innovation factory. IPA Slovakia, Žilina, 2012
[7] Grieves R: Virtually Perfect: Driving Innovative and Lean Products through Product Lifecycle Management. Space Coast Press, Florida, 2011, ISBN 978-0-9821380-1-4
[8] Brissaud D, Tichkiewitch S, Zwolinski P: Innovation in Life Cycle Engineering and Sustainable Development. Springer Verlag. 2006. ISBN-10 1-4020-4601-4
[9] Ernst and Young: Global Life Science Report 2012: Progressions.
[10] Huergo H, Jaumandreu, J: Firms 'age, process innovation and productivity growth. Journal of Industrial Organization, Elsevier, 2004, pp. 541-559
[11] Herrmann Ch: Ganzheitliches Life Cycle Management – Nachhaltigkeit und Lebenszyklusorientierung in Unternehmen. Springer Verlag, 2010. ISBN 978-3-642-01421-5

[12] Chromjakova F.: The Key Principles of Process Manager Motivation in Production and Administrative Processes in an Industrial Enterprise, Journal of Competitiveness, 2016, Vol. 8, Issue 1, pp. 95-110
[13] Dorst W: The Industrie 4.0 Platform – joint project from the Bitkom e.V., VDMA e.v. and ZVEI e.v. associations. Berlin, Kehrberg Druck Produktion Service, 2016
[14] Gentner S: Industry 4.0: Reality, Future or just Science Fiction? How to Convince Today's Management to Invest in Tomorrow's Future! Successful Strategies for Industry 4.0 and Manufacturing IT. CHIMIA, 2016, Volume 70, Issue 9, pp. 628-633, ISSN 0009-4293
[15] Gerlitz L: Design Management as a Domain of Smart and Sustainable Enterprise: Business Modelling for Innovation and Smart Growth in Industry 4.0. Entrepreneurship and Sustainability Issues, 2016, Volume 3, Issue 3, pp. 244-268, ISSN 2345-0282
[16] Gilchris 2016, ‚A: Industry 4.0:The Industrial Internet of Things. APRESS New York, ISBN 978-1-4842-2046-7
[17] Chromjakova F:: Flexible man-man motivation performance management system for Industry 4.0., International Journal of Management Excelence, Volume 7, No. 2, pp. 829-840, ISSN 2292-1648
[18] Chromjakova F, Bobak R, Hrusecka D and coll: RO project „"Parameters modelling for effective production and administrative processes in industrial companies based on Industry 4.0 concept". Tomas Bata University, Faculty of Management and Economics, Department of Industrial Engineering and Information Systems, 2016-2017
[19] Lin C C, Deng D J, Chen Z Y, Chen K C: Key Design of Driving Industry 4.0: Joint Energy Efficient Deployment and Scheduling in Group-Based Industrial Wireless Sensor Networks. IEEE Communications Magazine, 2016, Volume 54, Issue 10, pp. 46-52, ISSN 0163-6804
[20] Roblek V, Mesko M, Krapez A:: A Complex View of Industry 4.0. Sage Open, 2016, Volume 6,
Issue 2, ISSN 2158-2440
[21] Sandengen O Ch, Estensen L A, Rodseth H, Schjolberg P: High Performance Manufacturing An Innovative Contribution towards Industry 4.0. In: Proceedings of the 6th International Workshop of Advanced Manufacturing and Automation. November 10-11, 2016, Manchester, England
[22] Shafiw S I, Sanin C, Szczerbicki E, Toro C: Virtual Engineering Factory: Creating Experience Base for Industry 4.0. Cybernetics and Systems, 2016, Volume 47, Issue 1-2, pp. 32-47, ISSN 0196-9722
[23] Vuksanovic D, Ugarak J, Korčok D: Industry 4.0:The future concepts and new visions of factory of the future development. Conference Sinteza 2016, www.researchgate.net
[24] Wang S, Wan J, Zhang D, Li D, Zhang Ch: Towards smart factory for Industry 4.0: a self organized multi-agent system with big data based feedback and coordination. Computer Networks, Volume 101, pp. 158-168, ISSN 1389-1286
[25] Xing R: Development and Design of Investigation Platform Based on the Mobile Media. Scientific Research, Communications and Network, 2013, 5, 9-11
]26] European Parliament: Directorate-General for Internal Policies. Policy Department Economic and Scientific Policy.
http://www.europarl.europa.eu/RegData/etudes/STUD/2016/570007/IPOL_STU(2016)570007_EN.pdf
[27] Fraunhofer IPA Stuttgart:
http://www.ipt.fraunhofer.de/content/dam/ipt/en/documents/broschures/Industry%2040-Connected%20Adaptive%20Production.pdf
[28] Schroder Ch: The Challenges of Industry 4.0 for Small and Medium-sized Enterprises.
http://library.fes.de/pdf-files/wiso/12683.pdf
[29] McKinsey Company: Industray 4.0 at McKinsey's model factories. https://operations-excellence.mckinsey.com/files/downloads/2016/digital40modelfactoriesbrochure1.pdf

Multi-Criteria Decision Making For Determining A Simple Model of Supplier Selection

Harwati

Industrial Engineering Universitas Islam Indonesia, Yogyakarta, Indonesia

harwati@uii.ac.id

Abstract. Supplier selection is a decision with many criteria. Supplier selection model usually involves more than five main criteria and more than 10 sub-criteria. In fact many model includes more than 20 criteria. Too many criteria involved in supplier selection models sometimes make it difficult to apply in many companies. This research focuses on designing supplier selection that easy and simple to be applied in the company. Analytical Hierarchy Process (AHP) is used to weighting criteria. The analysis results there are four criteria that are easy and simple can be used to select suppliers: Price (weight 0.4) shipment (weight 0.3), quality (weight 0.2) and services (weight 0.1). A real case simulation shows that simple model provides the same decision with a more complex model.
Keywords: Supplier selection, simple model, criteria, AHP

1. Introduction

Supplier is a company and individuals who provide the resources needed by companies in both manufacturing and services to produce goods and services. Chen et al [3] stated that the problem of supplier selection is a very important issue in supporting the success of Supply Chain Management (SCM). Competitive opportunities in the philosophy of SCM can be achieved through a strategic collaboration with suppliers and other stakeholders [15]. It can be said that one key of SCM's success depends on the ability of companies to select the best suppliers [20]. The effective supplier means suppliers who can deliver the goods with the right amount at the right time with high quality [14]. Supplier selection is the early stages of procurement of raw materials for the production process. It starts from purchase through to delivery to the warehouse or production floor [8]. The evaluation of business processes from purchase to delivery is an important factor in choosing a supplier for the company to win the competition.

Supplier selection is a decision with many criteria starts from main criteria such as price, quality and delivery up to supporting criteria such communication, attitude and relationships [5]. Activities of supplier selection may take time, resources and costs but the success of selection supplier is potential to establish long term relationship that could determine the success of SCM. Ng [15] explains that the supplier selection process starting from the determination of supplier qualification and evaluation criteria for the selection of suppliers based on existing criteria. There are many alternative criteria that can be used to select suppliers as partners. The general criteria that widely used is the price [26], [23], [9], product quality [22], [11], [21], supplier services including lead time, after-sales service and etc [1], [7], and flexibility [26]. The issue of the environment has also become next important factor for supplier selection criteria today, such as technical efficiency, environmental efficiency and eco-efficiency [13]. Even more, some companies use ethical criteria as a factor in choosing the best [4].

Previous research showed that the model of supplier selection generally involves more than five main criteria and more than 10 sub-criteria. In fact, some studies use up to twelve criteria [16]. The proposed methods is also varied from quantitative to qualitative models. Analytical Hierarchy Process (AHP) is a widely used method in the model of supplier selection [6], [19]. The other method that widely used are Fuzzy DEMATEL [2], Fuzzy Multi-Objective Linear Model [25], VIKOR[24],Technique For Order Preference By Similarity To Ideal Solution (TOPSIS) [12], data envelopment analysis (DEA) [10], [20]. Problems that happened from these variations in criteria and methods in supplier selection is the lack of opportunity to apply the model in real cases. Companies practically need ways or solutions that are easy and quick to solve a problem. Engagement so many criteria and the use of a complicated calculation or method makes the decision-makers in companies prefer to use intuition based on experience to choose suppliers. This way certainly have a deficiency because it is not based on scientific calculations. The validity of intuition usually could not be accounted. This research purpose to designing an easy and simple model for supplier selection. Formed criteria consists of only the main criteria without sub-criteria. Because the different of interest among criteria, it is necessary to calculate the criteria weight to present the importance of each criterion. Calculation of the weight uses AHP as one of Multi Criteria Decision Method (MCDM) model. AHP is used because this method is able to produce weight by decomposition procedure from a complex systems into hierarchies of goals and criteria up to alternative [18]. Resulted model of supplier selection is tested on a real case in manufacturing company to prove that it can be used and applied to industry and providing a simple decision making tools in supplier selection.

2. Analytical Hierarchy Process

AHP is a decision support models developed by Thomas L. Saaty [17]. This method is used because it accommodated the need to compare weight of each criteria without alternatives. The decision maker compares each criterion with the other and determines the level of preference for each pair of criteria. Generally, steps using AHP follows these procedure:

a. Establish the hierarchy of the problem.
The resolved problem is decomposed into its elements: criteria, sub criteria and alternatives, then it is compiled into a hierarchical structure such as this following figure:

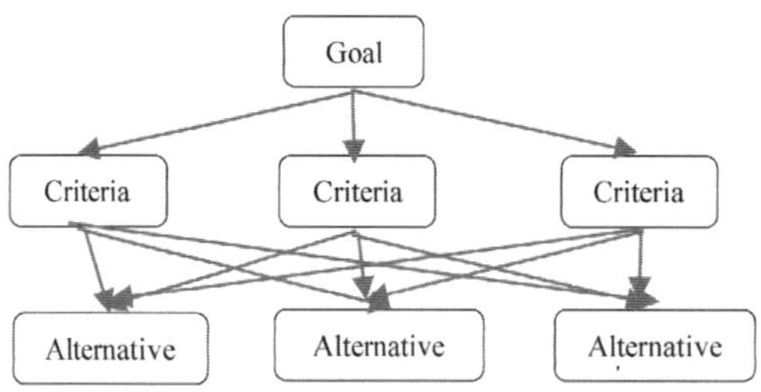

Fig 1. AHP Hierarchy Structure

b. Determine pairwise comparisons between criteria and sub-criteria.
Criteria and alternatives are assessed through pairwise comparisons. According to Saaty [17], for any kind of problems, a scale of 1 to 9 is the best scale in expressing opinions. Values and definition of qualitative opinion on a scale of Saaty can be seen in following table:

Table 1. AHP Rating

Scale	***Meaning***
1	Equal importance
3	Moderate importance
5	Strong importance
7	Demonstrate importance

IOP Conf. Series: Materials Science and Engineering **215** (2017) 012025 doi:10.1088/1757-899X/215/1/012025

9	Extreme importance
2,4,6	Intermediate importance

c. Calculate the consistency ratio

Consistency index (CI) in AHP is measure with formula CI = (λ-n)/(n-1) where λmax is the largest eigenvalue of A while A is n dimensional of AHP pairwise matrix. The degree of C.I measure by the formula CR=CI/RI. Where RI is a random index consistency from Saaty table. If the consistency ratio ≤ 0.1 means the comparison from expert is consistent.

d. Calculate priority weight

This step is processed to calculate the weight of every alternatives.

3. Method

3.1. Data Collection

Data is collected from previous research about supplier selection. All research was conducted in Indonesia specific in manufacture industry. Collected data covers some information as follow:

1. Company
2. Research
3. Criteria
4. Pairwise comparison among criteria/sub criteria and alternatives
5. Result

There are nine companies that were carried as a data source. Data description for each research is shown in the table below:

Table 2. Data Description

No	Research	Company	Criteria
1	Research 1	PT. Jogja Indo Global	Price, Quality, Delivery, Service, Payment
2	Research 2	Dii Leather	Price, Quality, Quantity, Capability
3	Research 3	CV. Bumi Multimedia	Price, Quality, Payment, Service
4	Research 4	Clothing Paradays	Price, Quality, Delivery, Service, Relationship
5	Research 5	CV. Jati Bhawono	Price, Delivery, Quantity, Capability
6	Research 6	Harmoni Rock Clothes	Price, Delivery, Quantity, Response, Flexibility, Capability
7	Research 7	CV. Prima Grafika	Price, Delivery, Quantity, Response, Flexibility
8	Research 8	PT. GKBI	Price, Delivery, Quality, Response, Relationship, Environment
9	Research 9	CV. Dita Kurnia	Price, Delivery, Quality, Response, Flexibility

The above data is the data of the main criteria used in each company. But in fact almost all companies use two-level hierarchy that includes sub-criteria for all of the criteria used. Recapitulation of eight common criteria used can be described in Pareto diagram as follow:

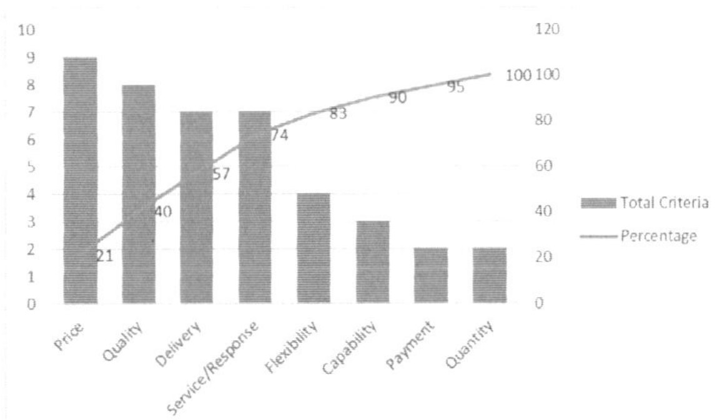

Fig 2. Pareto Diagram of Common Criteria

Using principle of Pareto 80:20, from the diagram above it can be concluded that there are four dominant criteria and they are used in all of company as consideration in supplier selection. Those criteria are: Price, Quality, Delivery, and Services.

4. Result

4.1. Data Processing Using AHP.

To obtain weight of each criteria is started by calculate the consistency index to convince the validity of the expert judgment. Calculation of consistency is applied for three expert from four company. Sample of calculation is shown in calculation of PT. Jogja Indo Global as follow:

- Pairwise Comparison Matrix

Table 3. Pairwise Comparison Matrix

Criteria	Price	Delivery	Quality	Service/Response
Price	1	½	9	7
Delivery	2	1	9	7
Quality	1/9	1/9	1	1/3
Service/Response	1/7	1/7	3	1

- Normalization Matrix

Table 4. Normalization Matrix

Criteria	Price	Delivery	Quality	Service/Response	Average
Price	0.307	0.285	0.409	0.456	0.365
Delivery	0.614	0.570	0.049	0.456	0.513
Quality	0.034	0.063	0.045	0.021	0.041
Service/Response	0.044	0.081	0.136	0.065	0.082

- Calculate Priority Weight

$$\begin{bmatrix} 1.000 & 0.500 & 9.000 & 7.000 \\ 2.000 & 1.000 & 9.000 & 7.000 \\ 0.111 & 0.111 & 1.000 & 0.333 \\ 0.143 & 0.143 & 3.000 & 1.000 \end{bmatrix} x \begin{bmatrix} 0.3645 \\ 0.5126 \\ 0.0412 \\ 0.0817 \end{bmatrix} = \begin{bmatrix} 1.563 \\ 2.183 \\ 0.330 \\ 0.017 \end{bmatrix}$$

$$Priority\ Weight\ (D) = \frac{0.563 \quad 2.183 \quad 0.330 \quad 0.017}{0.364 \quad 0.512 \quad 0.041 \quad 0.082} = 4.289 \quad 4.3 \quad 8.02 \quad 0.214$$

- Calculate λ maks

$$\lambda\text{maks} = \frac{4.2894 + 4.3 + 8.028 + 0.214}{4} = 4.19$$

- Calculate Consistency Index (CI)

$$\text{CI} = \frac{(\lambda\text{maks} - \text{n})}{(\text{n} - 1)} = \frac{(4.19 - 4)}{(4 - 1)} = 0.066$$

- Calculate Consistency Ratio (CR)

Using score for n=4 in Saaty's table is 0.9 so consistency ratio (CR) can be calculated as:

$$CR = \frac{0.66}{0.9} = 0.073$$

IOP Conf. Series: Materials Science and Engineering **215** (2017) 012025 doi:10.1088/1757-899X/215/1/012025

Since those score of consistency ratio is under 0.01 so pairwise comparison in the company can be said consistence and valid to be used for next process.

4.2. Geometric Mean

To calculate the weight of each of the selected criteria, entries adjustments are made for pairwise comparison matrix to accommodate opinion from three expert. Geometric mean is used to calculate average value of three rate with following formula:

$$\prod_{i=1}^{n} x_i = \sqrt[n]{x_1 . x_2 . x_3 \dots . x_n}$$

The geometric mean resulting from three expert calculation above are presented in the following table:

Table 5. Geometric Mean for Three Expert

	Price				Quality				Delivery				Service/Response			
Criteria	Expert 1	Expert 2	Expert 3	GM	Expert 1	Expert 2	Expert 3	GM	Expert 1	Expert 2	Expert 3	GM	Expert 1	Expert 2	Expert 3	GM
Price	1	1	1	1	0.5	3	1	1.145	9	1	1	2.08	7	3	3	3.979
Quality	2	0.333	1	0.874	1	1	1	1	9	0.333	1	1.442	7	1	1	1.912
Delivery	0.111	1	1	0.48	0.111	3	1	0.693	1	1	1	1	0.333	3	5	1.709
Service/Resp	0.143	0.333	0.333	0.251	0.143	1	1	0.523	3	0.333	0.5	0.794	1	1	1	1

4.3. Criteria Weight

The final pairwise comparison matrix for three expert is shown in the table below:

Table 6. Final Pairwise Comparison Matrix

Criteria	Price	Delivery	Quality	Service/Response
Price	1.000	1.145	2.080	3.979
Delivery	0.874	1.000	1.442	1.913
Quality	0.481	0.693	1.000	1.710
Service/Response	0.251	0.523	0.584	1.000

Normalization is done to know the weight for every criteria. The final result for criteria weight is shown in the table below:

Table 7. Final Result of Simple Criteria Weight

Criteria	Price	Delivery	Quality	Service/Response	Weight
Price	0.3837	0.3407	0.4074	0.4626	0.3986
Delivery	0.3354	0.2975	0.2824	0.2224	0.2844
Quality	0.1846	0.2062	0.1958	0.1988	0.1964
Service/Response	0.0963	0.1556	0.1144	0.1163	0.1206

From the calculation above it can be seen that the Price is still be the most important criteria for the supplier selection decision making with weight 0.4, and then followed by delivery (0.3), quality (0.2) and the last is service/response of the supplier (0.1).

4.4. Simulation

At this step, that simple model of supplier selection using four weighted criteria will be used to determine the rank of suppliers in a real case. Company in real case previously had calculated the rank of with using AHP-Promithee method. The company ranks three supplier using 5 criteria with 11 sub criteria. Score of each supplier based on AHP pairwise comparison is shown in the following table:

Table 8. Data Simulation

Criteria	Score				
	Price	Quality	Delivery	Flexibility	Service
Bobot	0.077	0.057	0.154	0.356	0.356
Supplier 1	0.362	0.328	0.353	0.327	0.327
Supplier 2	0.322	0.343	0.333	0.346	0.346
Supplier 3	0.316	0.328	0.314	0.327	0.327

By using more complicated formula of Promithee, the calculation obtains that the sequence supplier based on its criteria is supplier 1, supplier 2 and supplier 3. In the simple model, since every criterion already has weight from the previous calculation, so that the weight of every supplier can be simply calculated by the formula below:

$$V_n = X_n \,.\, R_i$$

Where Vn is the value of each supplier, Xv is the score of each supplier in each criterion and then Ri is the weight of every criterion. The result using the formula above are presented in the following table:

Table9. Simulation Result Using Simple Model

Supplier	Value
Supplier 1	0.346
Supplier 2	0.330
Supplier 3	0.320

5. Conclusion

Calculations with a simple model provides the best suppliers of the same series. Supplier 1 became the best supplier followed by Supplier 2 and Supplier 3. It is resulted both using complex model and simple model. From these results it can be concluded that the application of simple model and complex model did not give different results. But using simple model is more suggested because of its ease and effectiveness.

References

[1] Bala, A., Paco Muñoz, P., Rieradevall, J., Ysern, P., 2008. Experiences with greening suppliers. The Universitat Autònoma de Barcelona. Journal of Cleaner Production 16 (15), 1610-1619.
[2] Betty Chang, Chih-Wei Chang, Chih-Hung Wu (2011), Fuzzy DEMATEL Method For Developing Supplier Selection Criteria, *Expert Systems with Applications* 38, 1850–1858
[3] Chen, C.-T., Lin, C.-T. and Huang, S.-F. (2006), A Fuzzy Approach For Supplier Evaluation And Selection In Supply Chain Management, *International Journal of Production Economics,* Vol. 102 No. 2, pp. 289-301.
[4] Chen Jen-Yi, Baddam R Swathi (2015), The Effect Of Unethical Behavior And Learning On Strategic Supplier Selection, *Int. J.Production Economics* Vol:167 p:74–87
[5] Chen Yuh-Jen (2011), Structured Methodology For Supplier Selection And Evaluationin A Supply Chain, *Information Sciences* 181, 1651–1670
[6] Deng, X., Hu, Y., Deng, Y., & Mahadevan, S. (2014). Supplier Selection Using AHP Methodology Extended By D Numbers. *Expert Systems with Applications*, 41(1),156–167.
[7] Feyziogelu, O., Büyüközkan, G., 2010. Evaluation Of Green Suppliers Considering Decision Criteria Dependencies., Multiple Criteria Decision Making for Sustainable Energy and Transportation Systems, Part 2, pp. 145e154.
[8] Gencer Cevriye,Gurpinar Didem, (2007), Analytic network process in supplier selection: A Case Study In An Electronic Firm, *Applied Mathematical Modelling* 31, 2475– 2486
[9] Kuo, R.J., Wang, Y.C., Tien, F.C., 2010. Integration Of Artificial Neural Network And MADA Methods For Green Supplier Selection. *Journal of Cleaner Production* 18 (12), 1161e1170
[10] Kumar, A., Jain, V., & Kumar, S. (2014). A Comprehensive Environment Friendly Approach For Supplier Selection. *Omega*, 42(1), 109–123.
[11] Li, X., Zhao, C., 2009. Selection Of Suppliers Of Vehicle Components Based On Green Supply Chain., *International Conference on IE&EM, IEEE* 21e23-Oct, pp. 1588-1591
[12] Lima, F. R., Jr., Osiro, L., & Carpinetti, L. C. R. (2014). A comparison between fuzzy AHP and Fuzzy TOPSIS methods to supplier selection. *Applied Soft Computing*, 21, 194–209.
[13] Mahdiloo Mahdi, Saen Reza Farzipoor, LeeKi-Hoon, (2015), Technical, Environmental, and

eco-Efficiency Measurement for supplier Selection: Anextension and application of data envelopmentanalysis, *Int. J.ProductionEconomics* Vol:168 p:279–289

[14] Mwikali Ruth, Kavale Stanley, 2012, Factors Affecting the Selection of Optimal Suppliers in Procurement Management, *International Journal of Humanities and Social Science,* Vol. 2 No. 14 p:189-193

[15] Ng Wang Lun, 2008, An Efficient And Simple Model For Multiple Criteria Supplier Selection Problem, *European Journal Of Operational Research* Vol:186 p: 1059-1067

[16] Rezaei Jafar , Ortt Roland (2013), Multi-Criteria Supplier Segmentation Using A Fuzzy Preference Relations Based AHP, *European Journal of Operational Research* 225, 75–84.

[17] Saaty, T.L., 1993, *The Analytic Hierarchy Process,* University of Pittsburgh, Pittsburgh.

[18] Sharma, M. J., Moon, I. and Bae, H. (2008), Analytic Hierarchy Process To Assess And Optimize Distribution Network, *Applied Mathematics and Computation*, Vol. 202, pp. 256-265.

[19] Shaw, K., Shankar, R., Yadav, S. S., & Thakur, L. S. (2012). Supplier Selection Using Fuzzy AHP And Fuzzy Multi-Objective Linear Programming For Developing Low Carbon Supply Chain. *Expert Systems with Applications*, 39(9), 8182–8192.

[20] Toloo Mehdi, Nalchigar Soroosh, 2011, A New DEA Method For Supplier Selection In Presence Of Both Cardinal And Ordinal Data, *International Journal Of Expert Systems With Applications*, Vol: 38 p: 14726-14731.

[21] Wen, U.P., Chi, J.M., 2010. Developing green supplier selection procedure: a DEA approach. In: International Conference on IE&EM, IEEE 29e31 Oct, 79e74.

[22] Yan, G., 2009. Research On Green Suppliers' Evaluation Based On AHP & Genetic Algorithm. *International Conference on SPS, IEEE* 15e17 May, pp. 615e619.

[23] Yang, Y., Wu, L., 2008. Extension method for green supplier selection. *In: International* Conference on WiCom, IEEE 12-14 Oct, pp. 1e4

[24] You Xiao-Yue, You Jian-Xin, Liu Hu-Chen, Zhen Lu (2015), Group Multi-Criteria Supplier Selection Using An Extended VIKOR Method With Interval 2-Tuple Linguistic Information, *Expert Systems with Applications* 42, 1906–1916

[25] Yucel, Atakan, Guneri Fuat Ali, A Weighted Additive Fuzzy Programming Approach For Multi-Criteria Supplier Selection, Expert Systems with Applications 38 (2011) 6281–6286

[26] Zhang, H.C., Li, J., Merchant, M.E., 2003. Using Fuzzy Multi-Agent Decision-Making In Environmentally Conscious Supplier Management. CIRP Annals – Manufacturing Technology 52 (1), 385-388

IOP Conf. Series: Materials Science and Engineering **215** (2017) 012026 doi:10.1088/1757-899X/215/1/012026

Indonesia's Electricity Demand Dynamic Modelling

J Sulistio[1], A Wirabhuana[2], M G Wiratama[1]

[1] Industrial Engineering Department, Universitas Islam Indonesia, Indonesia
[2] Industrial Engineering Department, Universitas Islam Negeri Sunan Kalijaga, Indonesia

Abstract. Electricity Systems modelling is one of the emerging area in the Global Energy policy studies recently. System Dynamics approach and Computer Simulation has become one the common methods used in energy systems planning and evaluation in many conditions. On the other hand, Indonesia experiencing several major issues in Electricity system such as fossil fuel domination, demand – supply imbalances, distribution inefficiency, and bio-devastation. This paper aims to explain the development of System Dynamics modelling approaches and computer simulation techniques in representing and predicting electricity demand in Indonesia. In addition, this paper also described the typical characteristics and relationship of commercial business sector, industrial sector, and family/ domestic sector as electricity subsystems in Indonesia. Moreover, it will be also present direct structure, behavioural, and statistical test as model validation approach and ended by conclusions.

1. Introduction

The energy consumption continued to increase up to 7% each year, but in other hands, Indonesian fossil fuel resources and oil reserves experience a significant decline since the era of the 80s. Power shortage, fossil fuel dominated energy mix, environmental impact, electrification ratios, and energy loses in distribution channel, really are, the major issues of current Electricity system in Indonesia. Not to mention the electricity consumption of Indonesia was and still is showing a progressive increase over the years. This paper objective is to develop a dynamic model of electricity demand in Indonesia. Dynamic model suitable for complex system which is compliant with this case, it allow inclusion of external issues [1], contribute to question decision makers' mental models [2].

2. Methodology

The main purpose of the research is to develop dynamic model of Indonesia's electricity demand integrated with specific real system behaviour. Therefore, System Dynamics is fruitful in addressing such system. Data collected by observation as well as document and survey result from any related sources. Conceptual modelling based on causal loop and archetypes, cause and effect and influence diagram. Moreover, system analysis and design based on system dynamic based computer simulation, policy study framework, and multi-model ecologies perspective. On the other hand, structural, behavioural, and policy validation are the main approaches for model validity assessment. Nonetheless, due to the complexity of system, this research focus on developing model on demand side of the system.

Research stages are as follow. (1) Problem formulation. Comprehensive review of current condition of energy system in Indonesia, especially in electricity sector including Indonesian strategic policy for future energy. This review will also discuss the emerging sustainable modelling approach

IOP Conf. Series: Materials Science and Engineering **215** (2017) 012026 doi:10.1088/1757-899X/215/1/012026

for energy system and analysis. (2) Conceptual Model Development. This part explained the Conceptual or Framework Modelling of Electricity System for Socio-Economic and Environmental impact. Combination of causal loop and its generic archetypes, influence diagram, and circular-economy perspectives will applied as development tools and carried out as main approaches in this part [3]. (3) System Dynamic Computer Model Development. In this phase, system dynamic model will be developed and verified for Sustainable Electricity System in Indonesia. (4) Model Testing and Validation. After all aspects and scope of the system transformed into a computer simulation model, a validation is required. In addition, computer simulation conducted in this step either to analyze the existing policy or designing recommended policy for future sustainable energy system in Indonesia. (5) Policy Design, Analysis, and Report Writing.

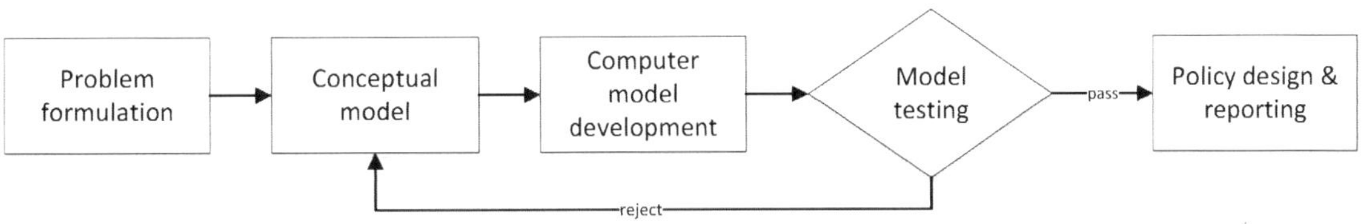

Figure 1 Research metholodogy

3. Dynamic modelling

The model consists of three sectors: (1) commercial business sector, (2) industrial sector, and (3) family/ domestic sector. Commercial business sector and industrial sector shows simple relations among its variables. A more complex dynamics exhibits in family sector whereas its relation to population sub-model. Population represented by level variable, which added by population growth in-rate caused by birth rate and immigration. On the other side, population decreased by population decline out-rate caused by death ratio and emigration.

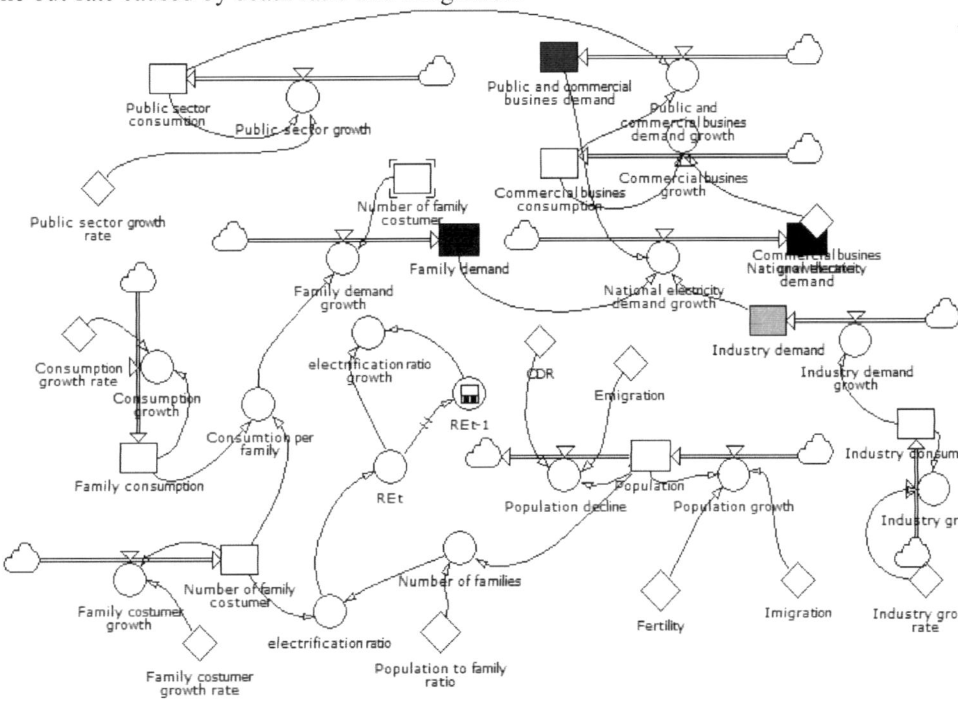

Figure 2. Flow diagram of Indonesia's Electricity Demand

IOP Conf. Series: Materials Science and Engineering **215** (2017) 012026 doi:10.1088/1757-899X/215/1/012026

3.1. Growth Rate

Population growth is the change in the number of residents in a certain area at a certain time compared with previous times, caused by fertility (births), mortality (deaths) and migration [4]. Indicators of population growth is very useful for predicting the population of a region or country in the future. By knowing the number of people, we also know the basic needs of this population, not only the social and economic fields but as well as the electrical energy needs in the future. Population grows geometrically and food items increased by arithmetically. The following equation calculate the rate of growth of customers using [5].

$$\text{Growth} = \frac{variable\ year(t) - variable\ previous\ year}{variable\ previous\ year} \times 100\% \quad (1)$$

Customer's growth rate equation above also used to calculate the growth rate of electrification ratio, growth in demand for the business sector and the industrial sector demand growth but it is also used to find the level of emigration and immigration rates. In the family sector, also affect the amount of electrification ratio or families that use electricity.

3.2. Electrification Ratio

The largest electrical energy consumers are the family sector. On one certain regional, electrical energy influenced by the amount of electricity users (electrification ratio) [6]. Electrification ratio defined as the number of families that have been electrified divided by the number of all families in Indonesia [7]. So that the electrification ratio calculations used the following formula:

$$\text{ER} = \frac{\text{number of families that have been electrified}}{\text{number of all families}} \quad (2)$$

The government has made great efforts to increase the electrification ratio in the area, including: instructed all regents / mayors to implement the accelerated development of rural electricity in their respective areas; allocate funds to provide power for at least 5 thousand families per year and PLN 50 thousand families per year. Rural electricity program expected to develop electricity grid villages to use the potential of renewable energy locals like power plants micro hydro (MHP), power plants wind / wind (PLTB) and solar power plants (PLTS) [8]. By using the electrification ratio, can be used to determine the needs and growth of electric energy in the family sector which will increase the demand for electrical energy overall.

3.3. Number of Family

Today, almost all classes of customers enjoy electricity subsidy, including the classes of customers that are highest economically. The subsidy recipient class is a class of small family 450 VA and 900 VA amounting to approximately 32.85% of the total subsidies in 2011. Another group that received subsidies are small family 1300 VA and 2200 VA, VA 3500-5500 medium families, medium industries > 200 kVA, the industry is 14-200 kVA, large industrial > 30 MVA, 6.6 to 200 kVA medium-sized businesses and public street lighting [8].

Table 1. Number of family growth *[8]*

Year		Year	Number of family
2000	52,008	2007	57,006
2002	55,041	2008	57,716
2003	56,623	2009	58,422
2004	58,253	2010	61,390
2005	55,119	2013	64,041
2006	55,942	2014	64,772

Seeing these data, the number of families would greatly affect the number of customers in the family sector that will determine the amount of the electricity demand in the family sector and the overall. The number of families estimated by the following formula [5].

$$\text{PHt} = \frac{Pt}{\text{RTt}} \tag{3}$$

Where:
RTt : The number of families in year t
Pt : The number of population in year t
Pht : The number of family occupants (assuming four person)

3.4. Growing electricity demand
Electrification ratio used to estimate the increase in electricity demand of family sector by the following formula [5].

$$Pel\ Rt = Ht\ x\ Ret\ x\ electricity\ consumption\ /\ customer \tag{4}$$

The improvement in the electrification ratio increased sales and the impact on national electricity consumption, especially in the family. Sales of electricity in the family sector in 2011 increased by 8.9% growth, followed by sales for social customer 8.1%, commercial business and 7.7%, industry 7.3% and sales of electricity for street lighting 3, 3%. Overall the total sales of the electricity in 2011 increased by 7.3% from 147.3 thousand GWh in 2010 to 157.9 thousand GWh in 2011. In addition to increase the electrification ratio, the improved reliability of electricity supply, especially in the Java-Bali electricity system also plays a role the increase in electricity sales during [8].

3.5. Number of Population
Number of population affected by changes in fertility (births), mortality (deaths) and migration [8]. The accumulation of the population of this will affect the number of families in Indonesia. Fertility influenced by many factors. Among the factors that affect the high and low fertility of the population is non-demographic factors and demographic factors. Factors that support and hinder the birth-rate in Indonesia are as follows:

a) Supporting birth (Pro birth-rate) (marry a young age, a view many children a lot of sustenance, a child of hope for parents as breadwinners, the child is a determinant of social status, the child is offspring especially boys) and
b) Factors inhibiting the birth (Anti birth-rate) (Implementation of the family planning program, delay the age of marriage because of completing education, a growing number of career women).

Discussing about various things regarding the cause of fertility and the impact on various aspects of life are more multi-disciplinary who studies the impact of the fertility rates of the various aspects of life, one of which is the growing demand for electricity from year to year. Mortality or death is one of the three components of demographic processes that affect the population. Factors that support the death (Pro mortality) is the lack of public awareness of the importance of health, inadequate health facilities, poor nutritional state of the population, the occurrence of natural disasters such as volcanic eruptions, earthquakes, floods, wars, disease and murder. While the death inhibiting factor (Anti mortality) is the increasing awareness of the population about the importance of health, inadequate health facilities, increasing the nutritional state of the population, multiply the medical personnel such as doctors and midwives as well as advances in medicine. Migration is one of the basic factors that influence the growth of population factors besides birth and death. Migration in question in this case is the migration permanently, that is to stay a lifetime in Indonesia or out leaving Indonesia lifetime. In developing countries, migration is regionally very important to study. Given the increase in population density rapidly in certain areas due to population distribution is uneven which of course it will affect other aspects of life that involves social and economic aspects is no exception will increase the demand for electrical energy [8]. The population may be decline due to emigration permanently affected and the crude death rate (CDR). CDR is a number that indicates the number of deaths per 1000 population within one year so depopulation calculated by the formula:

$$Depopulation = (CDR\ x\ Tot\ population/1000) + (Tot\ population\ x\ Lifetime\ emigration) \quad (5)$$

Increasing the number of people affected by the birth rate and immigration are still so used the formula:

$$Pop\ growth = (birth\ rate\ x\ Tot\ population) + (Tot\ population\ x\ lifetime\ immigration) \quad (6)$$

3.6. Model credibility

Validation is the latest stage in the development of a model to check the model by reviewing whether the model output in accordance with the real system, with a view of internal consistency, correspondence and representations [9]. Dynamic systems model validation assessed in several ways. Firstly, test structures directly (direct structure test) without operate (running) model. Secondly, the structural test model of behaviour (structure-oriented behaviour test) to operate the model. Lastly, comparing the behaviour of the model with real systems (quantitative behaviour pattern comparison) [10]. Model validation exercise by comparing the behaviour of the model with which to test the real system MAPE (Mean Absolute Percentage Error). MAPE (median absolute percentage error) is one measure of the relative involving percentage error. This test used to determine the suitability of the data resulting estimates with actual data.

$$MAPE = \frac{1}{n}\sum\frac{|X_m - X_d|}{X_d} \times 100\% \quad (7)$$

Where,
X_m : simulation data
X_d : actual data
n : period
MAPE < 5% : symmetric
5% < MAPE <10%: symmetric
MAPE>10% : asymmetric

Validation based on the model outlined in the flow chart and the variables used in the simulation. Validation results as shown by the graph and four tables above. From the table, the use of the business sector indicates that the accuracy of the model was not right with MAPE value amounted to 13.85916%. The public sector electricity consumption is very high degree of accuracy (very appropriate) with MAPE value of 0.53839%. The family sector and the use of the industrial sector has a level of accuracy of the model is very high as well (very appropriate), with a value of 2.0493505% MAPE and 3.0144276%. Based on this, it is safe to say that the model can describe the real condition.

Based on the MAPE test, known to no one model that is not appropriate (not describe the actual condition), so we need another test, for example by using a direct test structure, namely using direct extreme condition. Direct extreme condition done by changing the initial amount of electricity consumption in the business sector to zero. If there is no electricity consumption in business sector it is expected that, the stock usage in the business sector remain zero all the time, as well as the use of the public sector, families and industry. This is the same as that of the while researching the needs of antiretroviral drugs, i.e. by changing the initial number of HIV-positive population to 0 so that if there are no people who are HIV-positive, it is expected the stock HIV-positive population remains 0 all the time, as well with a population of AIDS [11].

4. Discussion

The simulation exhibits an exponential growth in national electricity demand because of cumulative growth in all existing sectors.

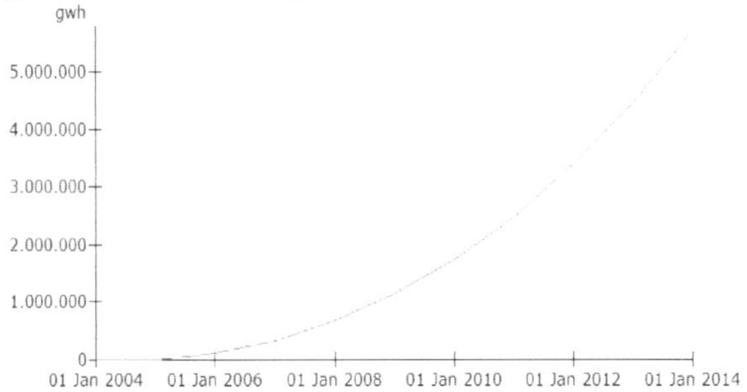

Figure 3. Simulation results of total Indonesia's Electricity Demand

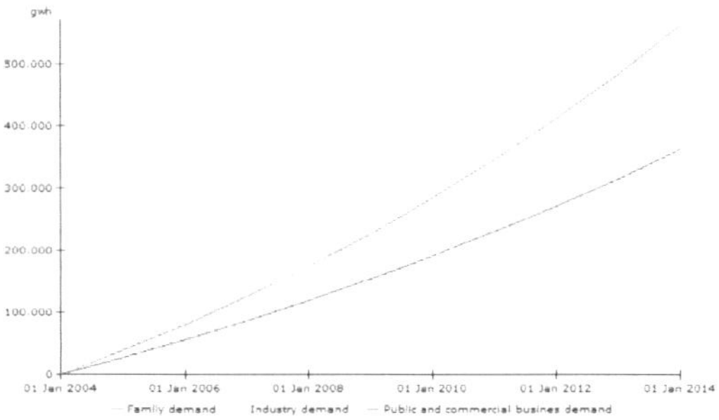

Figure 4. Simulation results of sectors effecting total electricity demand

Efforts on developing high degree accuracy models have been exercise. Correlation and regression analysis practiced to determine dominant variables effecting electricity demand in Indonesia. Average tariff and rate of capacity using per customers has given positive impact, and install capacity has given negative impact. Rate of capacity using per customer variable has significant influence to electricity demand of social sector but install capacity and average tariff has not significant influence to electricity demand of social sector [12]. Advance model such as Neural Network (NN) and Support Vector Machine (SVM) applied to predict the needs of electricity demand. Genetic Algorithm (GA) recommend as a method to optimize the value of NN and SVM parameters in predicting the demand of electrical energy [13]. Nonetheless, those models incapable to describe the dynamics and complexity of variables.

Similar to this research, a dynamics modelling approach employed in attempt to capture the complexity of electricity demand [14]. The research study the dynamics of electricity on both demand and supply side in East Java. The existing model shows that East Java will face energy crisis in 2025. In addition, the results of all design of experiment (pessimistic, optimistic, and most likely) are consistent. It concludes that no power plan supplying east java are adequate to meet future electricity demand in East Java.

The model's credibility will improve if at least two methods of validation applied. Those methods are (1) Structural validation, and (2) Behaviour validation. This research share the same method to validate initial model. Mean comparison and Error variance are applied to validate initial model. Mean comparison test objective is to assess the mean difference between two sets of data. The test's result, exclusively imply that there is no different between two sets of data.

5. Concluding remarks

From the description above, it can be concluded that the problem of policy planning and analysis of electricity systems is a significant issue in sustainable energy management in Indonesia. Furthermore, it can also be summarized, although still require various development, System Dynamics and computer simulations have contributed significantly in designing and modelling the demand side of Indonesia's Electricity System.

Next, throughout this work, it has been described that The Indonesia's Electricity System behaviour notably influenced by interdependency relation among commercial business, industrial, and domestic characteristic. Finally, from the description, it can be indicated that the research still has a sufficient opportunity to be developed specially to construct the integrated dynamics model of electricity system in Indonesia.

6. References

[1] D. W. A. Vega, "Modeling waste management in a bioethanol supply chain: A system dynamics approach," *DYNA*, vol. 83, no. 195, pp. 99-104, 2016.

[2] Anonymous, "Benefits of Entrepreneurship Education using an Interactive Learning Environment," in *ICSB World Conference Proceedings*, Washington, 2015.

[3] P. M. Senge, The Fifth Discpline: The Art and Practice of the Learning Organization, Doubleday , 2006.

[4] BKKBN, "Analisis Dampak Laju Pertumbuhan Penduduk Dan Kualitas Hidup Manusia Di Indonesia," Badan Kependudukan Dan Keluarga Berencana Nasional, Jakarta, 2012.

[5] A. A. Putri and S. Handoko, "Prakiraan Kebutuhan Beban dan Energi Listrik Kabupaten Kendal," Universitas Diponegoro, Semarang, 2013.

[6] R. Purnama, "Perkiraan Konsumsi Energi Listrik 2015 -2030 dan Potensi Pembangkit Listrik Energi baru Terbarukan di Aceh Tamiang," Universitas Gadjah Mada, Yogyakarta, 2015.

[7] PLN, "Statistik PLN 2013," PLN, Jakarta, 2014.

[8] Kementerian ESDM, "Kajian Supply Demand Energy," Kementrian Energi dan Sumber Daya MIneral, Jakarta, 2012.

[9] T. Simatupang, Pemodelan Sistem, Klaten: Penerbit Nindika, 2000.

[10] V. Daelan and W. Thissen , Dynamics Systems Modelling Continous Models, Faculteit Techniek, Bestuur en Management (TBFM). Technische Universiteit Delft, 2001.

[11] A. Sudrajar, "Perancangan Model Kebutuhan Obat ARV," FTI UI, Depok, 2010.

[12] A. Assagaf, "Analisis Faktor-faktor yang Memengaruhi Permintaan Tenaga Listrik Konsumen Sektor Sosial PT Perusahaan Listrik Negara (Persero)," *Jurnal Ekonomika,* pp. 39-44, 2011.

[13] O. Somantri and C. Supriyanto, "Algoritme Genetika untuk Peningkatan Prediksi Kebutuhan Permintaan Energi Listrik," *Jurnal Nasional Teknik Elektro dan Teknologi Informasi,* pp. 108-114, 2016.

[14] O. Axella and E. Suryani, "Aplikasi Model Sistem Dinamik untuk Menganalisis Permintaan dan Ketersediaan Listrik Sektor Industri (Studi Kasus : Jawa Timur)," *Jurnal Teknik ITS,* pp. 339-344, 2012.

Acknowledgments
This work is supported by Industrial Engineering Department of Universitas Islam Indonesia in collaboration with Universitas Islam Negeri Sunan Kalijaga and Delsim Laboratory. The authors are grateful for this support.

IOP Conf. Series: Materials Science and Engineering **215** (2017) 012027 doi:10.1088/1757-899X/215/1/012027

Operation Risk Mitigation On Halal Meat Supply Chain Management

Agus Mansur[1], Ayu Farida[1], Muhammad Ulil Albab[1]

[1] Universitas Islam Indonesia, Faculty of Industrial and Technology, Department of Industrial Engineering, Jalan Kaliurang KM. 14,5, Sleman, Yogyakarta, Indonesia

E-mail: agusmansur.am@gmail.com

Abstract. Business owner of perishable goods such as meats has barriers to meet the demand of worldwide shipping. One of potential risk is damage resulting in decreased quality product, while the product selection criteria not only focus on quality. As the development of health issues and needs, halal meats is not only considered by the Muslim consumer but also consumer globally. In addition, they have to pay attention to the possible risk that might occur during the logistic operations so as they prevent loss if they fail to deal with the problem and risk properly. Quality Function Development proposed to identify what consumer's need as well as to analyse the condition of market in the future. The study is conducted in the context of business owner's feelings based on their obstacles. It also aims to unravel the expectations and perceptions of owners with interrelated issues such as halal meats with each affecting on the other multidirectional ways. This study has revealed new insights for policy makers, logistics service providers, and practitioners whose decisions might impact the industry.

1. Introduction

In order to facilitate highly competitive business, a company should repair a system that never work properly. Business perishable goods like halal meats also have a barrier. One of another barrier is the customer needs. Customer orders with high product varieties in small quantities are often receive by the company, with request of customized value added service and time delivery. Therefore the company can't compete with competitors. Since the decision making process is one of the most complicated processes involved in the business, the fulfillment of customer order and varieties product are challenging as it necessary to satisfy increasing customer demand, and expand the marketing reach. In addition, due to the uncertainty and rapid changes in the business environment, the performance of company should be evaluate by the possible risk that might occur during business process.

Making a wrong decision will affecting on business. Bias and subjective judgment may result inaccurate result. The lack of a resource allocation strategy may mean that no suggestions regarding the allocation of resource are provided on how to handle the order when the risk occurs. This caused improper of result and loss of key customer, satisfaction. The purpose of this research is to evaluate the production process on halal meat using Quality Function Development (QFD) to identify the customer needs comparing with competitor's product and using Failure Mode and Effect Analysis (FMEA) to determine the problem that cause defect.

2. Literature Review

Growing number of Muslim population and the acceptance of non-Muslim toward halal product offer wider platform in the halal food market. The uniqueness of the halal industry is in the impression that holy values are driving from raw material to the final product. Halal, though the definition itself is anything that permissible. Hence, stringent requirement are bounded in complying halal. The global halal food market offers huge opportunity to be captured.

In halal food supply chain, the main goal is not only to ensure the satisfaction of the costumer is achieved, but also to ensure that the halal status of the food product remains intact throughout the whole process of supply chain. Lodhi (2009), Khan (2009) and Tieman (2011) believe that Halal integrity commitment is the main key factors in developing a certainty Halal food supply chain in the current complex and competitive environment. Deeply understanding about Halal food supply chain is needed based on those authors and it will protect the Halal integrity. One of process in supply chain is production, production also have an operational risk like another part of supply chain. To protect the integrity on operational production of halal meat, the company should mitigate the problem that caused halal product being non halal.

The integrity of halal meat must be protected by all means and all necessary step must be taken by all parties involved in the production to mitigate the consequences of risks and assure continuity of risk management in organization. Risk can be represented in a quantitative way by considering the relativity of risk and acceptability of risk. Dealing about concept risk management in organization scope is not effective which company only uses implicit knowledge and experience for risk management. It is happen because lack of sensitivity when the estimation of probability of possible outcomes which influenced by the risks. In some case organization of the company may tend to focus on performance than the risk. Therefore, the company should take focus on some risk that conducted in business operation process.

3. Methodology

Identify the reason of fail product is needed for guarantee to customer and also prevent losing customer key. Quality function development is a tools that used to identify the position of product with comparing competitor's product. To achieve this research goal, a comprehensive research methodology includes the research design and distribution. Primary data divided by 3 kind questionnaire and is collected through a survey from customer and owner. The research plan is shown in **figure. 1**.

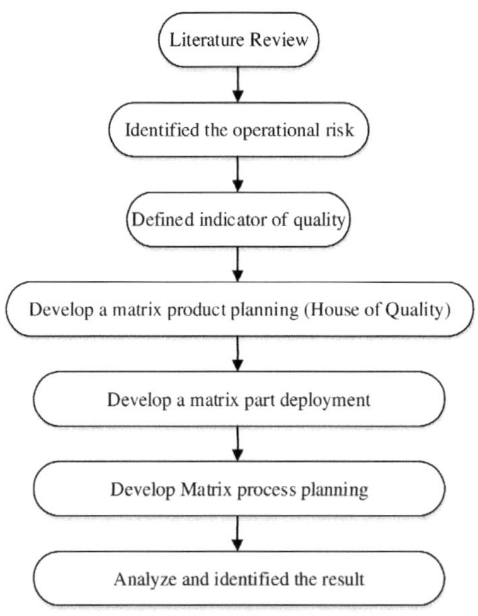

Figure 1. Research Plan

4. Case Study

Business process of perishable good like halal meats also have barrier, the meat must be assure that have same quality from production floor until arrive in the destination. Production process also have operation risk, this risk should be mitigate to ensure that the product still halal. To compete with another producer, we need to comparing the halal meat CV. Rapi Jaya and the competitors.

4.1. Discussion

We need to evaluate which one of operational processes that ineffective or in accurate. After the discussion session, we know that the product ever rejected because fail on production. So we only take focus on operational process in production. The five criteria has defined like halal, packaging, product varieties, how to order, and quality.

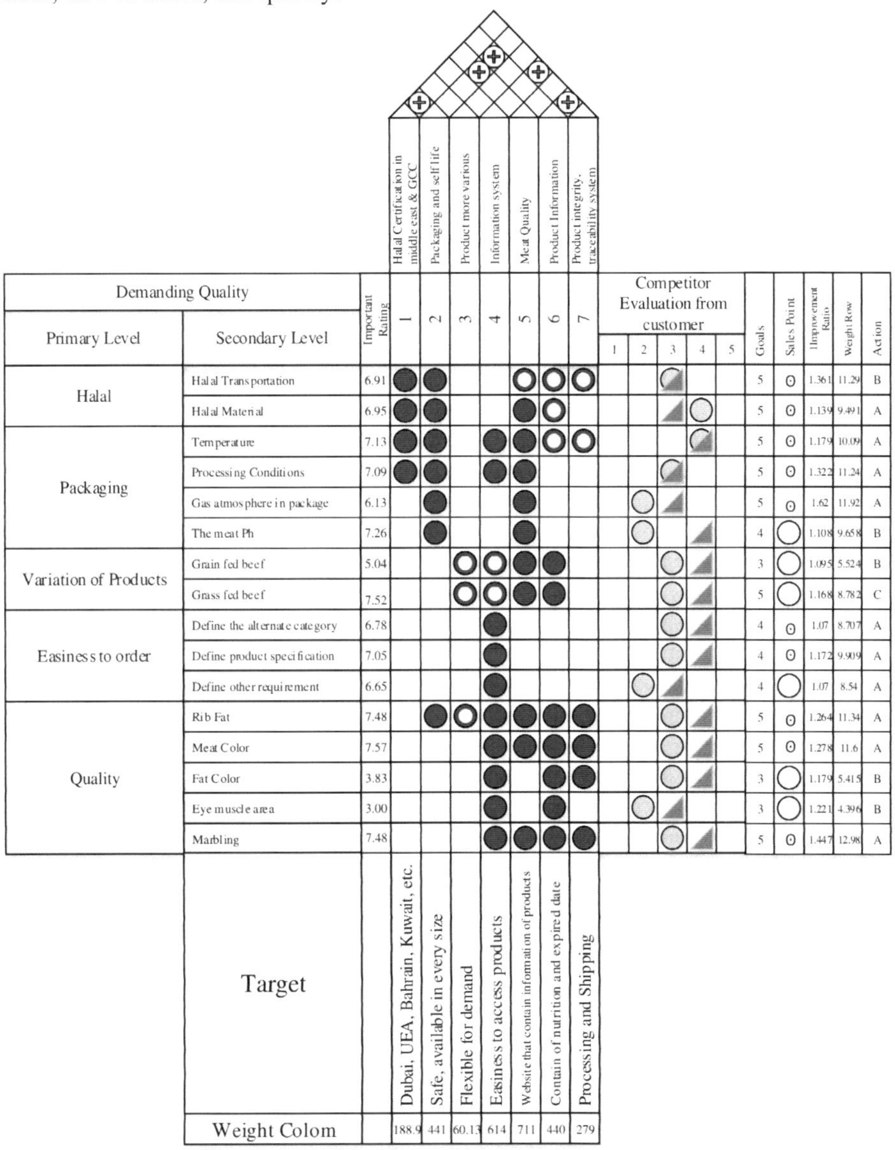

Figure 2. House of Quality

IOP Conf. Series: Materials Science and Engineering **215** (2017) 012027 doi:10.1088/1757-899X/215/1/012027

4.2. Result Analysis

Every criteria has a sub criteria that define the sub process. for example: packaging as a primary level has secondary level such as temperature, gas in atmosphere, processing condition, the meat pH Halal has secondary level like halal transportation and halal material, that so in another criteria. Every criteria, primary level and secondary level breakdown into technical requirement, target and also goal, with comparing competitor. After the subjective judgment on House of quality, we know that the company have a problem with product damage and product varieties.

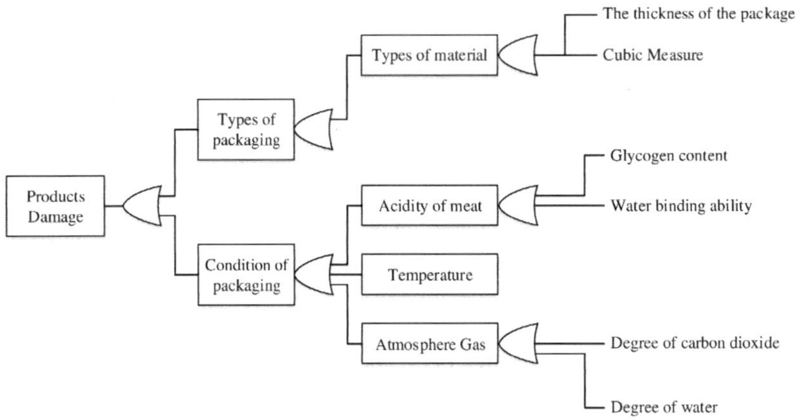

Figure 3. Risk of Damage Product

Product damage can happen with kind of packaging and packaging conditions. Wrong kind of packaging refers to the method that used like vacuum packaged (VP), layer packaged, ozone package and multi package. Packaging is very important to protect the meat's surface. Packaging product on the company only use cold box for 100kg halal meats without remove the gas inside the box. Gas that stay still would inhibits the growth of spoilage bacteria, while still allowing the natural tenderizing process of aging to continue. The pH meats caused will spoil more quickly than meat below pH 6, as some bacteria are able to survive in this high pH environment. Temperature caused Microbial growth rates at 32°F to 33.8°F are only about half those at 41°F. A storage temperature as low as 30.2°F without freezing the meat is the best policy to maximize shelf life.

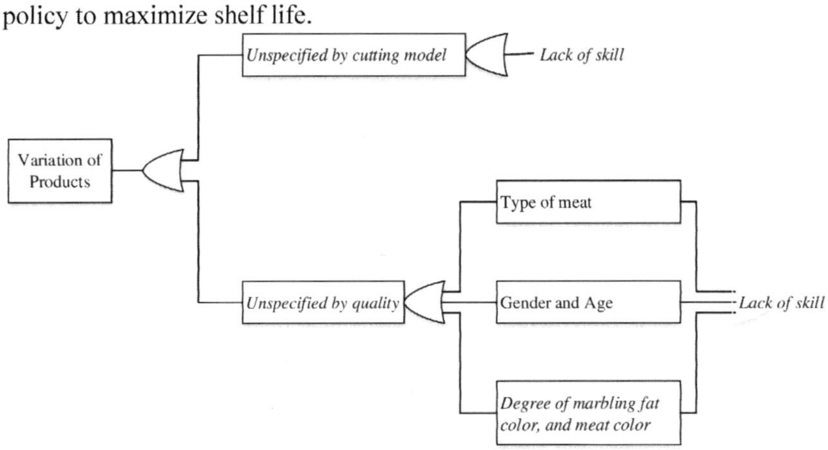

Figure 4. Risk of Variation Product

Another one biggest risk is varieties of product. The product should be easy to find, it can be define by the cutting model or specified by quality. It would help the customer to find product or alternative product. Product that define by quality such as meat color, fat color, certain age, sex, and also style of cutting. It would increasing customer satisfaction.

5. Conclusion

Halal integrity is the foundation of halal food industry. All parties in the supply chain must take responsibilities to protect the halal food product from being contaminated. One of the parties is production, to ensure the halal food still remain halal before it's travelled greater distance is minimize the damage product, that caused by wrong packaging and packaging condition. The company should change the packaging and minimize the size of packaging. The factor that caused this damage is lack of skill. The company can handle this with conducted a training for employee and work together with department of industry to evaluate the process. Employees is important to every company, it's an asset for continuity of the company.

To be able to survive in this era of business, the company need to innovate so as to overcome the saturation of the customer against product. The demand of varieties product of halal meats that's not only coming from end customer but also restaurant and hotel. This a big opportunity for company. The company should specified the product into different cutting model (tenderloin, striploin, brisket etc.) and also quality.

References

[1] Ageron, B., Gunasekaran, A., Spalanzani, A., (2012). Sustainable supply management: An Empirical Study. *Int., J. Prod. Econ* 140: 16-18.

[2] Alam, S.S. & Sayuti, N.M (2011). Applying the Theory of Planned Behavior (TPB) in Halal food purchasing. *International Journal of Commerce and Management*. 21: 8-20.

[3] Alba, J. W., & Hutchinson, W. (1987). Dimension of consumer expertise. *Journal of Consumer Research* 13: 411–454.

[4] Bahrudin, S.S.M., Illyas, M.I. & Desa, M.I. (2011). Tracking and tracing technology for Halal product integrity over the supply chain. *International Conference on Electrical Engineering and Informatics*

[5] Carter, C., Rogers, D.S. 2008. A framework of sustainable supply chain management: moving toward new theory. *International Journal of physical Distribution & Logistics Management* 38: 360-387.

[6] Cohen, L. (1995). Quality Function Development: How To Make QFD Work For You. Mussuchuttes: *Addition Wesley Publising Company.*

[7] Côté, R.P., Lopez, J., Marche, S., Perron, G.M., Wright, R., (2008). Influences, practices and opportunities for environmental supply chain management in Nova Scotia SMEs. *J. Cleaner Prod.* 16: 1561–1570.

[8] De Giovanni, P., Vinzi, V.E., (2012). Covariance versus component-based estimations of performance in green supply chain management. *Int. J. Prod. Econ.,* 135: 907–916

[9] Elkington, J., (1998). Partnerships from cannibals with forks: the triple bottom line of 21st-century business. *Environ. Qual. Manag.,* 8 (1): 37e51.

[10] Farouk, M. M. (2012). New Zealand meat industry must optimise halal market opportunities. *NZ Food Technology* 47(8): 9.

[11] Harland, C.M. 1996. Supply Chain management: relationships, chains and networks. *British journal of management* 7(1): 63-80.

[12] Jaafar, H.S., Endut, I.R., Faisol, N., Omar, E.N. (2011). Innovation in logistics services – Halal logistics. Paper presented at the 16th International Symposium om Logistics (ISL 2011), Berlin, Germany 10-13 July.

[13] Jamal, A. (2003). Marketing in a multicultural world: the interplay of marketing, ethnicity and consumption. *European Journal of Marketing* 37(11): 159-620.

[14] Kotler and amstrong (2006). Modern marketting. *Journal of management* 25: 115-120.

[15] Lam H.Y., Choy K.L., Ho G.T.S., Stephen W.Y., Cheng C.K.M. Lee., (2015). A Knowledge-based Logistics Operation Planning System for Mitigating Risk in Warehouse Fulfillment. *Int. J. Prod. Econ.,* 170: 763-779.

[16] Lodhi, A. H.,H.U.Ltd., (2010). Understanding Halal Food Supply Chain, HFRC UK Limited.

IOP Conf. Series: Materials Science and Engineering **215** (2017) 012027 doi:10.1088/1757-899X/215/1/012027

[17] Mena, C & Steve, G. (2010). Delivering Performance In Food Supply Chain: An Introduction. Cranfield University, UK.
[18] Rao, A. R., & Monroe, K. B. (1988). The moderating effect of prior knowledge on cue utilization in product evaluations. *Journal of Consumer Research* 15: 253–264.
[19] Regenstein, J. M., & Regenstein, C. E. (1991). Current issues in kosher foods. *Journal Of Food Science & Technology*, 50–54.
[20] Ronald G. Day. (1993). Quality Function Development. *ASQC Quality Press*. Milwaukee, Wiscounsin.
[21] Santoso, U. (2011). The Development of Halal Food in Indonesia: The 12th ASEAN FOOD CONFERENCE, 16-18.

IOP Conf. Series: Materials Science and Engineering **215** (2017) 012028 doi:10.1088/1757-899X/215/1/012028

Quality Control System using Simple Implementation of Seven Tools for Batik Textile Manufacturing

Muhammad Ragil Suryoputro[1], Muchamad Sugarindra[2], Hendy Erfaisalsyah[3]

[1,2,3]Department of Industrial Engineering, Faculty of Industrial Technology, Islamic University of Indonesia, Jl. Kaliurang Km 14.5 Yogyakarta, Indonesia
ragil.suryoputro@uii.ac.id[1], mragils@gmail.com[1], sugarindra@uii.ac.id[2], erfaisalsyah.hendy@yahoo.com[3]

Abstract. In order to produce better products and mitigate defect in products, every company must implement a quality control system. Company will find means to implement a quality control system that is capable and reliable. One of the methods is using the simple implementation of the seven tools in quality control defects. The case studied in this research was the level of disability xyz grey fabric on a shuttle loom 2 on the Batik manufacturing company. The seven tools that include: flowchart, check sheet, histogram, scatter diagram combined with control charts, Pareto diagrams and fishbone diagrams (causal diagram). Check sheet results obtained types of defects in the grey fabric was woven xyz is warp, double warp, the warp break, double warp, empty warp, warp tenuous, ugly edges, thick warp, and rust. Based on the analysis of control chart indicates that the process is out of control. This can be seen in the graph control where there is still a lot of outlier data. Based on a scatter diagram shows a positive correlation between the percentage of disability and the number of production. Based on Pareto diagram, repair needs priority is for the dominant type of defect is warp (44%) and based on double warp value histogram is also the highest with a value of 23635.11 m. In addition, based on the analysis of the factors causing defect by fishbone diagram double warp or other types of defects originating from the materials, methods, machines, measurements, man and environment. Thus the company can take to minimize the prevention and repair of defects and improve product quality.
Keywords: quality control, seven tools, batik manufacturing, defective products

1. Introduction

The growing global market introducing a competition between companies in providing high quality products in accordance with the wishes of the consumer. Due to the increasing consumer demand for certain products led to the company strives to meet the needs of consumers, thus demanding the company strives to improve the quality of its products. Judging from the views of consumers, subjectively people say the quality is something that is suited to the taste (fitness for use) [1]. With the good quality products, the company will benefit in terms of marketing and satisfaction of consumers.

Although the company has been striving hard to make a high quality product produced products must have disabilities. Defective product is one of the common problems shared by all companies. Usually a defective product contained in a product produced physical changes that are not in accordance with prescribed standards due to causes in the production process as well as other causes. Disability products may be due to the factors of machines, humans or the environment. Product defects can also cause loss of the company such as cost and production time. Moreover, if many defective products produced, it will increase the cost for repairs and other expenses. Not only that, defective products also reduce the selling price. In order to cope with product defects each company must have a quality control department that is in charge of checking whether there is a defect of the product before it is marketed.

This company is engaged in textiles also experienced defective products. Products produced by this company is grey cloth (calico cloth), and Batik fabric (fabric for batik material). The company also has a quality policy for orientation in its business activities. To respond with the often discovery of defective fabric products therefore it is necessary to control the quality of the existing products so that later can reduce defective products. One basic way in quality control of products is to make corrections to the elements concerned with the production and evaluation of the results. So with the correct quality control company will know how to deal with the problem there and get a result that is useful for subsequent excretion process. One of the tool that can be used is quality control system using simple implementation of seven tools. This seven tools includes flowchart, check sheet, histogram, scatter diagram combined with control charts, Pareto diagrams and fishbone diagrams (causal diagram).

From the above purposes, for reducing and preventing of defects, researchers will conduct research on quality control in cloth/textile product in Batik Company. Hence the title that will be raised is "Quality Control System using Simple Implementation of Seven Tools for Batik Textile Manufacturing".

2. Literature review

2.1. Quality Control

Quality consists of a number of product features, either directly or privilege attractive feature that meets the needs of customers and thus provide satisfaction on the use of the product [2]. Quality can also be defined as the number of attributes or properties as described in the product (and services) are concerned with [3]. Quality definition is also sourced from two sides, producers and consumers. Manufacturers determine the requirements or specifications of quality, while consumers determine the needs and desires [4]. Quality control is an activity to ascertain whether the policy in quality (standard) can be reflected in the final results, or in other words the effort to maintain the quality or the quality of goods produced to fit the specifications of the products that have been determined by policy leaders [5]. The purpose of quality control [5]: 1. that manufactured goods can reach the quality standards that have been required, 2. arranged for inspection costs to be as small as possible, 3. improve the cost estimation of the design of products and processes with the use of certain production quality to be as small as possible. And 4. provide production costs as low as possible. In general, the factors that can affect the quality [1] can be classified as follows: 1. The operating facility such as the physical condition of the building, 2. equipment and supplies (tools and equipment), 3. raw materials or materials, and 4. employment or staff organization.

2.2. Statistical Quality Control

There are many methods and tools used to perform statistical quality control but famous is using seven tools or some call the magnificent seven tools. According, Paliska, G, Pavletic, D and Sokovic, M. (2007) in every stage of the production, Quality tools could be applied, from the start of manufacturing of the products until the sale of these products [6]. Seven tools can also be used to identify the problems that exist within an enterprise. Where seven tools are as follows:

- Flowchart
  Flow chart one effective method for defining a process. Flow chart is a simple drawing of a process [7]. It could be said processing from start to finish the process in a system.
- Check sheet
  check sheet is a form, in which the items to be inspected has been printed on the form, with the intention that can be collected easily and concisely [9].
- Histogram
  Histogram is one tool that helps to find variations, and histogram is one method to make a summary of the data so that the data is analysed, which present the data graphically on how often elements in the process of emerging [9] and [7].

- Scatter Diagram
  Scatter diagram is a useful tool to clarify whether there is a relationship between two variables, and whether the relationship is positive or negative [7].
- Control chart
  Control chart is a tool in the form of process control charts to determine the upper control limit (upper limit control) and lower control limit (lower limit control) process performance [4].

  In the quality control of products, often used Quality Control Chart. With chart or a control chart can be known immediately whether the quality of the goods being processed under or out of control. Control chart or map control is a graphic comparison of data performance process (test results/observations of the nature of the product), for calculating limit control which is described as a limit line on the map. The main objective of Control Chart is to determine whether there uniformities that can be avoided in the process. The types of control charts, there are 5 types:
  - X - Chart,a chartis that used for an average price, and chart thisis the most frequently used.
  - R - Chart, used alongside the X - Chart, intended to show the magnitude of the variation of each subgroup.
  - P - Chart, used to indicate the value of the defect (percent defective).
  - Pn - Chart, is used to control the number of defects with sample values fixed size.
  - C - Chart, used to indicate the number of times the value of the actual defect (not in percent) per piece or per unit.
- Pareto diagram
  Charts Pareto are used to classify problems according to cause and symptoms [8]. Problems were depicted by priority or importance, using a bar graph format, where 100% indicates a total loss. The underlying principle of this diagram is the '80 -20 rule 'which states that "80% of the trouble comes from 20% of the problems'.

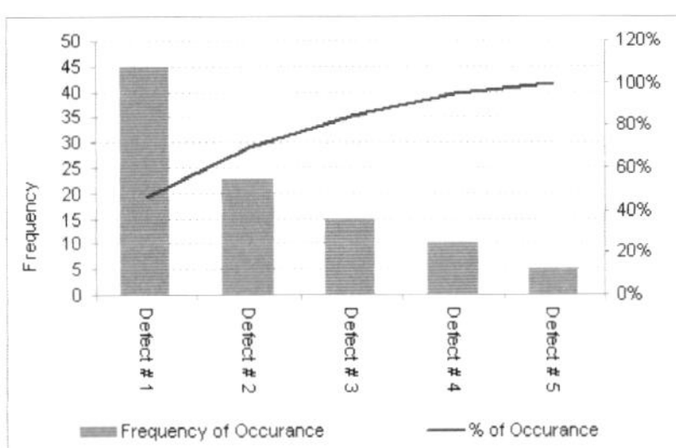

Fig 1. Pareto diagram

- Fishbone Chart
  Fishbone diagram (fishbone diagram - because it is shaped like a fish bone) is often called Cause-and-Effect Diagram. This diagram was developed in 1950 by a Japanese quality expert, ie Kaoru Ishikawa and is one of the seven basic quality tools (7 basic quality tools). Fishbone diagrams are used to identify and analyse a process or situation and find possible causes of an issue / problem that occurred [8]. As Figure 4.3 above common causes of a problem that is used in the fishbone diagram is 5M 1E or materials (raw materials), methods (method), machines (machines), measurements (measurement), man (people) (human), and environment (environment) which is used as a factor for manufacturing the initial fishbone diagram. As for the manufacture of end-use fishbone diagram 5W 1H or W1 (What), W2 (Who), W3 (When), W4 (Where), W5 (Rev), and H (How). 5W 1H basically is a method used to carry out investigation and research on the problems that occur in the production process. The addition of5 Why this question and answer simple technique to investigate the causal relationship is at the root of a problem. [10] In essence this diagram can be used for the needs of the following:
  - Helps identify the root cause of a problem.
  - Helps generate ideas for solutions to problems.

- Assist in the investigation or fact-finding further.

3. Methods

This study was conducted in Batik Textile Manufacturing Company. The resulting product is a yarn clothes, grey cotton, cambric (bleaching, mercerized, Sanforized).

The first stage of the research is done in the field observations, in the shop floor. In this observation phase researchers will conduct direct observation to see all production activities in the company.

After making observations the next step is to identify the problem, researchers will know the problems-problems that occur in the company. In this research focus is the problem identification shuttle loom two departments that have problems in production on disability production so it needs to be quality control of products.

After looking for theories that could be used as a support and research support to match the existing theory and workable, next is the data collection. The required data are flawed product data for the month of a certain month in Shuttle loom 2. Data have been collected covering total production, the type of defect and the number of defects cheek sheet then combined in a useful order for the data neatly and concisely. After it makes a comparison chart between the amount of production and the number of defects in order to determine the level of the comparison.

The next step is data processing, At this stage there are a few sequences in the data processing that is the first to calculate and determine the value of upper and lower control limits of the product are used to determine whether the process that has been known to be in a controlled or uncontrolled. The second is to create a scatter diagram to see the positive and negative relationship between two variables: the amount of production and the number of defects. The third is to identify the highest type of disability through the histogram. The fourth is to create Pareto diagram is used to classify problems according to cause and symptoms. The fifth is to identify the most influential defects and identify the cause. Sixth, once known cause then create fishbone diagrams or cause-effect diagram which is useful as a tool to identify problems and provide solutions to improvement or change in the object studied.

In the discussions, the researcher will discuss and analyse the overall results of the data processing method of the seven tools. From this analysis, researchers will find out what the cause of the defect can cause problems in production.

Lastly is the conclusions and suggestions, which conclusion is the result of that will be missed from the formulation of the problem has been determined. While the advice is input and solutions to related parties are expected to be useful for future evaluation.

4. Results and Discussion

From the research conducted, it is effective to use seven tools for implementing and analysing quality control system. These tools combined would be a beneficial for research conducted in quality system [11], [12], [13] and [14]. The table below is a defective product data type xyz grey cloth during the month of May 2015 on a shuttle loom 2. Following the results of the data collection check sheet:

Table 1. check sheet defect and production data

Day number	Length produced (m)	Defect type									Total
		Unwoven weft	Double warp	warp breaking	Double woven	Empty woven	warp tenuous	Ugly edge	Thick warp	Rust	
2	6669	25,11	0	166,73	968,97	81,60	12,90	0	65,2839	0	1320,59
3	4984	141,21	354,55	436,16	654,35	53,08	111,24	0	0	0	1750,59
4	6236	8,84	357,34	141,37	1114,38	0	73,91	0	164,11	0	1859,95
5	4948	45,16	384,51	98,96	746,51	0	30,69	0	134,25	0	1440,08
6	5491	80,70	281,20	422,08	788,46	25,93	306,54	161,50	161,50	73,09	2300,99
7	6458	53,36	0	671,46	734,15	0	434,08	0	0	0	1893,05
8	4700	139,30	329,18	537,54	743,43	44,91	68,39	85,35	0	0	1948,10

Day number	Length produced (m)	Defect type									Total
		Unwoven weft	Double warp	warp breaking	Double woven	Empty woven	warp tenuous	Ugly edge	Thick warp	Rust	
9	5391	45,38	163,36	87,52	1174,41	73,59	6,85	0	174,92	0	1726,02
10	6149	0	154,32	332,38	332,38	0	321,89	0	0	0	1140,97
11	5341	213,74	0	164,91	1222,53	0	719,21	0	166,91	0	2487,30
12	6116	55,24	38,77	164,21	813,62	102,18	343,46	0	165,30	0	1682,78
13	6188	147,16	0	133,76	799,10	0	136,36	0	157,09	0	1373,48
14	5128	86,22	165,42	836,45	423,10	15,13	496,26	0	0	0	2022,57
17	6377	110,66	589,11	654,05	830,63	119,98	113,28	0	65,60	0	2483,31
18	3990	34,47	388,87	69,84	599,95	0	0	0	61,72	0	1154,86
19	7326	432,45	666,00	241,73	1653,04	0	348,62	266,68	0	0	3608,52
20	4411	34,84	278,03	51,05	251,32	163,37	265,57	0	0	0	1044,19
21	6267	0	548,15	553,42	464,60	0	419,55	164,92	0	0	2150,64
22	6046	53,08	163,41	755,01	1320,78	0	322,72	0	0	0	2614,99
23	5105	43,13	329,35	248,46	525,83	0	330,94	0	0	0	1477,71
24	5739	0	272,55	544,07	764,95	77,78	238,79	163,97	0	0	2062,10
25	6154	23,90	166,32	460,13	788,62	0	332,65	0	0	0	1771,62
26	6278	135,70	66,28	0	800,59	61,26	566,99	0	0	0	1630,82
27	5310	14,43	402,55	165,94	830,37	51,29	165,94	165,94	0	0	1796,46
28	5910	61,68	328,33	265,04	645,96	0	292,76	0	0	138,40	1732,17
29	7273	14,76	409,79	195,80	1845,10	31,48	169,24	0	0	0	2666,17
30	4559	0	117,87	487,08	980,42	115,87	14,97	0	0	0	1716,21
31	5677	70,56	491,97	746,31	817,57	0	589,36	0	0	0	2715,77
Total	**160221**	**2071,08**	**7447,25**	**9631,43**	**23635,11**	**1017,46**	**7233,16**	**1008,36**	**1316,68**	**211,48**	**53572**

From the data on known types of defects are often found in xyz grey cloth that has been set by the company, the type of defect, respectively are:

- Unwoven weft weaving is a type of defect in which there is no pitch thread woven perfectly due to several reasons.
- Double wrap is a type of defect in which yarn that goes double numbering resulting in woven wicker is not perfect.
- Wrap breaking out is a type of defect in which there are broken threads in woven wicker resulting imperfect.
- Double woven is a type of defect in which there is a thread that is woven thick double that grounds exist on the surface of the fabric.
- Empty woven is a type of defect in which there is a part of the fabric that is discharged from the turn of the pallet is not perfect during the weaving process.
- Warp sprawl is a type of defect which occurs apart on the fabric.
- An ugly edge where the defect type on the edge of the fabric is not woven well.
- Warp thick is a type of defect in which there is a thread that is woven with thick, so there is a thick line on the surface of the fabric.
- Karat is a type of defect in which there are rust stains on the fabric, this is due to rust obtained from the engine when the fabric weaving process.
- Here is a graph illustrating the amount of production and the number of defective products in the grey cloth xyz starting on day 2 to day 31, is as follows

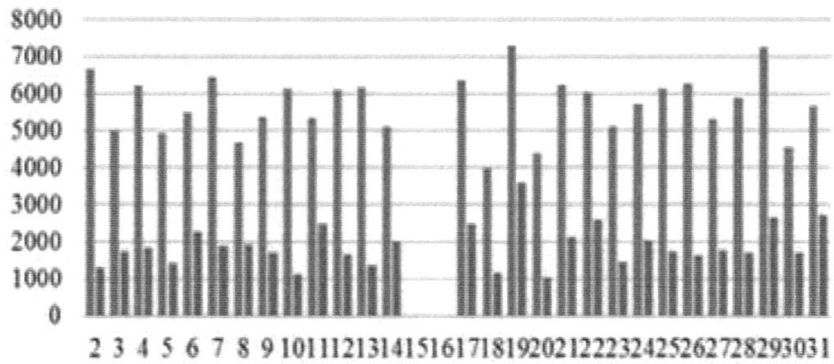

Fig 2. total production vs total defect

- The flowchart to understand the process in the company is as follows (fig 3.). This process was obtained by observation and discussion with the company.

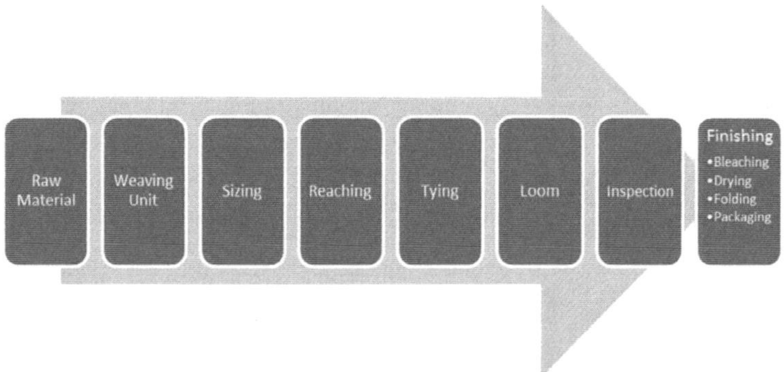

Fig 3. Production process (form observation)

4.1. C-Chart

The C Chart control data will be shown on the next table.

Table 2. C-Chart control data

	Total production number	Total Cacat	UCL	CL	LCL
	160221	**53572,02**	2044,51	1913,29	1782,06
Average	**5722,17**	**1913,29**			

Once known the value of the upper limit (UCL), a middle value (CL) and the lower control limit (LCL), later described the results of calculation of a control chart that has been created as the following chart

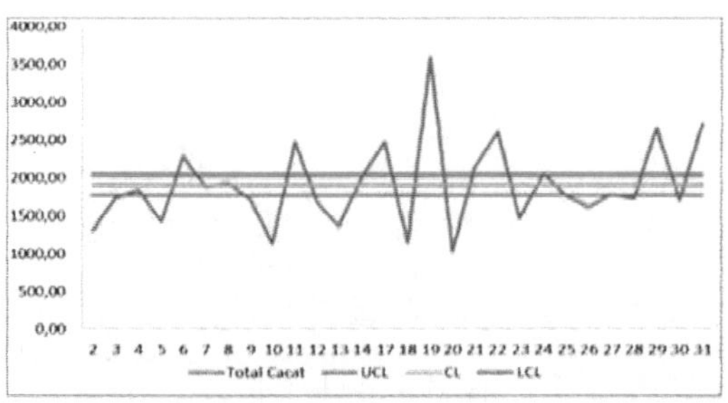

Fig 4. C-Chart

4.2. Histogram

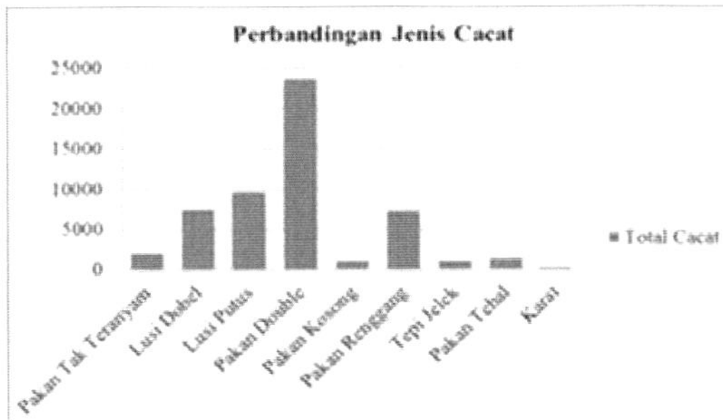

Fig 5. Comparison of defect

4.3. Scatter Diagram

Table 3. Data Calculation for Scatter Diagram

Length of Production/Jumlah Produksi (x)	Defect in Total/Total Cacat (y)	x^2	y^2	xy
160221	**53572,02027**	**93488412 5**	**111258420,6 0**	**312742634,4 1**

So the correlation coefficient between the total production and total production defects is 0.4923 means that the two variables have a fairly close relationship because if the value of r> 0 means that there is a positive relationship. Close to 1 means a strong relationship. If made into a scatter diagram of the shape as shown below:

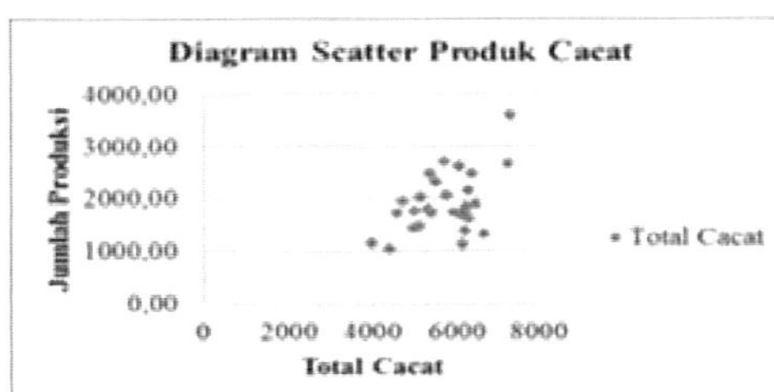

Fig 6. Scatter Diagram

4.4. Pareto diagram

Table 4. Defect Type Percentage

Defect Type	Frequency	Percentage from Total	Cumulative Percentage
Double Warp/Pakan Double	23635,11	44%	44%
Warp Break/Lusi Putus	9631,43	18%	62%
Double WarpLusi Double	7447,25	14%	76%
Warp tunous/Pakan Renggang	7233,16	14%	89%
Unwoven Warp/Pakan Tak Teranyam	2071,08	4%	93%
Thick Warp/Pakan Tebal	1316,68	2%	96%
Empty Warp/Pakan Kosong	1017,46	2%	98%

Ugly Edges/Tepi Jelek	1008,36	2%	100%
Rust/Karat/Oli	211,48	0%	100%
Total	53572,02	100%	

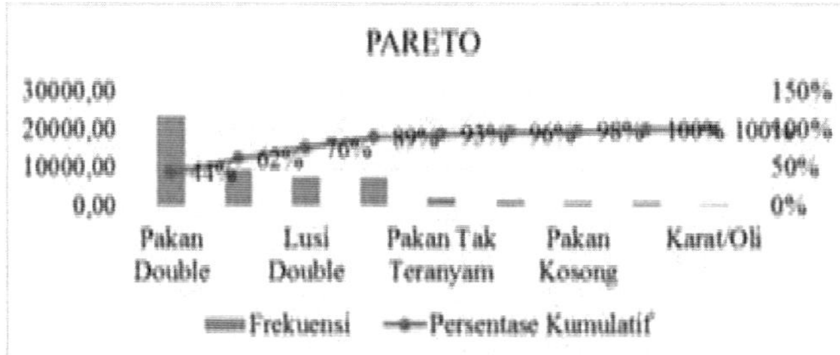

Fig 7. Pareto Diagram

4.5. Fishbone diagram

After an inventory of the problems and unknown factors cause of the defect under 5M 1E, then do manufacture fishbone diagram, that also had been combined with 5 W (what, why, when, who and where), 1 how and 5 whys.

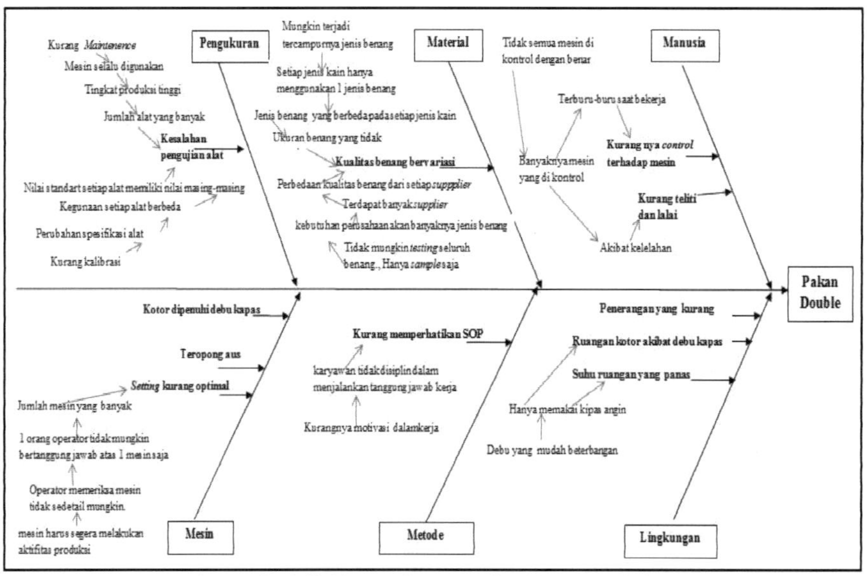

Fig 8. Final fishbone diagram double warp defect

The Pareto analysis and fishbone diagram would enhance company to explore the root cause of the problem identified as shown in the fig 8. This result is similar for building remediation contractor for elimination problem [15].

5. Conclusions and Recommendations

5.1. Conclusions

- The types of defects that occur in grey cloth warp xyz is not woven, warp double, dropping out of warp, weft double, empty warp, warp tenuous, ugly edges, thick fodder, and rust. And of the various types of disabilities that are categorized by the company, the most common defect is a defect warp double the percentage value of 44%.

- Based on the analysis of the fishbone diagram above, it can be seen that there are several factors causing the defect types that double warp material, human, machine, and environment management. Hence there is need for improvement and evaluation of the company.
- Efforts need to be made by the company in order to reduce defects in the fabric of gray xyz types are
  - In terms of the company's material should be tightened and more thoroughly in attention and choose the quality of the raw material fabric
  - n terms of human or workers in the company will have to give directions or training to them in order to improve performance and improve the regulation and strict supervision over again so that the workers could work properly
  - In terms of engine actions that must be done is to improve the care and cleaning of the machine regularly
  - In terms of its working methods should be the elements of the company's more obedient again and adhere to the SOP that has been applied
  - In terms of the measurement of the company should be more attention to measurement of the testing equipment so as to meet the standards in the company so as to produce a desired output quality
  - In terms of the environment should the company further improve the implementation of 5S workplace culture that covers everything about the company's environmental conditions.

5.2. Recommendations

- The result shown that the seven tools could be effectively used in this company, and therefore, seven tools could be implemented in other garment industry and duplicated the same method.
- Companies should improve the quality of raw materials according to the standard specified
- Companies should improve control over the performance of its employees in order to avoid deterioration in performance and negligence at work
- Perform routine engine maintenance, and pay attention to cleanliness of the engine so that the engine is not damaged or distractions while walking
- Follow standard operating procedures that have been defined
- Perform calibration routine within 6 months-12 months
- Evaluate the working environment such as lighting and 5s for workplace [10].

6. Reference list:

[1] Yamit Z. Manajemen Produksi dan Operasi. Ekonisia, Yogyakarta, 2011
[2] Gaspersz, V. 2003. Total Quality Management (TQM). Jakarta: PT. Gramedia Pustaka Utama.
[3] Ahyari, A. 1987. Manajemen Produksi Pengendalian Produksi. Yogyakarta: BPFE-Yogyakarta.
[4] Purnama, N. 2006. Manajemen Kualitas Perspektif Global. Yogyakarta: Ekonisia.
[5] Assauri, Sofjan. 2004. Manajemen Produksi dan Operasi, Edisi Revisi. Jakarta: Lembaga Penerbit Fakultas Ekonomi Universitas Indonesia.
[6] Paliska, G., D. Pavletic., and M. Sokovic, 2007. 'Quality tools-systematic use in process industry".Journal of Achievments in Materials and Manufacturing Engineering, Volume 25, Issue1, November.
[7] Yamit, Z, 2010. Manajemen Kualitas Produk dan Jasa, Yogyakarta : Ekonisia.
[8] Tjiptono, F. & Diana, D. 2001. Total Quality Management. Yogyakarta: Andi.
[9] Gaspersz, V. 1998. Statistical Process Control : Penerapan Teknik-Teknik Statistikal Dalam Manajemen Bisnis Total. Jakarta: PT. Gramedia Pustaka Utama.
[10] Suma'mur. 1989. Ergonomi Untuk Produktivitas Kerja. Jakartta : CV Haji Masagung

IOP Conf. Series: Materials Science and Engineering **215** (2017) 012028 doi:10.1088/1757-899X/215/1/012028

[11] Magar, V.M. and Shinde, V.B., 2014. Application of 7 Quality Control (7 QC) Tools for Continuous Improvement of Manufacturing Processes. International Journal of Engineering Research and General Science Volume 2, Issue 4, June-July, 2014

[12] Sokovic, M., Jovanovic, J., Krivokapic, Z., and Vujovic, A., 2009. Basic Quality Tools in Continuous Improvement Process. Strojniški vestnik - Journal of Mechanical Engineering 55(2009)5.

[13] Pavletic, D., Sokovic, M., and Paliska, G., 2008. PRACTICAL APPLICATION OF QUALITY TOOLS. 2nd International Quality Conference, Kragujevac.

[14] Ograjensek, Irena, 2002. Applying Statistical Tools to Improve Quality in the Service Sector. Developments in Social Science Methodology. Metodološki zvezki, 1 8, Lj ublj ana: FDV, 2002.

[15] Panthi, K., and Ahmed, S.M., 2008. Application of Quality Tools to Improve Quality: A Case Study of a Building Remediation Contractor. In ASC Proceedings of the 44th Annual Conference, Auburn University, April 2-5.

Hazard Identification and Risk Assessment of Health and Safety Approach JSA (Job Safety Analysis) in Plantation Company

Muchamad Sugarindra[1], Muhammad Ragil Suryoputro[2], Adi Tiya Novitasari[3]

[1,2,3] Department of Industrial Engineering, Faculty of Industrial Technology, Universitas Islam Indonesia Km 14.5 Jalan Kaliurang Sleman, Yogyakarta 55184.
sugarindra@uii.ac.id[1], mragils@gmail.com[2], adityanovitasari@gmail.com[3]

Abstract. Plantation company needed to identify hazard and perform risk assessment as an Identification of Hazard and Risk Assessment Crime and Safety which was approached by using JSA (Job Safety Analysis). The identification was aimed to identify the potential hazards that might be the risk of workplace accidents so that preventive action could be taken to minimize the accidents. The data was collected by direct observation to the workers concerned and the results were recorded on a Job Safety Analysis form. The data were as forklift operator, macerator worker, worker's creeper, shredder worker, workers' workshop, mechanical line worker, trolley cleaning workers and workers' crepe decline. The result showed that shredder worker value was 30 and had the working level with extreme risk with the risk value range was above 20. So to minimize the accidents could provide Personal Protective Equipment (PPE) which were appropriate, information about health and safety, the company should have watched the activities of workers, and rewards for the workers who obey the rules that applied in the plantation.
Keywords: OSH, Health and Safety, JSA,

1. Introduction

In an effort to produce a high quality product and as meeting the needs of consumers, company demanding to use advanced technology in the form of modern tools. The use of advanced tools of course will inevitably bring danger to the operator or to other employees as well as a negative impact on the environment and surrounding communities. This is where the management of occupational safety and health job is expected to minimize the negative impact of a production process, so that the business efficiency maximized and increased productivity of the company can be implemented.
Potential accidents are common in the workplace and working environment resulting in losses from the company, employees and the surrounding community. Efforts should be made to prevent these and to implement a concept of Occupational Health and Safety (OSH). Occupational accidents besides causing direct losses also causing indirect losses which are losses in damage to machinery and work equipment, the stop of the production process, environmental damage and others [1].
In the application management system of OSH one of the steps that must be taken is to identify the risk for accidents that may occur in the work area. Risk identification aims to identify potential hazards that pose a risk of workplace accidents. It can be determined so precautions initiatives will be performed to minimize accidents. Accidents that occur in the workplace largely caused by the carelessness of labour. It can be caused by lack of knowledge of the labour force, inadequate skills in carrying out its work. In addition, especially when faced with new technology or new equipment which is not in accordance with the size of anthropometry of Indonesian workers and also safety was not the focus of workers while

IOP Conf. Series: Materials Science and Engineering **215** (2017) 012029 doi:10.1088/1757-899X/215/1/012029

performing their work [2]. To control the work accidents and occupational illness, it is needed the efforts to identify the factors/sources of hazards in the workplace and to evaluate the risks and suggests an adequate control. In the field of OSH there are ways to identify, analyse and evaluate the factors that creates danger in the workplace. One way to identify hazards is the safety analysis, or better known as Job Safety Analysis-JSA [3].

Making Job Safety Analysis, in addition to providing remedial action for potential hazards can also provide other benefits to management. With the implementation of Job Safety Analysis, a supervisor can provide training of safe and efficient procedures for workers, making it easier to give instructions to new workers who will carry out the work and hazards inherent in the work, and can be used to assess or relearn when accidents happen. With the Job Safety Analysis, workers can work in safe and efficient, knowing the dangers inherent in the work and control the action, and to improve the knowledge and awareness of the importance of Occupational Safety and Health [4]

This plantation is one company that produces rubber Low Grade (LG) called SIR 20 and its production processes use a lot of machines that have the potential hazards. On the production process there are many potential hazards and factors that cause workplace accidents or occupational illness. From the data report accidents that previously occurred in the company in 2011 to 2015 there has been a 32 accidents. Therefore, to minimize the risk aspect is committed to protecting the workforce through the implementation of OSH program. This study aims to identify the hazard and risk assessment of health and safety with the approach of Job Safety Analysis (JSA) in order to find the right solutions to minimize workplace accidents on the plantation.

2. Literature

2.1. Occupational Health and Safety (OSH)

Occupational health and safety (OSH) is an instrument that protects workers, the company, the environment and the surrounding community from harm due to workplace accidents. Such protection is a fundamental right that must be met by companies [1]. According to the America Society of Safety and Engineering (ASSE), OSH is defined as the field of activities aimed to prevent all kinds of accidents that are related to the environment and work situation.

In general, work safety can be considered as a science and its application relating to machines, appliances, tools, materials and processing, foundation work and the work environment and how to perform the work to ensure the safety of workers and the company's assets in order to avoid accidents and other losses [1]. Safety also includes the provision of Personal Protective Equipment (PPE), machine maintenance and humane working hours' arrangements.

In OSH also known terms of Occupational Health, which is a specialization in health sciences as well as practice aimed at keeping workers or community workers obtain health standard as high as both physical, mental, social, with business ventures preventive and curative, against illness or health problems caused by the factors of work and working environment, as well as against common illness [1].

2.2. Work Accident

The main definition describing accident combined form OHSAS 18001:2007 and BS 8800:2400 is the undesired event resulting in injury, sickness or death. [5]. So, accident is defined as an unexpected event, originally undesirable that disrupt the process that has been set up from an activity and can lead to losses for both human and or property. While understanding of occupational accidents according to Ministry Regulation number 03/Men/1994 [6] regarding the program of Jamsostek, is an accident in connection with a work activity, including disorders arising from the employment relationship as well as accidents that happen in the course set out from home to work and back home through road ordinary or reasonable impassable.

Basically, the background of the accident is influenced by two factors, namely:

a. *unsafe condition*

Where accidents occur due to unsafe working conditions, as a result of the following ways:

- Machinery, Equipment, Materials, etc.
- Work Environment
- Work Process
- Nature of Work
- How it Works

b. *Unsafe Action*

Where accidents occur because the act / acts that are not safe, as a result of the following ways:

- Lack of knowledge and skills
- Physical characteristics
- Work Process
- Psychological mental characteristics
- The unsafe attitude and behaviour

2.3. Risk assessment

The definition of risk assessment is the process for determining the priority of controlling the level of risk of accidents or occupational diseases [7].
Methods in risk assessment include:

- Determine Opportunity accident
  Determine probability incident that occurred in the workplace, we can use a scale based on the level of potential.
- Determine Accident Consequences
  To determine the consequences, we must make provision for the severity of its potential
- The level of each risk
  level or levels of risk is determined by the relationship between the value of the identification of hazards and consequences.
  Risk Value (NR)= Value Exposure (E) x Opportunities accident (L) x Consequences of accidents (K)

2.4. Job Safety Analysis

JSA often called Job Safety Analysis is one attempt to analyze the tasks and procedures that exist in the industry. JSA is defined as a method of studying a job to identify hazards and potential incidents associated with each step, to develop a solution that can eliminate and control hazards and incidents [4] When the hazard has been identified, it can be done the control measures in the form of physical changes or repair work procedures to reduce occupational hazards. In practice, the safety analysis procedures require training, supervision and writing job descriptions, known as the JSA to facilitate understanding of the working procedures of employees

3. Research method

The research method used in this issue by using primary data through observation, which is direct observation of events happened, and interview methods, which is questioned directly by the parties concerned. In addition, this also uses secondary data, obtained by the method of internal data that is on the books or reports that are available in the company, and external data that is based on the literature and other references that are outside the company.
In identifying the hazards to the job, workers were selected based on the selection of work in the Job Safety Analysis proposed by OSHA 3071 [7], namely:

- Jobs that have a high accident rate
- The work that has the potential hazards, accidents and injury, although there is no record of accidents on the job.
- The work may cause accidents or serious injury as a result of serious human error is simple.
- The new work, the work is not routine, or work experience.
- Works complex / complicated.

Then the selected workers and are analysed as follows:

- Forklift Operator
- Macerator Worker
- Crepe Workers
- Shredder Worker
- Workshop Worker
- Line engineering Worker
- Workers Cleaning Trolley
- Workers decline Crepe in Pre drying

4. Results and discussion

4.1. Results for Job Safety Analysis

Job Safety Analysis in the data collection is performed directly by the direct observation to the workers concerned and the results are recorded on a form Job Safety Analysis. The results from Job Safety Analysis of some of the work that was observed in the plantation.

- Forklift operator

Table 1. Results of Job Safety Analysis Forklift Operator

Job Steps	Hazards Type	Impacts	E	L	K	NR
Running Forklift	Crashing worker / work area	Forklift damage, injury to operators, workers and work area damage	10	0.3	5	15
The process of hiring and laying load	Overload cargo	Forklift toppled , operator injury, the fall of the product	10	0.3	5	15
	Altitude removal exceed the limit	Load override Forklift, Forklift Damage, operator injury	3	0.05	5	0.75

- Workers macerator

Table 2. Results of Job Safety Analysis of Macerator Worker

Work Steps	Hazards Type	Impacts	E	L	K	NR
Incorporate pieces of rubber into tanks machine	working posture is not good	CTD / MSD	3	0.3	1	0.9
	Strangulated machine	Fractures, lacerations on fingers or hands	2	0.3	10	6
	Slipping	Bruises, sprains, fractures lightweight	2	0.1	5	1
	engine Noise	hearing loss	10	0.3	5	15
	Exposure to dust comes from rubber	respiratory problems	10	0.3	5	15

Based on the results analysis using JSA, it is obtained several risks/hazards that are owned by the macerator worker which has a risk value is 0.9, 6, 1, 15 and 15 where the risk value 15 included in the high risk value so that the necessary risk control to minimize the possibility of accidents needed.

- Crepper *Worker*

Table 3. Results of Job Safety Analysis Crepper Worker

Work Steps	Hazards Type	Impacts	E	L	K	NR
	Strangulated machine	Hands or fingers broke,	3	0.3	20	18

Work Steps	Hazards Type	Impacts	E	L	K	NR
		broken bones, lacerations				
Incorporate into the rubber sheet machine	engine noise	Hearing impairment	10	0, 3	5	15
	pieces of rubber dust	respiratory problems, asthma	10	0.3	5	15
	working posture is not good	CTD /MSD	3	0.3	1	0.9

Based on the analysis results using the JSA, it is obtained several risks/hazards that are owned by Crepe worker that have a risk value, as 18, 15, 15 and 0.9 where the risk value 15 included in the high risk value so that it is a necessary risk control to minimize the likelihood of accidents.

- Shredder Workers

Table 4. Results of Job Safety Analysis shredder worker

Work Steps	Hazards Type	Impacts	E	L	K	NR
Incorporate rubber that had dried into the engine	Strangulated machine	Hands or fingers broke, broken bones, lacerations	3	0.3	20	18
	noise machine	Hearing loss	10	0.3	10	30
	Exposure to dust rubber flakes	respiratory problems, asthma	10	0.3	5	15
	working posture is not good	CTD /MSD	3	0.3	1	0.9

Based on the analysis results using the *JSA* obtained several risks / hazards are owned by *shredder* workers that has a value of risk that is 18, 30, 15, 15 and 0.9 where the risk value 30 entered in the value of the extreme risk that necessary risk control to minimize the possibility of accidents extreme *badly needed*.

- Workshop Worker

Table 5. Results of Job Safety Analysis workshop worker

Job Steps	Hazards Type	Impacts	E	L	K	NR
Welding	Eye exposed to sparks	Damage the vision senses	3	0.3	10	9
	work posture is not good	CTM / MSD	3	0.6	1	1.8
	Hand exposed welding sparks	burns, blistering	2	0.6	2	2.4
Sharpening	Hands affected grinding	scratch wounds, fingers cut off	2	0.3	2	1.2
	Noisy equipment / machinery grinding	damage of auditory	10	0.3	5	15
	working posture is not good	CTM /MSD	3	0.6	1	1 8

Based on the analysis results using the *JSA* obtained several risks / hazards to workers workshop that has a value of risk that is 9, 18, 2.4, 1.2, 15 and 1.8 where the risk value 15 included in the high risk value so necessary for risk control minimizing the possibility of accidents *needed*.

- Line Engineering Worker

Table 6. Results of *Job Safety Analysis* Line Engineering Worker

Working step	Hazard Type	Impact	E	L	K	NR
Maintenance Engineering	Exposure to sharp	Cut objects hand dropping	2	0.3	2	1.2
	Slipped	Bruises, sprains, fractures lightweight	2	0, 3	1	0.6
Dismantling machine	Wedged	Bruises, fractures lightweight	2	0.3	2	1.2
	machinery part fall	'Bruises, fractures	2	0.3	2	1.2

Based on the analysis results using the JSA obtained several risks / hazards that are owned by the worker's job line technique that has a risk value is 1.2, 0.6, 1.2 where the risk value of 1.2 included in the value of low risk but remain vigilant in doing any work to prevent work accidents.

- Cleaning Trolley Workers

Table 7. Results Job Safety Analysis cleaning workers Trolley

Working step	Hazard Type	Impact	E	L	K	NR
Submerge Trolley with chemicals	Exposure to chemicals caustic soda	Skin irritation	2	0.3	5	3
Cleaning Trolley	Strangulated Trolley	Cut, scraped	1	0, 3	2	0.6

Based on the analysis results using the *JSA* obtained several risks / hazards that are owned by the worker's job cleaning trolley that has the risk value is 3 and 0.6 where the risk value 3 included in the value of intermediate risk but remain vigilant in every work order *to* work *accident* do not occur.

- Decline Crepe Worker

Table 8. Results of Job Safety Analysis Decline Crepe Worker

Working step	Hazard Type	Impact	E	L	K	NR
Decrease crepe	Falling from height	injuries, fractures, sprains	1	0.3	20	6
	Crushed crepe	Bruises, wounds on the skin surface	2	0.3	2	1.2
	Exposure to dust from crepe	respiratory problems	6	0.3	5	9
Crepe Transport	working posture is not good	CTD / MSD	3	0.6	1	1.8
	Overload cargo	falling, injury operator	2	0.3	2	1.2

Based on the analysis results using the JSA obtained several risks / hazards to workers decline crepe that has a risk value is 6, 1.2, 9, 1.8 and 1.2 where the risk value 9 included in the value of intermediate risk but remain vigilant in doing any work to prevent work accidents.

4.2. Proposed Risk Management

Following this is the result of the Job Safety Analysis of some of the work that was observed in the company:

- Forklift Operator

Table 9. Proposed Risk Management on Operator Forklift Operator

Department / Working step	Type Hazards	Impact of	Proposed Risk Control
Running Forklift	Bumping workers / work area	Damage Forklift, injury to operators, workers and damage to the work area	Determination of special lanes for Forklift and transport
Processes of appointment and the laying of a charge	Overload charge	Downfall product, Forklift falling, injury operator	Determination of the maximum Forklift load, extension and operator training, use of personal protective equipment such as helmets and boots
	Altitude removal exceed the limit	Load override Forklift, damage, Forklift injury operator	

- Macerator *Worker*

Table 10. Proposed Risk Management in macerator worker

Department / Working step	Type Hazards	Impact of	Proposed Risk Management
Incorporate pieces of rubber into troughs machine	posture work is not good	CTD / MSD	Redesign workstations, using the tools of work
	Strangulated machine	Fractures, lacerations on fingers or hands	Running SOP correctly and using tools work
	Slipping	Bruises, sprains, fractures light	use of PPE in the form of boots and helmets and cleaning floor
	Noise machine	hearing loss	use of PPE in the form of earmuff or earplug
	Exposure to dust comes from rubber	respiratory problems	use PPE in the form of masks

- Crepper *Worker*

Table 11. Proposed Risk Management in Crepper Worker

Department / Working step	Type Hazards	Impact of	Proposed Risk Control
Incorporate rubber sheets into the machine	Strangulated machine	Hands or fingers broken, broken bones, lacerations	Running SOP correctly and using the tools of labor
	noise machine	Disorders hearing, deaf	Use of PPE in the form of earplug or earmuff
	Flakes Dust rubber	respiratory problems, asthma	Use of PPE in the form of a mask
	work posture is not good	CTD / MSD	Redesign workstations

IOP Conf. Series: Materials Science and Engineering **215** (2017) 012029 doi:10.1088/1757-899X/215/1/012029

- Shredder Workers

Table 12. Proposed Risk Control at Full Shredder

Department / Working step	Type Hazards	Impact of	Proposed Risk Control
Incorporate rubber that had dried into the machine	Strangulated machine	Hands or fingers broken, broken bones, lacerations	Running SOP correctly, and use work tool
	engine noise	hearing loss	use of PPE in the form of earplug or ear muff, use noise-cancelling machinery
	Exposure to Rubber dust flakes	respiratory problems, asthma	use of PPE in the form of a mask
	work posture is not good	CTD / MSD	Redesign workstations

- Full workshop

Table 13. Proposed Risk Control at Full workshop

Department / Working step	Type Hazards	Impact of	Proposed Risk Management
welding	Currency splash fire	damage vision senses	use of PPE in the form of a face shield, goggles
	posture work is not good	CTD / MSD	Redesign workstations
	Hand exposed to sparks welding	Luka burn, scald	use of PPE such as gloves heat resistant and scratch resistant
grinding	hands exposed to grinding	Luka scratch, finger cut	use of PPE such as gloves
	Noisy equipment / machinery grinding	damage of auditory	use of PPE in the form of earplug or earmuff
	posture work that is not good	CTM / MSD	Redesign workstation.

- *Line Engineering Worker*

Table 14. Proposed Risk Management in Line Engineering Worker

Ministry / Working step	Type Hazards	Impact of	Proposed Risk Control
Engine Maintenance	Exposed sharp objects	Cut, hand dropping	to Use of PPE in the form of gloves anti-scratch, run the SOP with secure
	Slipped	Bruises, sprains, a mild fracture	Using of PPE in the form of boots and helmets, carry out the cleaning of floors
	Strangulated	Bruises, broken bones light	SOP Running with secure
Demolition machine	machinery part fall	Bruises, fractures	Running SOP correctly, and use of PPE such as helmets

- *Cleaning Trolley Workers*

Table 15. Proposed Risk Control at Cleaning Trolley Workers

Department / Working step	Type Hazards	Impact of	Proposed Risk Control
Submerge Trolley with chemicals	Exposure to chemicals caustic soda	Skin irritation	Use of PPE in the form of special gloves
Cleaning Trolley	Strangulated Trolley	Cut, scraped	Running SOP correctly

- *Decline Crepe Worker*

Table 16. Proposed Risk Control on Decline Crepe Worker

Department / Working step	Type Hazards	Impact of	Proposed Risk Control
Lose crepe	Falling from height	injuries, fractures, sprains	Use of PPE in the form of a helmet
	Crushed crepe	Bruises, wounds on the skin surface	to Use of PPE in the form of a helmet
	Affected dust from crepe	respiratory problems	use of PPE form of a special mask
Crepe Transport	working posture is not good	CTD /MSD	Redesign work stations, Extension of MMH (manual material handling) is good and right.
	Overload charge	toppled, operator injury	Load adjustment of Trolley

4.3. Table Value Risk of Overall Results JSA

Table 17. NR overall analysis JSA

Job Type	Value Risk Largest	Category NR
Been Forklift	15	High-Risk
Workers macerator	15	High-Risk
Workers Crepper	18	High-Risk
Workers Shredder	30	Risk of Extreme
Workers Workshop	15	High-Risk
Workers Mechanical Line	1, 2	Low risk
Care workers Trolley	3	Medium risk
Decreased workers Crepe	9	Medium risk

Risk seen from the table value (NR) the overall results of JSA shredder worker included in the category of the highest NR or an extreme value risk, because it is in the range of NR> 20.

Description [3] Value risk > 20: Extreme, Values risk 10-20: High Risk, Risk Value 3-10: Risk Medium, risk value<3: Low Risk

This model is predicted to be able for implementation in same characteristic plantation.

IOP Conf. Series: Materials Science and Engineering **215** (2017) 012029 doi:10.1088/1757-899X/215/1/012029

5. Conclusions and Recommendations

5.1. Conclusion
Based on the result, there are some conclusions as follows:

- Management system for OSH in company has been good. On the other hand, in the implementation of health and safety works were not enough because of his own workers still paid less attention and did not realize the importance of health and safety of their work and therefore there are still accidents were due to the carelessness of workers themselves.
- Based on the analysis using method Job Safety Analysis can be concluded that the worker's job shredder that is worth 30 and has a working level with extreme risk with the risk value range is above 20. Moreover, that can be done to minimize accidents that can use Personal Protective Equipment (PPE) appropriately, counselling on health and safety at work, the company must monitor every activity of employees, and give rewards to workers who obey the rules that apply in the company.

5.2. Suggestions
As for suggestions from the author that if it can help companies to be better are:

- Controlling workplace accidents carefully so that control can run optimally, therefore occupational accidents can be minimized.
- Companies are advised to raise awareness of the employees will be essential and mandatory use of PPE to avoid accidents.
- Companies are expected to conduct regular training to employees about the meaning and importance of an effective OSH program so that compliance with the OSH program become a habit in the workplace.

6. References

[1] Suma'mur PK 1996. *Keselamatan Kerja dan Pencegahan Kecelakaan*. Jakarta: PT. Gunung Agung.
[2] International Labor Organization (ILO) Jakarta, Keselamatan dan Kesehatan Kerja (Sarana untuk Produktivitas), 2013.
[3] Rudi Suardi. 2007.Sistem Manajemen OSH. Jakarta: PPM :69-106
[4] NOSA, 1999. Keselamatan kerja dan *Job safety analysis* dalam lingkungan kerja di Indonesia
[5] Ceyhan Cumhur, 2012. Occupational Health and Safety Hazard Identification, Risk Assessment, Determining Controls: Case Study On Cut-And-Cover Underground Stations and Tunnel Construction, Master of Science in Civil Engineering Department, Middle East Technical University
[6] Ministry regulation.03/Men/1994 About JAMSOSTEK Program
[7] Occupational Safety and Health Administration. 2002. Job Hazard Analysis OSHA3071. US: Department of Labour

IOP Conf. Series: Materials Science and Engineering **215** (2017) 012030 doi:10.1088/1757-899X/215/1/012030

Temperature-dependent photoluminescence characteristics of $CH_3NH_3PbI_{3-x}Cl_x$ perovskite thin film

F. J. Cheng[1], S. Y. Hu[2,*], L. Y. Chen[3], Y. C. Lee[4,], G. Z. Yin[5], K. K. Tiong[5], and J. L. Shen[6]**

[1] Department of Electrical Engineering, Tungnan University, Shenkeng, New Taipei City, Taiwan
[2] Department of Digital Technology Design, Tungfang Design Institute, Kaohsiung, Taiwan
[3] Department of Chemical Engineering, National Taiwan University of Science and Technology, Taipei, Taiwan
[4] Department of Electronic Engineering, Tungnan University, Shenkeng, New Taipei City, Taiwan
[5] Department of Electrical Engineering, National Taiwan Ocean University, Keelung, Taiwan
[6] Department of Physics, Chung Yuan Christian University, Chungli, Taiwan

Corresponding author's e-mail: * *shenghu2729@gmail.com*
****jacklee@mail.tnu.edu.tw*

Abstract. Recently, organometal halide perovskites are widely considered as promising materials in light emitting and lasing devices due to their potential optical properties. The luminescence characteristics of $CH_3NH_3PbI_{3-x}Cl_x$ perovskite material have investigated by temperature-dependent PL measurements. We suggest that the tetragonal inclusions in the orthorhombic phase of $CH_3NH_3PbI_{3-x}Cl_x$ thin film at low temperature result in the 12 K PL spectrum with a broad emission band. In addition, the PL peak of $CH_3NH_3PbI_{3-x}Cl_x$ thin film as a function temperature shows a tendency of buleshift, which is contrary to the typical tendency of redshift for traditional semiconductor materials.

1. Introduction

Methylammonium lead iodide perovskite material $CH_3NH_3PbI_3$ have attached much attention, because it has unique optical of a direct bandgap with high absorption coefficients over the visible to near-infrared range and small exciton binding energies for the application of solar cells [1-3]. Comparing to the widely used $CH_3NH_3PbI_3$ material, the mixed halide $CH_3NH_3PbI_{3-x}Cl_x$ material has further been used as the active layer of the highly efficient solar cells owing to its improved optical properties [4-5]. The wavelength emission of $CH_3NH_3PbI_{3-x}Cl_x$ material can be tunable from ultraviolet to near infrared by substituting iodine with chlorine and bromine, and these perovskite films can possess significant photoluminescence (PL) characteristics, which make them potential candidates for use in light-emitting devices [5-6]. However, only a few studies of the temperature dependence of the photo-physics of organic–inorganic perovskites present. Temperature-dependent steady state PL measurements allow the depiction of a complex photoexcitation performance, and the low-temperature PL spectra are also used to elucidate the basic photophysical properties of the perovskite materials. In this work, we investigate the luminescence properties of solution-processed $CH_3NH_3PbI_{3-x}Cl_x$ material from temperatures ranging from 10 to 300 K using PL measurement with temperature dependence.

2. Experimental

The $CH_3NH_3PbI_{3-x}Cl_x$ perovskite material was synthesized with a mixture of $PbCl_2$ and CH_3NH_3I (1:3 molar ratio) dissolved in dimethylformamide (DMF) as the precursor solution. The synthesized $CH_3NH_3PbI_{3-x}Cl_x$ perovskite material was spin-coated on the quartz substrates and heated to form a thin film. The temperature-dependent PL spectra were measured by using a diode laser with an excitation wavelength of 686 nm (PicoQuant). The diode laser produces light pulses with 50 ps duration and a repetition rate of 1 MHz. PL measurement was setup using a spectrometer (Zolix omni–λ 500) with a grating of 1200 grooves/mm, and detected using a photomultiplier tube (PMT). Janis Research Model CCS-150 and LakeShore Model 321 temperature controller were used to carry out the temperature-dependent PL measurement.

3. Results and discussion

Figures 1(a) shows the 12 K PL spectrum with a broad emission band, implying the existence of multi-band structures. By fitting with Gaussian function, the 12 K PL spectrum can be further decomposed into two peaks at 1.67 and 1.60 eV. According to the previous literatures [7-8], the higher peak at 1.67 eV and the lower peak at 1.60 eV are attributed to the low-temperature orthorhombic and the high-temperature tetragonal phase, respectively. It explains that the complex performance of PL spectrum in the $CH_3NH_3PbI_{3-x}Cl_x$ thin film at 12 K could be due to the coexistence of two structure phases.

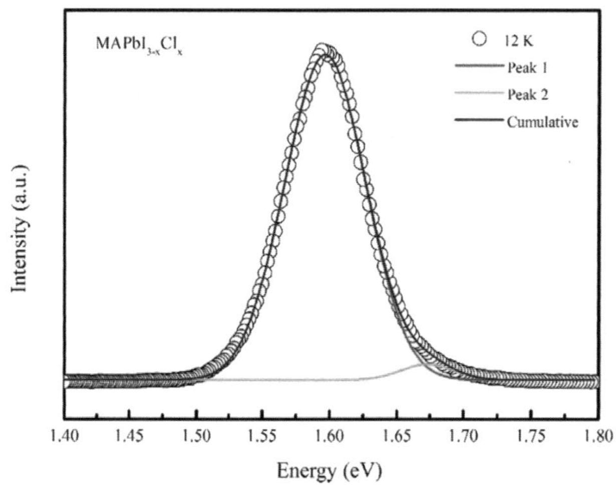

Figure 1. PL spectrum at 10 K for the $CH_3NH_3PbI_{3-x}Cl_x$ thin film. The spectrum is decomposed into two Gaussian peaks.

The PL spectra of $CH_3NH_3PbI_{3-x}Cl_x$ thin film taken at selected temperatures of 12, 100, 140, 160, 220, and 300 K are displayed in Figure 2. It is clear that the PL peak energy exhibits a blue-shifted tendency with increasing temperatures. The variations in the temperature-dependent PL peak of MAPbICl perovskite material is contrary to that of typical semiconductors, which shows a red-shifted PL peak with increasing temperature.

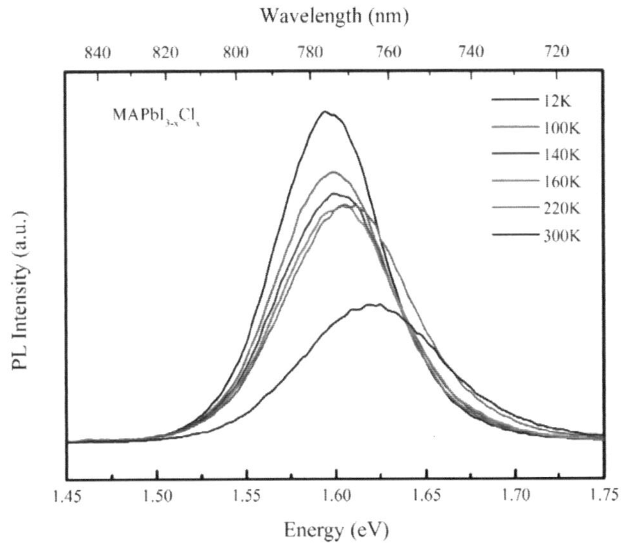

Figure 2. PL spectra of $CH_3NH_3PbI_{3-x}Cl_x$ thin film at selected temperatures.

Figure 3 display the PL peak positions as a function of temperature. The variations in the PL peak position could be discussed in three distinct regions separated by about 80 and 160K. Below 80 K, It can be observed that the PL peak position towards blueshift with increasing temperature to 50 K. In the temperature region of 80~160 K, the variations in the PL peak position shows a significant v-like shape. The performance could be attributed to the phase change from the orthorhombic to the tetragonal structure [9-10]. Above 160 K, the PL signal originated form the near-band-edge (NBE) transition in the $CH_3NH_3PbI_{3-x}Cl_x$ with tetragonal phase exhibits a continuous blueshift as the temperature increased to 300 K.

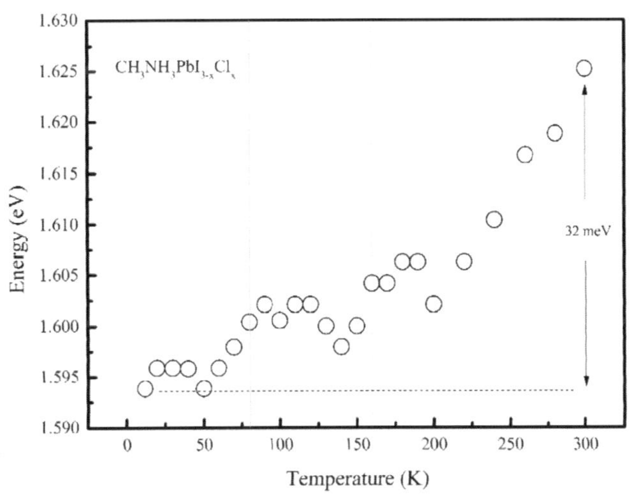

Figure 3. Temperature dependence of PL peak position for the $CH_3NH_3PbI_{3-x}Cl_x$ thin film.

IOP Conf. Series: Materials Science and Engineering **215** (2017) 012030 doi:10.1088/1757-899X/215/1/012030

4. Summary

In this work, we have discussed the luminescence behavior of $CH_3NH_3PbI_{3-x}Cl_x$ perovskite material by temperature-dependent PL spectra. We suggests that the broad PL shapes are originated from the excitons transitions between the complication of orthorhombic and tetragonal phases in $CH_3NH_3PbI_{3-x}Cl_x$ thin film. The tendency of PL peak as a function temperature of $CH_3NH_3PbI_{3-x}Cl_x$ perovskite material presents a buleshift, which is contrary to that of traditional semiconductor materials belonging to redshift. Our results provide useful luminescence properties of hybrid perovskites to be applied in advanced optoelectronic devices.

5. References

[1] Sum T C, and Mathews N 2014 *Energy Environ. Sci.* **7** 2518.

[2] Hsiao Y C, Wu T, Li M, Liu Q, Qin W, and Hu B 2015 *Mater. Chem. A* **3** 15372.

[3] Sum T C, Chen S, Xing G, Liu X, and Wu B 2015 *Nanotechnology* **26** 342001.

[4] Liu D, Wu L, Li C, Ren S, Zhang J, Li W and Feng L 2015 *ACS Appl. Mater. Interfaces* **7** 16330.

[5] Wu K, Bera A, Ma C, Du Y, Yang Y, Li L and Wu T 2014 *Phys. Chem. Chem. Phys.* **16** 22476.

[6] Xing G, Mathews N, Lim S S, Yantara N, Liu X, Sabba D, Grätzel M, Mhaisalkar S and Sum T C 2014 Nat. Mater. **13** 476.

[7] Milot R L, Eperon G E, Snaith H J, Johnston M B and Herz L M 2015 *Adv. Funct. Mater.* **25** 6218.

[8] Tahara H, Endo M, Wakamiya A and Kanemitsu Y, 2016 *J. Phys. Chem. C* **120** 5347.

[9] Wehrenfennig C, Liu M, Snaith H J, Johnston M B and Herza L M 2014 APL Mater. **2** 081513.

[10] Yamada Y, Nakamura T, Endo M, Wakamiya A and Kanemitsu Y 2014 *Appl. Phys. Express* **7** 032302.

Usability Testing For Android Based Application "Jogja Smart Tourism"

Harwati[1], Imam Djati Widodo[2]

[1,2]Industrial Engineering Department, Universitas Islam Indonesia

[1]harwati@uii.ac.id, [2]imamdjati@uii.ac.id

Abstract. The android based application "Jogja Smart Tourism (JST)" is designed to help everyone who visited Yogyakarta to enjoy their travel. As new application, it is need to be tested for its usability before launched. Usability testing will show how easy user interfaces are to used. The objective of this research is to demonstrate the result of usability testing for application JST based on five characteristics: learnability, effectiveness, memorability, errors, and satisfaction. About 30 respondents were involved to test the usability of this application. Learnability and effectiveness is calculated from some task that should be finished by respondents, and the rest aspects are calculated from questionnaires that should be answered after simulation. There are 14 functions bound in this usability testing. The result shows total usability level is in 81.75%. Learnability testing shows that 98.8% of respondent could finish the task successfully with 87.5% in efficiency. The memorability level of respondents is good (84.5%) where their ability to fix the errors is 71.5%. And the last for satisfaction level of application interface is 66.25%. Low level of satisfaction occurred because most of respondent felt uncomfortable with landscape interface of application because they should turn their mobile phone while using JST application and also it happened because the lack of using picture and colour inside the application. Both of these becomes important note for the improvement of further applications where the interface in a portrait version is more comfort the use and also utilization of colour and the image will be the main focus to improve customer satisfaction.
Keywords: usability; android; application; tourism; Yogyakarta

1. Introduction

Tourism is a sector with strong growth around the world and has gradually changed from a secondary requirement (luxuries) becomes a primary need [1]. The growth of the tourism industry certainly requires special strategies how tourism services could offset the desire of the consumer. So with Yogyakarta area. Yogyakarta is one of the very promising tourism destinations in Indonesia. This is because Yogyakarta has many aspects that support as a potential natural beauty, the area is rich in ancient sites and cultural distinctiveness. Besides, Yogyakarta also has a lot of souvenirs and handicrafts such as batik, silver and pottery with competitive quality. With these forces Yogyakarta has a chance to become one of the important tourist destinations, nationally, regionally, and internationally. With tourism potential is quite large, it should be able to attract tourists Yogyakarta larger. There was lot of effort and strategy to market the tourism object in Yogyakarta. One way is to utilize information technology. The development of communication technology has made it possible for someone to be able to access content via internet wherever they are via smartphones. Most people today are active users of smart phone and get benefit from many applications that are built on software [2]. The use of smart tourism technologies such as travel-related websites, social media, and smartphones in travel planning has been pervasive and growing [3]. Smart tourism is a tourism information service for tourists [4]. It is a facility that can be utilized in supporting the ease of travel. Development of tourism information system is needed to support the ease of potential tourists to conduct its activities. Specialized in designing travel application still has not been done and is still limited to the application that contains information about the tourist areas along the navigation menu to discover the attractions sought. From the comparison of the previous application, it can be seen that the entire application was only based on presenting static information about the object and tourist facilities in Yogyakarta. There are no applications to provide recommendations tour packages according to the behavior and conditions of tourists who come to Yogyakarta. At the past research it has designed a tourism application to help tourists enjoy the city of Yogyakarta. This travel application focuses on the fulfillment of the essential attributes in a single application based on consumer needs, they arey:

- The list of restaurants and hotels and resorts
- The recommendation sites,
- The recommendation package,
- The estimated cost of travel

This system was designed for use on Android mobile phone using the Java programming language and to assist in the process of programming using the Eclipse IDE Indigo. This application provides proposed packages and travel destination adjusted with the tourism personal profile and it is completed with appropriate cost estimation. The interface of this application is shown in the following figure

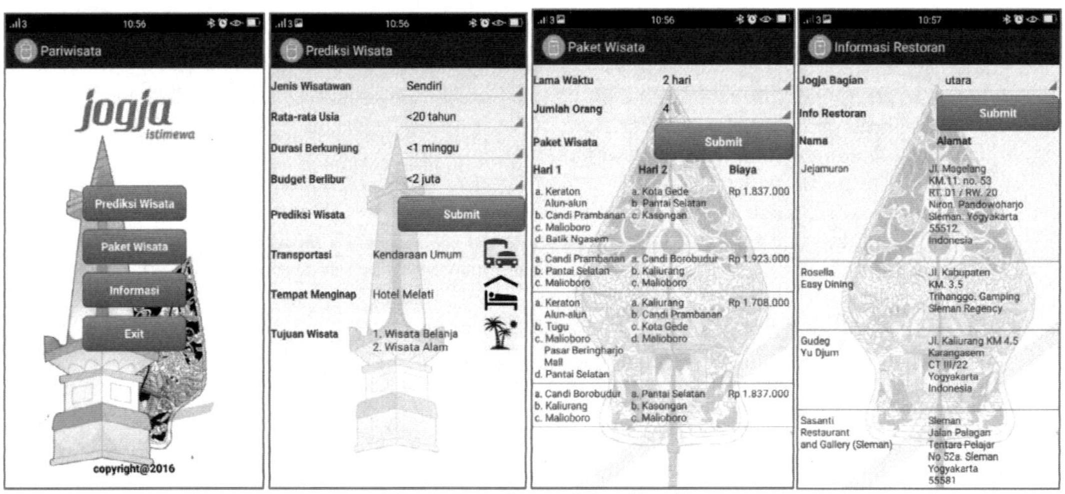

Fig 1. Interface of Application

1.1 Research Question

Questions to be solved in this study is how to test the usability for new applications JST based on five aspects learnability, effectiveness, memorability, errors, and satisfaction.

1.2 Research Purpose

As new application, it is need to be tested for its usability before launched. Usability testing will show how easy user interfaces are to used. The objective of this research is to demonstrate the result of usability testing for application JST based on five characteristics: learnability, effectiveness, memorability, errors, and satisfaction.

2. Literature Review

Usability basically is the quality level of the system (software) that is easy to learn, easy to use, and encourages users to use these devices. Usability is defined as "the extent to which a product can be used by specified users to achieve specified goals with effectiveness, efficiency and satisfaction in a specific context of use" [5]. Usability is also associated with the learnability of the system, its efficiency of use, its easiness to remember, its ability to prevent and recover from errors, and user satisfaction [6]. Usability issues need to be considered early during the software development phase to avoid design rework [7]. In general, there are five requirements or characteristics that must be fulfilled in order for an application reaches the ideal level of usability, they are:

- Learnability: explain the degree of user convenience in studying the applications to meet the basic tasks when first using the application.
- Efficiency: to explain the rate of speed in completing the tasks after studying the application.
- Memorability: explain the degree of user convenience in using the application properly, after long use.
- Errors: explain how the number of mistakes made by users, and how users correct mistakes easily.
- Satisfaction: explain the degree of user satisfaction in using the application.

Research on usability has been widely applied to various areas. Applications in the healthcare industry using usability testing before being released to consumers [8], [9], [10]. In other areas, usability testing is also used in other fields such as academic applications [11], [12], [13] government website [13], [15], [16] and many more. An application with low usability can lead to wastage of time, user frustration, increase frustration and tendency to not want to go back to visit or use existing applications.

3. Method

Usability testing was processed using a questionnaire which involves five aspects: Learnability, efficiency, memorability, errors and satisfaction. Respondents cover 30 people with ages 20-40 years consist of 47% women and 53% men. Overall respondents are familiar with and accustomed to using mobile devices with the operating system based on Android. Preliminary observations to the respondents also results assumptions that all of them have the ability to *define*: the ability to recognize, for example the ability to identify the icon in the application, *access*: the ability to collect the necessary information in the digital environment, such as the ability to open the browser and open a search engine, Usability test was conducted by following a flowchart as follows:

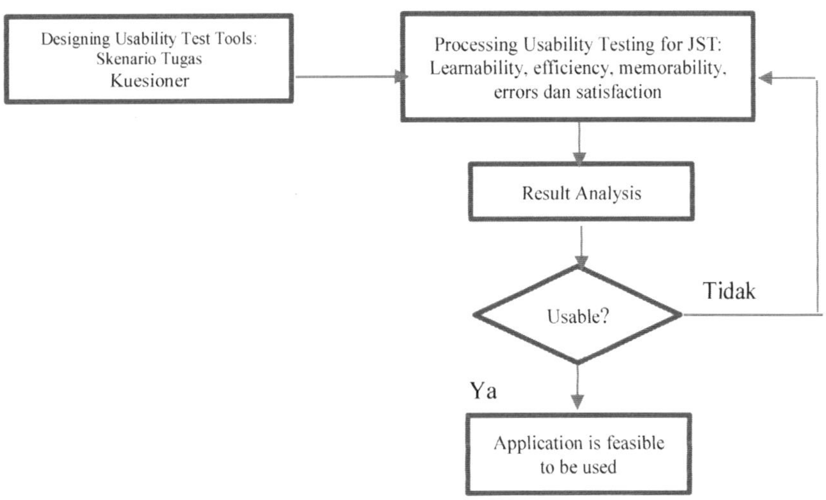

Fig. 2 Usability Testing Method Flowchart

IOP Conf. Series: Materials Science and Engineering **215** (2017) 012031 doi:10.1088/1757-899X/215/1/012031

Usability testing instrument was designed to measure five aspects: learnability, efficiency, memorability, errors and satisfaction. Learnability and efficiency was measured using observation sheet after the respondent perform certain tasks. Memorability, errors and satisfaction was measured using a questionnaire *post task.* Function 1 through 6 is a function to measure learnability and efficiency, function 7 to 10 are used to measure satisfaction, functions 11 and 12 are used to measure the errors and the last function 13, and 14 are used to measure aspects of memorability. The following table show plot of usability testing posed to respondents:

Table 1. Aspects of Usability Measurement Plot

No of Functions	Questions	*L*	Ef	M	Er	S
1	Is there a travel prediction in the application?					
2	Is there an application menu in the travel package?					
3	Is there a selection of travel forecast budget visit in application menu?					
4	Is there a selection of travel packages duration in application menu?					
5	Looking for travel predictions					
6	Looking for travel packages					
7	What is your opinion regarding the organization of menus in applications					
8	What is your opinion regarding the terms used in the application					
9	What is your opinion about the language used in the application					
10	What is your opinion regarding the design of the application					
11	Do you fix errors during the use of the system with ease?					
12	Do you easily feel 'lost' when using this application?					
13	Do you remember back in the menus and the look of your application easily?					
14	Do you think you can use these applications easily in the future?					

To measure aspects of learnability and efficiency it was required a scenario that support the functions of 1 to 6. The scenario tasks that must be done for each respondent is presented in the following table:

Table 2. Scenario Task

No.	Task A must Done	Yes / No	Level of Ease
1	Find menu prediction of travel		1 2 3 4 5
2	Find menu travel packages		1 2 3 4 5
3	Find the option to visit the prediction budget travel		1 2 3 4 5
4	Finding long been in menu selection travel packages		1 2 3 4 5
5	Looking for travel predictions by category: Type of traveler : Families Average of age : 30-40 years Length : 1-2 weeks Budget : 2-5 million		1 2 3 4 5
6	Looking for travel packages by category: Length of time : 2 days Number of people : 5 people		1 2 3 4 5

4. Result

Data taken from each respondent was declared valid if these respondents completing and filling the whole of tasks and questions. Next is example of stuffing recapitulation from the respondent:

Table 3. Summary of Sample Respondents Entry

	Learnability						Efficiency						Satisfacton				Errors		Mem	
Responden	F1	F2	F3	F4	F5	F6	F1	F2	F3	F4	F5	F6	F7	F8	F9	F10	F11	F12	F13	F14
1	1	1	1	1	1	1	4	4	4	3	4	4	4	4	4	5	3	5	4	4
2	1	1	1	1	1	1	4	5	4	4	4	4	3	4	4	4	4	4	3	4
3	1	1	1	1	1	1	4	4	5	4	3	4	3	4	3	3	3	2	3	4
4	1	1	1	1	1	1	4	3	2	4	2	3	3	4	5	4	3	3	4	2
5	1	1	1	1	1	1	3	3	2	3	3	4	3	2	4	4	3	4	3	5
6	1	1	1	1	1	1	5	5	5	5	5	5	4	4	4	5	4	4	5	4
7	1	1	1	1	1	1	5	5	4	3	4	4	4	4	5	3	2	4	5	4
8	1	1	1	1	1	1	3	3	4	4	3	3	4	3	4	2	3	4	2	4
9	1	1	1	1	1	1	3	4	4	4	4	4	4	3	4	3	4	2	4	4
10	1	1	1	1	1	1	5	5	5	5	5	5	4	4	5	3	3	3	5	1

The level of reliability is calculated using the formula the user's success rate. It is the percentage of respondents successfully completed the task correctly.

$$Learability\ (\%) = \frac{\sum_{i=1}^{n} X_i}{n} x100\% \quad \ldots\ldots\ldots\ldots \quad (1)$$

Where X_i is the value of the success of the task by respondent i, Xi = {0,1}

$$Learnability\ (\%) = \frac{237}{240} x100\% = 98{,}8\%$$

The efficiency is calculated from the ratio of speed for the respondents in doing the task i (in a Likert scale) to maximum multiple weight and the number of respondents.

$$Efficiency\ (\%) = \frac{\sum_{i=1}^{n} X_i}{5xn} x100\% \quad \ldots\ldots\ldots\ldots \quad (2)$$

$$Efficiency(\%) = \frac{1050}{1200} x100\% = 87{,}5\%$$

Memorability is calculated from the comparison grades contents of respondents for F13 and F14 (on a Likert scale) to maximum multiple value by the number of respondents.

$$Errors\ (\%) = \frac{\sum_{i=1}^{n} X_i}{5xn} x100\% \quad \ldots\ldots\ldots\ldots \quad (3)$$

$$Erors\ (\%) = \frac{286}{400} x100\% = 71.4\%$$

Satisfaction is calculated from the ratio between respondent satisfaction grades (on a Likert scale) to maximum multiple value by the number of respondents.

$$Satisfaction\ (\%) = \frac{\sum_{i=1}^{n} X_i}{5xn} x100\% \quad \ldots\ldots\ldots\ldots \quad (4)$$

$$Satisfaction\ (\%) = \frac{636}{960} x100\% = 66.3\%$$

Thus for overall usability value can be calculated using the following formula:

$$Usability\ (\%) = \frac{Learnability + Eff + morability + errors + Satisfaction}{5} x100\% \quad \text{..........} (5)$$

$$Usability\ (\%) = \frac{98.8 + 87.4 + 84.5 + 71.4 + 66.3}{5} x100\%$$

$$Usability(\%) = \frac{408.8}{5} x100\% = 81.75$$

The result showed that from six tasks almost all tasks can be done by all respondents (98.8%). It can be concluded that the respondents did not experience difficulties in operating the menus of your application. For testing the efficiency, the overall speed of the respondents to finish the task is very good (87.5%) where the average fastest time achieved to perform the first task (FI, 91.5%), while the average time the longest used to perform tasks to 3 (F3, 80.5%). The level of remembrance of respondents in using the application is at a good level (84.5%). That means respondents are very good at remembering menu functions that exist and sure can use the application without a guide in the future. For an error rate of respondents in using the application is not a lot. It was proved by achieving level of usability for errors at 71.5%. The ability of respondents to correct errors when using an application is 72% of respondents, while the ability to browse the menus on the application resides on the achievement of 71%.

The level of satisfaction with the overall application at the level of 66.25%. The results were not good enough. This happens because of in general of respondents are still not satisfied by the application. Utilities for design or display application only reached 63.5%, while for the organization, language and terminology used is sufficient to satisfy of respondents (84.8%). From the open questionnaire it is known that many respondents felt less satisfied with the display of the menu which made in landscape format. It is considered that respondents feel not comfortable in using the application having to rotate the phone while used. Besides, the use of background colors and images in applications is still very poor thus affecting the comfort of respondents in using the application. Both of these things be important for the improvement of further applications where the application display in a portrait version as well as the use of color and images become a major focus to improve customer satisfaction so that the application is feasible to use.

5. Conclusion

From the overall calculation it can be concluded that the level of reusability JST currently on the percentage of 81.75% which there are two things that become the focus of improvement which display the menu and the lack of use of images and colour. This research can be continued in the future by improving the design of existing applications and perform usability tests again to see the increase in user satisfaction levels against the application.

References

[1] Kastarlak, B. I., & Barber, B. (2012). Fundamentals of Planning & developing tourism. Upper Saddle River, NJ: Pearson-Prentice Hall.

[2] Nyembo Salama, Christian Bach, 2013, Usability Testing for Android and Apple Smart Phone, *International Journal of Emerging Technology and Advanced Engineering,* Volume 3, Issue 7, p: 273:277.

[3] C. Derrick Huanga, Jahyun Gooa, Kichan Namb, Chul Woo Yoo, 2017, Smart Tourism Technologies In Travel Planning: The Role Of Exploration And Exploitation, Information &

Management
[4] Yunpeng Li, Clark Hu, Chao Huang, Liqiong Duan, 2017, The Concept Of Smart Tourism In The Context Of Tourism Information Services Tourism Management, Vol: 58, p: 293-300
[5] ISO 9241-11. (1998). Guidelines for specifying and measuring usability.
[6] Nielsen, J. (1993). Usability engineering. Boston, MA: AP Professional
[7] Cigdem Altin Gummusoy (2016), Usability Guideline For Banking Software Design, *Computers in Human Behavior*, Vol 62, p: 277-285
[8] Yu-Kuang Wua, Hsin-Yi Liua, Annmarie Kelleher, Jonathan Pearlmana,b , Rory A. Cooper, 2016, Evaluating The Usability Of A Smartphone Virtual Seating Coach Application For Powered Wheelchair Users, *Medical Engineering and Physics*, Vol 38, p: 569-579
[9] Dorian Davis, Steven Jiang, 2016, Usability Testing Of Existing Type 2 Diabetes Mellitus Websites, *International Journal of Medical Informatics*, Vol 92, p: 62–72
[10] Wiard Jorritsmaa, Fokie Cnossenb, Rudi A. Dierckxa, Matthijs Oudkerka, Peter M.A. Van Ooijen, 2015, Post-Deployment Usability Evaluation Of A Radiology Workstation, *International Journal of Medical Informatics* Vol 85, p: 28–35
[11] Sharmistha Roy, Prasant Kumar Pattnaik, Rajib Mall,2014, A Quantitative Approach To Evaluate Usability Of Academic Websites Based On Human Perception, *Egyptian Informatics Journal* Vol 15, p: 159–167
[12] Mona Masood, Amiza Musman, 2015, The Usability and its Influence of an e-Learning System on Student Participation, *Procedia - Social and Behavioral Sciences* Vol 197 , p:2325 – 2330
[13] Reese Hoi Yin Fung, Dickson K.W. Chiu, Eddie H.T. Ko, Kevin K.W. Ho, Patrick Lo, 2016, Heuristic Usability Evaluation of University of Hong Kong Libraries' Mobile Website, *The Journal of Academic Librarianship* Vol 42, p:581–594
[14] Bridgett A. King, Norman E. Youngblood, 2016, E-Government In Alabama: An Analysis Of County Voting And Election Website Content, Usability, Accessibility, And Mobile Readiness, *Government Information Quarterly* Vol: 33, p:715–726
[15] Viswanath Venkatesh, Hartmut Hoehle, Ruba Aljafari, A Usability Evaluation Of The, 2014, Obamacare Website, *Government Information Quarterly* Vol: 31, p:669–680
[16] Zhao Huang, Morad Benyoucef, 2014, Usability And Credibility Of E-Government Websites, *Government Information Quarterly* Vol: 31, p: 584–595

Implementation of 5S Method for Ergonomic Laboratory

Amarria Dila Sari[1)], Fety Ilma Rahmillah[2)], Bagus Prabowo Aji[3)]

[1,2,3)] Department of Industrial Engineering, Faculty of Industrial Technology, Islamic University of Indonesia, Jl. Kaliurang Km 14.5 Yogyakarta, Indonesia

amarria@uii.ac.id [1)], fety.rahmillah@uii.ac.id[2)], ajikgoez@gmail.com[3)]

Abstract. This article discusses 5S implementation in Work System Design and Ergonomic Laboratory, Department of Industrial Engineering, Islamic University of Indonesia. There are some problems related to equipment settings for activity involving students such as files which is accumulated over the previous year practicum, as well as the movement of waste in the form of time due to the placement of goods that do not fit. Therefore, this study aims to apply the 5S method in DSK & E laboratory to facilitate the work processes and reduce waste. The project is performed by laboratory management using 5S methods in response to continuous improvement (*Kaizen*). Moreover, some strategy and suggestions are promoted to impose 5S system within the laboratory. As a result, the tidiness and cleanliness can be achieved that lead to the great performance of laboratory users. Score assessment before implementing 5S DSKE laboratory is at 64 (2.56) while the score after implementation is 32 (1.28) and shows an improvement of 50%.This has implications for better use in the laboratory area, save time when looking for tools and materials due to its location and good visual control, as well as improving the culture and spirit of '5S' on staff regarding better working environment

Keywords: 5S, laboratory, management, waste

1. Introduction

5S system is commonly used by several laboratories in the global scope [1][2]. Implementation of 5S is part of a continuous improvement (Kaizen), aims to improve quality and work systems within an organization. Scholars [3] state that 5S is a management method of the work area as a consequence of the application of kaizen culture. Implementation of 5S method has been applied to the various work areas such as in manufacturing companies [4], metal SMEs [5], hotel service industry [6], and education especially in university laboratories. Some researches related to the implementation of 5S in the laboratory Coast University [2], laboratory Quality Assurance [7], Pharmaceutical laboratories [1], laboratory of Graphic Communications Management at the University of Wisconsin-Stout [8] and NDE laboratory in India [9].

A neat and clean workplace can be a modal of the good work culture. It gives a reflection of service quality and provides image of excellent to the institution. Laboratory as one of the component supporting university services should have a neat and clean workplace. DSK & E is one of six laboratories in Department of Industrial Engineering, Islamic University of Indonesia. The activities include teaching and learning/tutorials, research and administrative activities. Tutorial activities performed for subjects Physiology and work measurement and work system engineering and ergonomics. Tutorial activities carried out more use of practical tools such as antropometer, lux meter, stopwatch, sound level meters and other lab tools such as toys, plug, and resistor. Besides the practical activities, the business processes in the laboratory are administrative activities such as lending tools, attendance and research activities.

There are some problems related to equipment settings for activity involving students such as filing which is accumulated over the previous year practicum, as well as the movement of waste in the form of time due to the placement of goods that do not fit, inappropriate storage space for bin, broken printer, old computer and lab tools such as toys and resistors. In addition, wasting time in searching files, documents, books and lab tools due to workplace management and unnecessary material as well as lack of visual control for SOP (Standard Operational Procedures). However, the major issue in the laboratory management is the shortage of system which can determine recent arrangement and disproportionate participation of all staff in workplace management. Therefore, this research aims to apply the 5S method in DSK & E laboratory in order to respond continuous improvement in several aspects such as equipment setting, work processes, and reduce waste.

2. Literature review

5S is commonly known as a simple housekeeping method which can be used to arrange equipment or household, but actually the main concept of 5S is on the details and continuous working [2]. This method is also implemented by Japanese companies at the administration field which infiltrate in the ideology and technique management [10]. Furthermore, 5S methodology is believed can reduce the waste, improve safety and work process.

Basically, 5S method consists of 5 stages representing the meaning of five Japanese words and all of the levels are integrated in a whole working process [11]:

1. ***Seiri*/sort**, it means removing the waste and unused items from the working area. By sorting the unnecessary tools employee will be able to prioritize the equipment whether it is needed or not to be placed in their workplace.
2. ***Seiton*/set in order**, it means arranging tools in specific place and keeping it into the right position. Therefore, the items can be found easily by the workers when it is needed.
3. ***Seiso*/shine**, it means cleaning all working space and necessary tools. Therefore, the workers will have convenient place to work, and the equipment will always be ready to be used for other employee.
4. ***Seiketsu*/standardize**, it means ensuring all setup and equipment which has been arranged and cleaned in the right condition by promoting visible and modest rules.
5. ***Shitsuke*/sustain**, it means obeying the rules and procedures which has been issued as a respond of the first four phases.

In practice 5S methodology is used to build and keep the quality of environment within an organization [12]. Toyota Production System (TPS) is one of the obvious model of 5S implemented at the firm [13]. In addition, 5S methodology had been implemented in various manufacturing company such ceramics [14], sheet metal [15], and luggage bag [16]. Scholar [17] point out that 5S implementation creates domino effects which influence personal or company performance as depicted in Fig.1. 5S environment as a result of the 5S implementation will affect working behaviour of a person or culture in a company. Then, it will boost the personal skills and knowledge as well as company's operational excellence. Further, the skills will build the mindset which affect positive thinking of person and care optimism in a firm. Next, the mindset will lead the person and company to build identity in order to create personal excellence and top brand. Finally, the goal can be achieved in the term of company growth or a better personal life.

Fig.1 5S domino effect

3. Methods

This study was conducted in Work System Design and Ergonomic Laboratory. In this project, there are four strategy practiced by Industrial Engineering Department. First, 5S training is given to the laboratory personnels in order to raise awareness about the importance of 5S implementation and to provide the 5S basic knowledge. The next step is building the 5S team organization structure, including designating the referees, operation area and team leader. Furthermore, project scheduling and sharing responsibilty are also performed during this session. Third, 5S action is performed with prior preparation namely, providing 5S check list, identification form, as well as identifying the laboratory arrangement problem (laboratory tools and furniture). Then 5S method is implemented sequentially by 5S team in accordance with roles and duties. Moreover, 5S board indicator and poster are provided in order to foster and to assert 5S method in the laboratory. Finally, periodical inspection are performed by selected judge to keep the laboratory arrangement in 5S condition. In addition, the management also give further 5S training for its personnel, and conduct 5S contest among the laboratories. This process is briefly described in Table. 1

Table. 1 5S Implementation Strategy and Suggestion

Strategy	Suggestion
- 5S Training	• Evoking 5S consciousness to the laboratory management. • Equipping the laboratory organization with 5S knowledge.
- Creating 5S team and dividing the tasks	• Assigning the 5S implementation leader. • Creating 5S organization structure and defining personnel responsibility • Pointing 5S judges in order to review 5S implementation at the laboratory
- 5S action	• Providing 5S check list and identification form for each laboratory • Identifying the problem of laboratory arrangements

IOP Conf. Series: Materials Science and Engineering **215** (2017) 012032 doi:10.1088/1757-899X/215/1/012032

Strategy	Suggestion
- Continuous Improvement	• Performing 5S method : *Seiri, Seiton, Seiso, Seiketsu, Shitsuke* • Presenting 5S board indicator and poster inside the laboratories for educating and enforcing 5S method. • Conducting periodical inspection • Establishing further 5S training • Conducting 5S Competition

4. Results and Discussion

4.1 *Seiri/Sort*

In this phase, sorting necessary or unnecessary items is practiced. All non-essential items located inside the laboratory are given red tag. This step is useful to identify whether unimportant papers and books from previous practicum, lab tools (i.e resistors, toys), equipment (i.e chairs, boxes, sticks) will be discarded or stored at the shed (Fig.2).

Fig.2 Sorted item and Red tagging

4.2 *Seiton/Set in Order*

This step is important to put essential equipments and furnitures remain in its proper place. Moreover, lab items are labelled, and colored tapes are sticked for defining the boundary (Fig.3).

Before 5S Implementation	After 5S Implementation

Fig.3 Sticked colored tape, labelled and arranged items

4.3 Seiso/Shine

It is done by arranging the goods to be easy to use, easy to find and easy to revert as always done in cleaning work area and any objects that have been used must be returned to the same place to reduce the risk of missing items and hard to find.

In maintaining the cleanliness, neatness and comfort of Work System Design and Ergonomic laboratory and sustainability of 5S which has been implemented, therefore each teaching assistant has been required after the projector is turned off again, trimmed projector cables, and AC is turned off. And also, before leaving the lab assistants are required to clean up first in the area, especially at room assistants such as washing cutlery and drinks that have been used, reposition the seat into place, smoothing the goods there on the table, and taking out the trash if there is rubbish in assistant room area. In addition, before going home assistant is required to ensure that no return bins in every room and turn off all electricity devices which still alive such as air conditioning, projector or computer and turn off all the lights. Whereas, for the cleanliness of the general's own lab DSK & E coordinate with the cleaning service to assist in cleaning up any room assistants in lab DSK & E such as sweeping, mopping, etc.

In this stage, 5S organizational structure is created. It contains team members duties within their operation area. Although all academics community must be responsible in the use of laboratory, there should be person in charge to ensure 5S method is practiced in the lab (Fig.4).

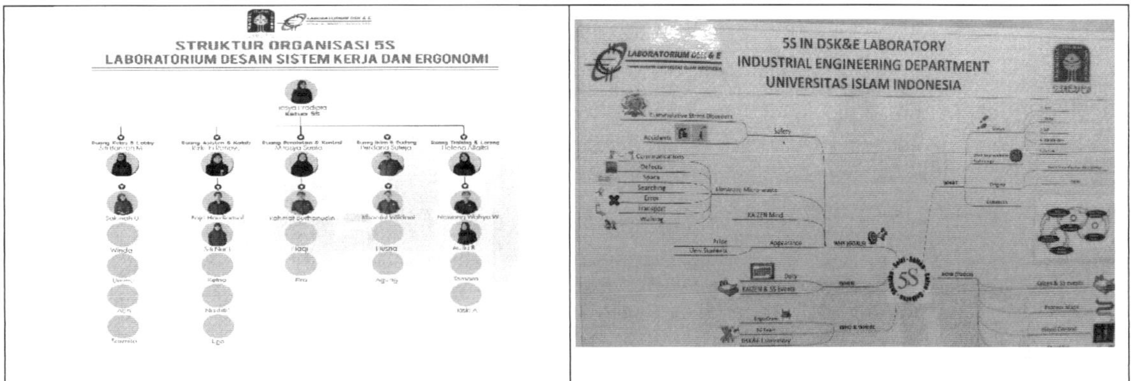

Fig.4 5S organizational structure

4.4 *Seiketsu/Standardize*

This stage aims to standardize 5S setting by providing visual control attached in the 5S board indicator. Therefore, every irregularity happening in the laboratory can be quickly identified and adjusted (Fig.5).

Standart Condition	Picture
Books should be in a good arrangement The book shelf should be in clean condition	
Table and chair should be inside the boundary (colored tape) Room must be in spotless condition	

Standart Condition	Picture
Table and chair should be arranged neatly Table surface should be free of paper, pen, etc.	

Fig. 5 Visual control of 5S

4.5 *Shitsuke/Sustain*

In this stage, the management urges all lab users to maintain 5S condition and to obey the laboratory standard which has been organized by installing 5S poster as depicted in Fig.6

Fig. 6 Poster of 5S

Visual Board of 5S is one way to make the staffs and all stakeholders understand the concept of 5S. It is created with the concept of 5W + 1H, What filled with what the 5S concept, and second W is filled with the list of waste that can be reduced in the current of 5S, the third W is When, it is the execution time of 5S periodically, the fourth W is Where, it is the work area of 5S, the fifth W is Who, who was responsible for 5S, and H is How to measure 5S DSKE lab. The detailed explanation is available on the picture (Fig.7).

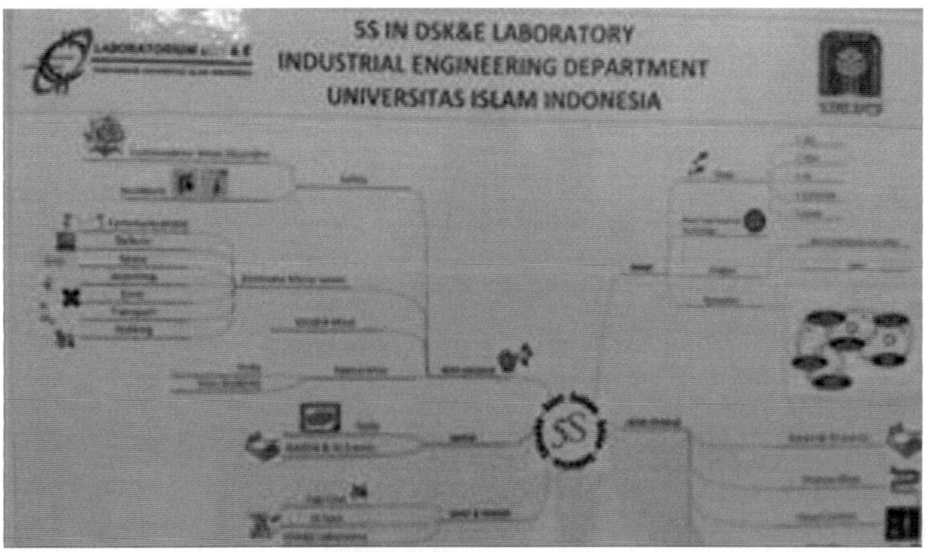

Fig. 7 Visual Board of 5S

4.6 '5S' Radar Chart

The radar map system is a result of the implementation of 5S performance indicators that have been run. Radar chart based on the results of the audit of the rating table 5S checklist consisting of five check items / questions on each of the 'S', so that the total check items are 25 check items which include signage and standards, general work area, information areas and documentation, visual controls, and habitual behavior. By using a 0-4 rating, with the conditions of observation 0 = very bad (above 7 problems), 3 = Bad (5-6 problems), 2 = ok (3-4 problems), 1 = good (1-2 problems), 0 = very good (perfect conditions). Results score check items before the implementation of 5S (audit in May) is 64 obtained from the sum score for sort/*seiri* (16), store/*seiton* (12), shine/*seiso* (11), standardize/*seiketsu* (12), and sustain/*shitsuke* (13) with an average of 25 items is 2.56 means the application of 5S in DSK&E Laboratory is good enough (3-4 problems). Fig. 8 is a radar chart before the implementation of 5S. While the score of the check items after the implementation of 5S (audit in September) was 32 obtained from the sum score for sort/*seiri* (6), store/*seiton* (8), shine/*seiso* (6), standardize/*seiketsu* (8), and sustain/*shitsuke* (4) with an average of 25 items is 1.28 means good (1-2 problems). Fig.9 shows a radar chart after the implementation of 5S. As a result, there is an improvement of 50% from 2.56 became 1.28. This has implications for better use in the laboratory area, save time when looking for tools and materials due to its location and good visual control, as well as improving the culture and spirit of '5S' on staff regarding better working environment.

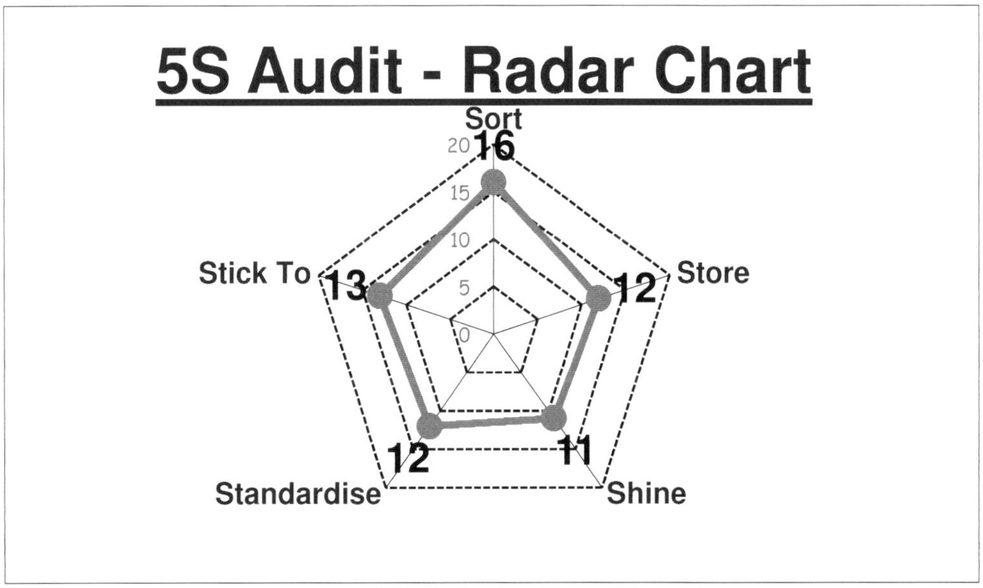

Fig 8. Radar Chart 5S on May

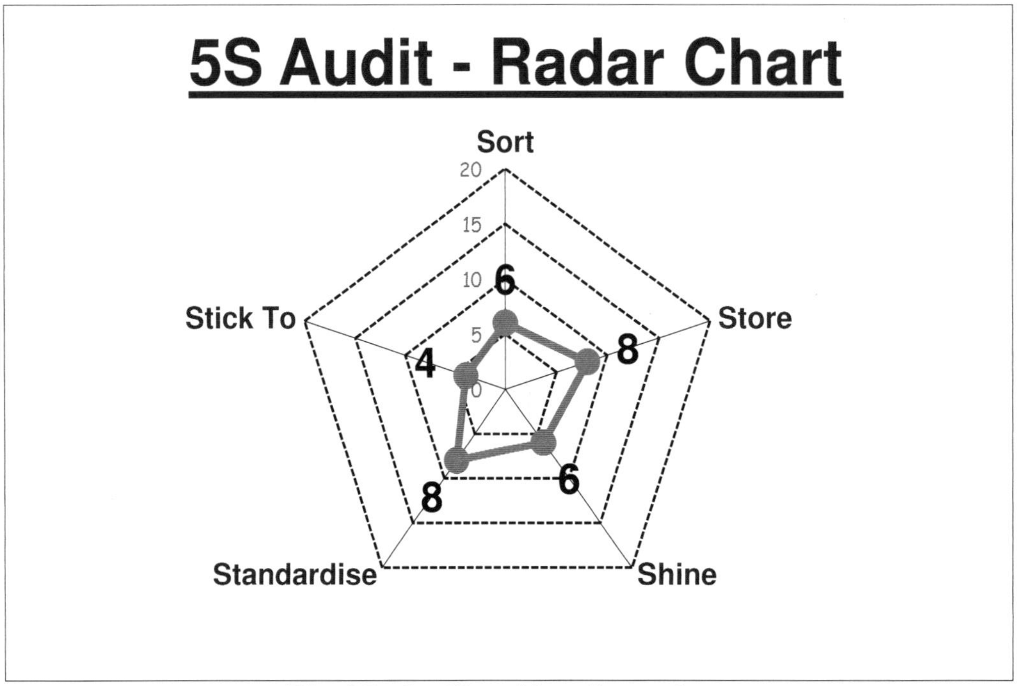

Fig. 9 Radar Chart 5S on September

5. Conclusions and Recommendations

Score assessment before implementing 5S DSKE laboratory is at 64 (2.56) while the score after implementation is 32 (1.28) and shows an improvement of 50%. This has implications for better use in the laboratory area, save time when looking for tools and materials due to its location and good visual control, as well as improving the culture and spirit of '5S' on staff due to better working environment

As a recommendation, it is necessary to have continuous improvement, provide more visualization regarding the number of minimum / maximum objects arranged in a box, as well as the consistent '5S' audit / regular self assessment. Further research studies can be done by calculating the time before and after the implementation process, and research into the 6S (Seiri, Seiton, Seiso, Seiketsu, Shitsuke and Safety).This research also can be implemented in other laboratory.

6. References

[1] Mallick, A., Kaur, A. and Patra, M. (2013) Implementation of 5S in pharmaceutical laboratory. IJPRBS 2 (1), 96–103.

[2] Pentti, O. (2014) Applying the Lean 5S Method to Laboratories and Prototype Workshops. Bachelor's Thesis. Turku University of Applied Sciences. Finland

[3] Jiménez, M., Romero, L., Domínguez, M., & del Mar Espinosa, M. (2015). 5S methodology implementation in the laboratories of an industrial engineering university school. *Safety science*, 78, 163-172

[4] Ginting, P., Matondang, R., & Buchari, B. (2014). Analisis Program Keselamatan Dan Kesehatan Kerja Di Bagian Produksi Dengan 5s Dalam Konsep Kaizen Sebagai Upaya Pencegahan Kecelakaan Kerja Di Pt. Xyz. *Jurnal Teknik Industri USU*, 3(5).

[5] Sitompul, Z. A. (2015). Pengukuran Nilai Efisiensi Penerapan 5s Pada IKM Logam Alumunium (Studi Kasus Di Km Alumunium, Yogyakarta) (Doctoral Dissertation, UIN Sunan Kalijaga Yogyakarta).

[6] Gürel, D. A. (2013). A conceptual evaluation of 5S model in hotels. *African Journal of Business Management,* 7(30), 3035.

[7] Chitre, A. (2010). Implementing the 5S Methodology for Lab Management in the Quality Assurance Lab of a Flexible Packaging Converter (Doctoral dissertation, University of Wisconsin-Stout).

[8] Maharjan, S. (2011). Implementing the 5S Methodology for the Graphic Communications Management Laboratory at the University of Wisconsin-Stout (Doctoral dissertation, University of Wisconsin--Stout).

[9] Ananthanarayanan, K. R. M. (2006). Application of 5S management system in NDE laboratory. In National seminar on non-destructive evaluation.

[10] Gapp, R., Fisher, R. and Kobayashi, K. (2008) Implementing 5S within a Japanese context: an integrated management system. *Manage. Decis.* 46 (4), 565–579.

[11] Kobayashi, K. (2005) What is 5S? A Content Analysis of Japanese Management Approach. Unpublished Master's Thesis, Griffith University, Southport.

[12] Khamis, N., Abrahman, M.N. and Jamaludin, K.R. et al. (2009) Development of 5S practice checklist for manufacturing industry. In: *Proceedings of the World Congress on Engineering, vol.* 1, pp. 978–988.

[13] Monden, Y. (2012) Toyota Production System: An Integrated Approach to Just-In-Time, fourth edition. Institute of Industrial Engineers.

[14] Patel V.C. and Thakkar, H., (2014). A Case Study: 5s Implementation in Ceramics Manufacturing Company, *Bonfring International Journal of Industrial Engineering and Management Science,* 4(3)

[15] Kakkar, V., Dalal, V.S., Choraria, V., Pareta, A.S., Bhati, A. (2015) Implementation Of 5S Quality Tool In Manufacturing Company: A Case Study, *International Journal Of Scientific & Technology Research,* 4(2), pp. 208-213

[16] Deshpande1, S.P., Damle, V.V., Patel, M.L., and Kholamkar, A.B. (2015) Implementation Of '5S' Technique In A Manufacturing Organization: A Case Study, IJRET: International Journal of Research in Engineering and Technology, 4(1), pp. 136-148

[17] Firdaus, F. (2016) 5S Implementation. 5S Training module, Indonesia Islamic University.

IOP Conf. Series: Materials Science and Engineering **215** (2017) 012033 doi:10.1088/1757-899X/215/1/012033

The Analysis of Thermal Comfort in Kitchen

Fety Ilma Rahmillah[1], Agustina Hotma Uli Tumanggor[2], Amarria Dila Sari[3]

[1,3]Department of Industrial Engineering, Faculty of Industrial Technology, Islamic University of Indonesia

[2]Department of Mechanical Engineering, Faculty of Engineering, Politechnique of Hasnur

fety.rahmillah@uii.ac.id[1], agustina.hotma@gmail.com[2], amarria@uii.ac.id[3]

Abstract

Human also has a thermoreceptor which is a non-specialized sensory receptor that has relative changes in temperature. Thermal comfort is a very important element for human body. Kitchen as an important part of a home is often forgotten. Cooking in the kitchen is a routine activity which is done from the morning until the evening; begin with preparing breakfast, lunch and dinner. The problem in this study was the occurance of heat when cooking in the kitchen without air conditioning in tropical countries. This research analyzes thermal comfort while doing cooking activities in conventional kitchen with gas stoves in tropical dry season. Two residential kitchens are observed by measuring the temperature and humidity as well as analyze other possible factors. Psychometric chart is used to assess the comfort zone in the kitchen. This research is using Predicted Mean Vote (PMV) Index and Predicted Percentage Dissatisfied (PPD) Index. By using online psychometric chart, the sensation is in warm condition with the range value of PMV between 1.73 up to 2.36 and PPD 63% untill 90%. However, 71% respondents perceived morning kitchen thermal as comfortable.

Keywords— Thermal comfort, PPD index, PMV index, psychometric chart.

1. Introduction

Thermal comfort has a significant influence towards human body. It is a heat balance between human and environment occupied. Human also has a thermoreceptor which is a non-specialized sensory receptor that has relative changes in temperature. This sensory acts to convey information to the brain in which the brain will give order to maintain human body's temperature stabile at 37 ± 0.5 °C under different climatic conditions. It is necessary so that the organs can perform its functions well. Some factors affect human thermal comfort are clothing worn, climate condition, and physical activity [1].

Kitchen is an important part of a home that is often forgotten. It is one of the few rooms that must exist in the residence. It is not just a place to prepare food, but it also can be used for recreation, family communication or even working for certain job. People are too focused on areas such as private rooms, a family room, and even bathroom. However, kitchen is part of the largest contributors to discomfort, because it often causes odor when cooking and increase the heat surrounding the room. Cooking is a routine activity which is done from the morning until the evening; begin with preparing breakfast, lunch until dinner.

Relatively few studies on thermal comfort in kitchen than that in house or outdoor environment in general. Many of the available studies regarding house have focused mainly on country which has four seasons [2][3][4]. However, the studies actually does not cover residential kitchen. Only scholars [5][6] analyzes conventional kitchen in Taiwan, while another scholar [7] studies commercial kitchen in United States. Scholar [5] states that the highest average concentrations of Carbon monoxide (CO) in residential kitchens in Taiwan during the period of cooking with gas stoves (24 hours on average) varied from 0.1 to 13.9 ppm. While research [8] concluded that the use of LPG stoves produce CO emissions of about 5.83 ppm which is smaller than charcoal-fueled stoves or kerosene. There are many problems in kitchen such as discussed by [9] regarding older people problems in kitchen and its design solution and proper clothing worn in institusional and restaurant kitchen [10].

Even though the cooking process of residential and commercial kitchen are similar, but at some points, they are so different such as in exhaust airflow rate and installation height. Residential kitchen is intermittent operation and no replacement air requirement for ventilation. Moreover, it used to be more conservative. A residence also has a much lower background ventilation rate, so escaping contaminant is more persistent. This situation presents a different challenge because it is not possible to overcome problems by simply increasing the ventilation rate at the cooking process. Most residential kitchens are ventilated by wall-mounted, conventional range hoods so that it depends almost entirely on buoyancy capture [11]. Moreover, very limited research carried out in non air-conditioned spaces.

This research analyzes thermal comfort while doing cooking activities in conventional kitchen with gas stoves in tropical season by considering air temperature, humidity and clothing.

2. Literature Review

Thermal comfort is a psychological concept that relies on the preferred physiological state (uncomfortable through comfortable). While British Standard BS EN ISO 7730 specify it as 'that condition of mind which expresses satisfaction with the thermal environment.'

The most widely used indicator of thermal comfort is air temperature because every body can easily measure it. However, if it is lonely used, it will be not capable to measure thermal comfort. Some of environmental factors should be considered are radiant temperature, relative humidity and air velocity while the personal factors are activity and clothing [2].

In term of activity, the work rate has influence toward heat produced by human body. In addition, the effect of metabolic rate on thermal comfort is very critical. Metabolic equivalent (MET) can be used to measure physical activity in which expressing energy expenditure. While clothing worn also give influence toward thermal comfort due to clothing insulation which is measured in Clo units (Icl). One clo is the amount of thermal resistance which is significant to maintain thermal comfort for subject' activities in a normally ventilated room (air movement 20 ft/min or 10 cm/sec) with temperature of 70°F (21.1°C) and RH<50%. It is necessary to understand how the fabric contributes toward thermal comfort or discomfort [1].

Predicted Percentage Dissatisfied (PPD) index can be used to quantify thermal comfort at specific environment. While the average thermal sensation which is measured by using ASHRAE thermal sensation scale usually called the Predicted Mean Vote (PMV). The value ranges from +3 (hot), +2 (warm), +1 (slightly warm), 0 (neutral), -1 (slightly cool), -2 (cool), and -3 (cold).

Research relates to residential thermal comfort had been done in Belgium [4], Nigeria [2], Singapore [12], and Malaysia [13]. Scholar [6] and [12] study the effect of ventilation in indoor environment. None research done in Indonesia yet especially related with kitchen thermal comfort.

3. Research Method

The study is conducted in two kitchens located in Sleman, Yogyakarta. The type of kitchen studied is residential kitchen with the dimension of Kitchen A is 3 x 3 meters as shown in Figure 1 while Figure 2 visualizes Kitchen B with the dimension of 4x6 meters.

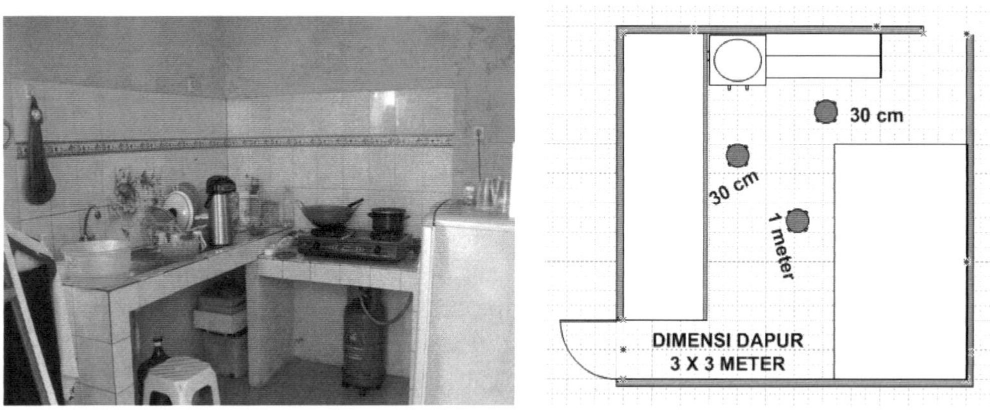

Fig 1. Kitchen A

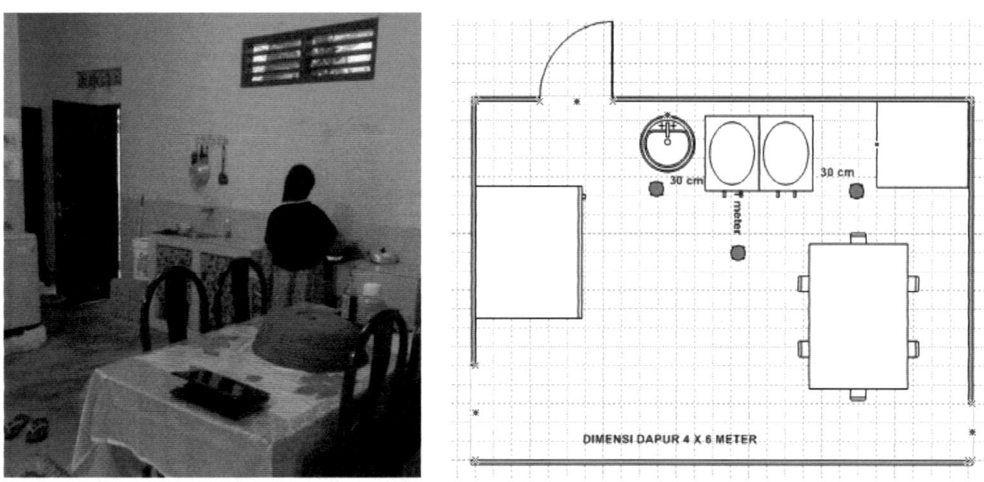

Fig 2. Kitchen B

The tools used in the study are Multienvirometer to measure temperature (^{0}C) and humidity (%RH), Graph thermal comfort ASHRAE-55, Psychometric Chart and camera. Method that is used in this study are observation and survey. The data of temperature and humidity were taken during cooking time for four days in dry season with six times measurement in every morning (5.30 - 7 AM), afternoon (11 AM - 12.30 PM) and evening (4 - 5.30 PM). In addition, survey to twenty women was conducted to find the preferences of comfort material for clothing worn while cooking activity conducted.

4. Results and Discussion

Factors affecting thermal comfort

The temperature and relative humidity varies greatly throughout the day. Based on Ministry of Health of Republic Indonesia No 1405/MENKES/SK/XI/2002 concerning the requirements of the work environment, comfort temperature is between 18–28^{0}C. Other literature stated that thermal comfort for tropic can be divided into: cool comfortable (20,5^{0}C–22,8^{0}C); optimal comfortable (22,8^{0}C–25,8^{0}C); and warm comfortable (25,8^{0}C–27,1^{0}C). That's why, the kitchen temperature while cooking in the morning and evening is categorized as warm comfortable (Figure 3).

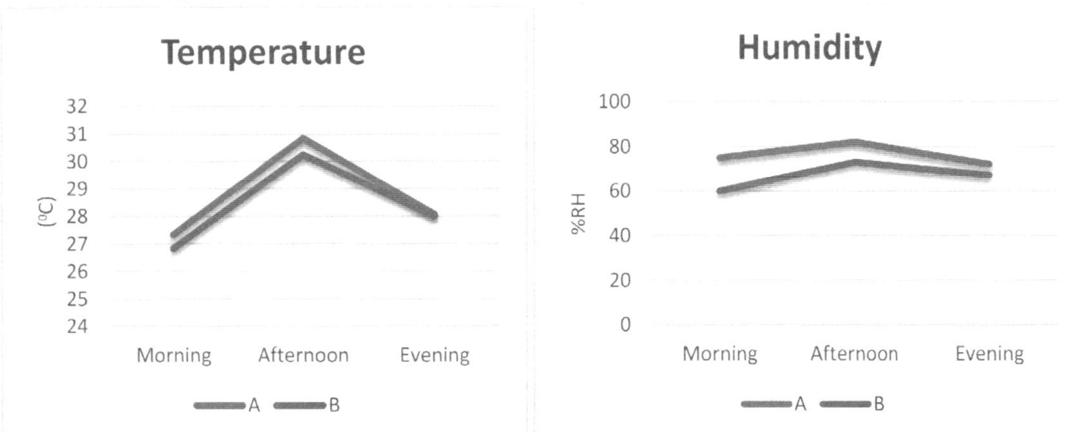

Fig 3. Graph of Temperature and Humidity in Kitchen A and B

The kitchen temperature during morning and day tend to be higher in kitchen A than that in B. It may due to the use of ceiling in kitchen A in which it could trap more heat remain in the kitchen than the roof tile did in kitchen B. In addition, the position of kitchen A got more sunlight than kitchen B which was located in the middle of the house. The temperature during afternoon in kitchen B was lower than A. It is probably due to the influence of air insulation coming inside from the left door and ventilation at the top of the cooking area. Whereas the temperature of kitchen A slightly decreases. Moreover, there is no ventilation inside the kitchen A which makes the heat remain in the kitchen.

According to SNI 03-6572-2001: the natural ventilation provided should consists of permanent opening, window, door or others that can be opened, with the number of ventilation openings not less than 5% against the floor area. Based on calculation, the ventilation for kitchen B should around 1.2 m^2. It considered to be fulfilled because there is

IOP Conf. Series: Materials Science and Engineering **215** (2017) 012033 doi:10.1088/1757-899X/215/1/012033

1.2 m^2 ventilation in the top of cooking area (0.4 x 1 m) and 2 m^2 from the left door (1 x 2 m). In addition, it is also completed by five big windows in directly opposite the kitchen. Whereas an opening needed in kitchen A is as much as 0.45 m^2 (which comes from 0.05 x (3x3m)). It is not enough since the recent door is covered by a heavy gorden. Applying natural ventilation also had been applicated in Singapore, as a result, comfortable indoor environment can be provided [12].

Recommended humidity for tropic area is between 40-60% along the year, but it should be 55-60% for meeting room in which many people involved. The average of humidity in this study is 71.53% (the detail can be seen in Figure 3) which considers as comfort since research in Malaysia which is also a humid tropic nation stated that the mean relative humidity corresponding to the optimum comfort temperature was close to 73% [13].

Activities will also influence thermal conduction. Based on Table MET for various of activities, cooking activity has MET value of 1,6 to 2.0. People with this activity has metabolic heat generation around 90-115 W/m^2 [14]. Thermal insulation in clothing is an important parameter of thermal comfort [15]. Based on a survey to twenty women; 87.5% prefer wearing cotton for clothing while cooking, whereas 12.5% choose jersey material because of familiarity. Cotton fabric is a textile with a soft and porous character that makes the skin easier to 'breathe'. It is made from cotton plant so that it is very comfortable to wear as well as become the most widely used in the clothing industry [16].

The influence of clothing and operative temperature in the morning can be seen in Figure 4. The clothes that are used: the maximum limit of 1.4 clo, standard: 1 clo and the bottom limit is around 0.75 clo. According to the graph, the condition of the kitchen is not including into comfortable because it is over the line limit. But the relative humidity is recognized as 50% while the air velocity is 0.15 m/second. The cooking activities varies from 1.6 to 2.0 MET while the operative temperature is 27.075^{0}C and to be comfort at that area, it is necessary to wear clothing which less than 0.1 clo.

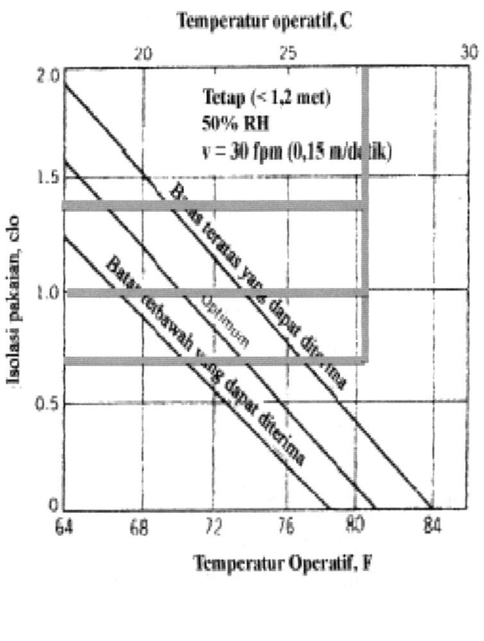

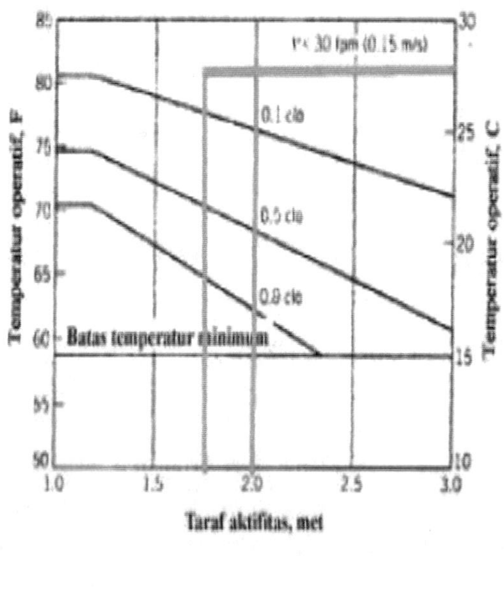

Fig 4. The Effect of Clothing and Operative Temperature in the Morning

Psychometric Chart

Online psychometric chart is used to get PMV, PPD, and comfort zone area. The input data of operative temperature, humidity, metabolic rate, and clothing are determined based on the research results. The comfort zone can be seen in the blue area, whereas the operative temperature is symbolized by red circle point. The complete results for comfort zone in the morning, afternoon, and evening can be seen in Figure 5.

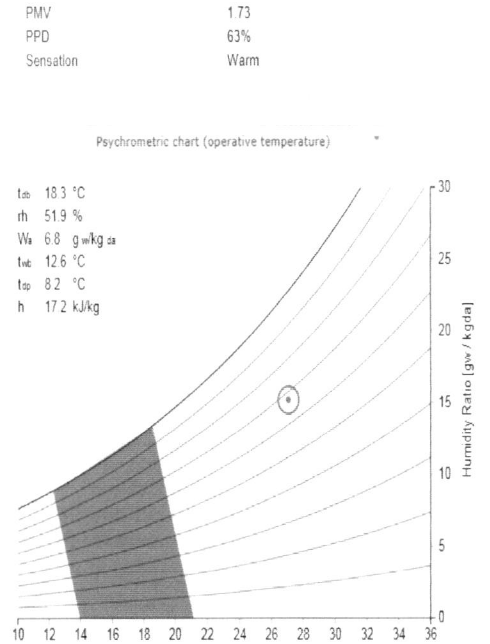

(a)

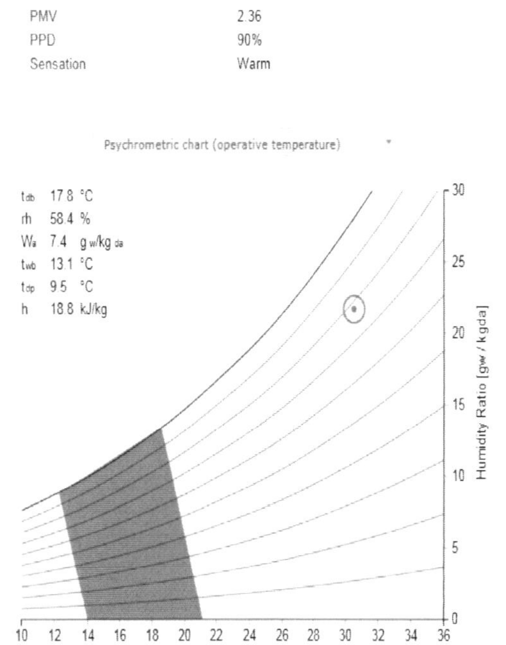

(b)

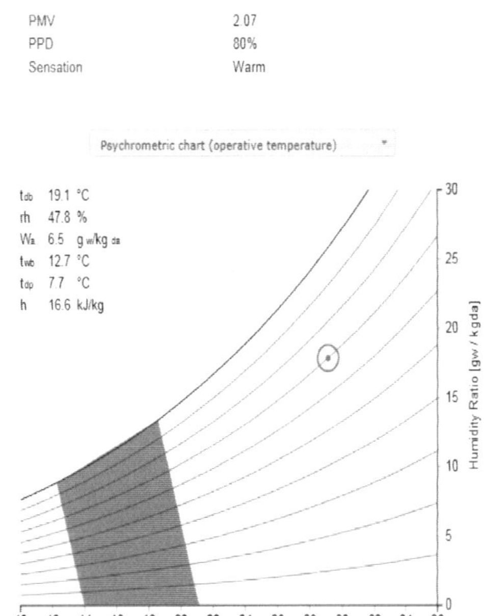

(c

Fig 5. Psychometric Chart in the Morning (a), Afternoon (b), and Evening (c)

Based on the calculation of online psychometric chart, the value of PMV lies in the range of 1.73 up to 2.36 which means in warm sensation. While the PPD range from 63 until 90%. Eventhough 63% people feel dissatisfied with the thermal comfort in the morning, but based on the survey conducted, most women prefer to cook in the morning (71%) because of comfortable weather. While only 29% want to cook in the evening, and none choose afternoon to do cooking activity. Scholar [7] also states that PMV/PPD may not suitable for commercial kitchen which has characteristic of high activity level and hot temperature. Moreover, research by [3] showed that people live in tropical countries such as Ghana may have higher perceived heat tolerance. Eventhough the standard is between 26-28^{0}C (at summer), but the interviewees accept higher temperature as comfort.

5. Conclusion and Recommendation

According to the discussion in the previous, the sensation of thermal comfort in residential kitchen is in warm condition with the range value of PMV between 1.73 up to 2.36 and PPD 63% untill 90%. However, 71% respondents perceived morning kitchen thermal as comfortable.

Recommendation

Future studies should be practiced in countries that has four seasons.

6. References

[1] Layton J.M., (2001). The science of clothing comfort, *The textile institute*, Textile progress, 31 (1/2). UK.

[2] Adaji, M., & Adler, P. G. (n.d.). an Investigation into thermal comfort in residential buildings in the hot humid climate of Sub-Saharan Africa : a field study in Abuja-Nigeria.

[3] Appah-Dankyi, J., & Koranteng, C. (2012). An assessment of thermal comfort in a warm and humid school building at Accra, Ghana. *Adv. Appl. Sci. Res.,* 3(1), 535–547.

[4] Singh, M., Mahapatra, S., & Teller, J. (2013). Study on indoor thermal comfort in the residential buildings of Liege, Belgium. Cisbat, 481–486. Retrieved from http://orbi.ulg.ac.be/handle/2268/155634

[5] Chiang, C. M., Lai, C. M., Chou, P. C., & Li, Y. Y. (2000). The influence of an architectural design alternative (transoms) on indoor air environment in conventional kitchens in Taiwan. *Building and Environment.* Retrieved from https://doi.org/10.1016/S0360-1323(99)00029-3

[6] Kosonen, R. (2007). The effect of supply air systems on the efficiency of a ventilated ceiling. *Building and Environment*, 42(4), 1613–1623. https://doi.org/10.1016/j.buildenv.2005.09.030

[7] Simone, A., & Olesen, B. W. (2013). Thermal environment evaluation in commercial kitchens of United States, 1–4.

[8] Hidayat, A., Huboyo, H. S., & Budisulistiorini, S. H. (n.d.). Analisis emisi karbon monoksida (CO) dan distribusi aliran gas pembakaran pada udara dalam ruang dapur rumah tangga menggunakan model computational fluid dynamics (CFD).

[9] Maguire, M., Peace, S., Nicolle, C., Marshall, R., Sims, R., Percival, J., and Lawton, C. (2014). Kitchen living in later life: exploring ergonomic problems, coping strategies and design solution, *International Journal of Design,* 8(1)

[10] Zhang, H., McQueen, R.H., Batcheller, J.C., Paskaluk, S.A., and Murtaza, G. (2015). Clothing in the kitchen: Evaluation of fabric performance for protection against hot surface contact, hot liquid and low-pressure steam burns. *Textile Research Journal*, 85(20), 2136–2146

[11] ASHRAE Application Handbooks. (1999). Kitchen ventilation. California Building Energy Efficiency Standards, (Chapter 30).

[12] Liping, W., & Hien, W. N. (2007). Applying natural ventilation for thermal comfort in residential buildings in Singapore. *Architectural Science Review*, *50*(3), 224–233. https://doi.org/10.3763/asre.2007.5028

[13] Djamila, H., Chu, C-M., Kumaresan, S., (2014). Effect of humidity on thermal comfort in the humid tropics, *Journal of Building Construction and Planning Research,* 2, 109-117

[14] Butera F.M., (1988). Chapter 3: 'Principles of thermal comfort', *Renewable & Sustainable energy reviews,* 2, p. 39-66.

[15] Oğulata, R.T., (2007). The effect of thermal insulation of clothing on human thermal comfort., *Fibres & Textiles in Eastern Europe*., 15(2), 67-72

[16] Polhemus, T. (2005). What to wear in the global village? *In J. Brand & J. Teunissen (Eds.),* Global fashion local tradition: On the globalisation of fashion (87-101). Arnhem: Uitgeverij Terra Lannoo.

IOP Conf. Series: Materials Science and Engineering **215** (2017) 012034 doi:10.1088/1757-899X/215/1/012034

Design of Smart Home Systems Prototype Using MyRIO

Dwi Ana Ratna Wati[1], Dika Abadianto[2]

[1,2]Department of Electrical Engineering, Universitas Islam Indonesia, Jl. Kaliurang Km 14,5 Yogyakarta Indonesia

dwi_ana@uii.ac.id

Abstract. This paper presents the design of smart home systems prototype. It applies MyRIO 1900 embedded device as the main controller of the smart home systems. It includes wireless monitoring systems, email based notifications and data logging. The simulated sensors such as temperature sensor, push button as proximity sensor, and keypad, as well as simulated actuators namely buzzer as alarm system, LED as light and LCD are applied in the prototype. Based on the test and analysis, the smart home systems prototype as well as the wireless monitoring systems have real time responses when input signals are available. The performance of MyRIO controller is excellent and it results in a stable system.

1. Introduction

Smart home systems are designed for a variety of purposes and reasons including safety, security, comfort, and energy efficiency. The smart home systems communicate with the user, whether the user is at home or outside via email [1] as well as web applications. Wireless network plays an important role in the development of smart home systems [2]. By using camera, image processing, and email services which are applied in an embedded controller, the home security can be monitored remotely by the homeowners [1]. In addition to home automation, research on smart home systems has been well developed in the field of medical applications, such as health monitoring systems [3], as well as to assist users with disabilities [4]. Management of power consumption in the smart home systems can be realized by scheduling, such that provides solutions to optimize the use of electrical equipment [5]. As an example is the room temperature control which increases energy efficiency in the home automation system while provides the comfort at home. It applies intelligent controller, which automatically adjusts the temperature set point according to user behaviour as well as environmental conditions. Neural network as temperature controller has a stable performance and is able to adapt to the various environmental conditions [6]. The equipment at home already functions like smart devices, which along with the ability to set up operations when the presence or absence of the user and also learn from the operation of the scheduling pattern that has been received [7].

As the increase in complexity and applications of smart home systems as well as the need for speed, and easy access, this research designed a prototype of smart home systems using NI MyRIO 1900. It has onboard FPGA and microprocessor for ensure the speed of computation processes. It is equipped with a WiFi device that enables communication between devices in the home as well as data communication throught the internet services. The goal of this study is to design the concept of smart home systems using MyRIO 1900 and test the performance while providing the basis for further advanced smart home systems development.

2. Methodology

This research consists some steps as shown in the Figure 1. The smart home systems prototype used MyRIO 1900 as information processing and controller. Besides functioning to monitor the state of the house, the system is designed with the aim to improve the security of the residence. Some features of the smart home system are as follows.

- Remote Monitoring : the state of the home include temperature, state of the lights, electronic equipment and the presence or absence of people in the house can be monitored from a distance as long as the homeowner is connected to the Internet.
- PC monitoring (available at home) which is connected to the smart home systems via a wifi network.
- Security system :
    - Homeowners are asked to enter a password to unlock the gate. If the password is wrong, then the alarm will light and the camera will take a picture of the face of the opening of the gate. Meanwhile, if the password is correct, then the gate will open.
    - If the house is being abandoned by the owner, then there are foreigners who are found in the house, the alarm is on and the system will send a notice to the owner of the house that there was a stranger in the house.
- Data logging: all data (temperature, lighting conditions, the presence or absence of people in the house and faces photograph of the person who opens the gate) to be stored and delivered to the homeowners.

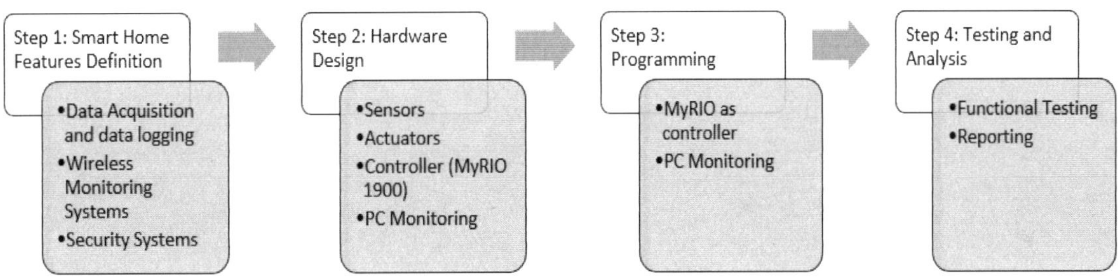

Figure 1. Research methodology

2.1. Hardware Design

Figure 2 shows the block diagram of smart home systems prototype. The input to NI MyRIO are temperature sensors, switches/pushbutton assumed as a proximity sensor, keypad, and camera. MyRIO has outputs namely the buzzer as the alarm, the LCD as a viewer, the LED lights which are assumed as lights control systems.

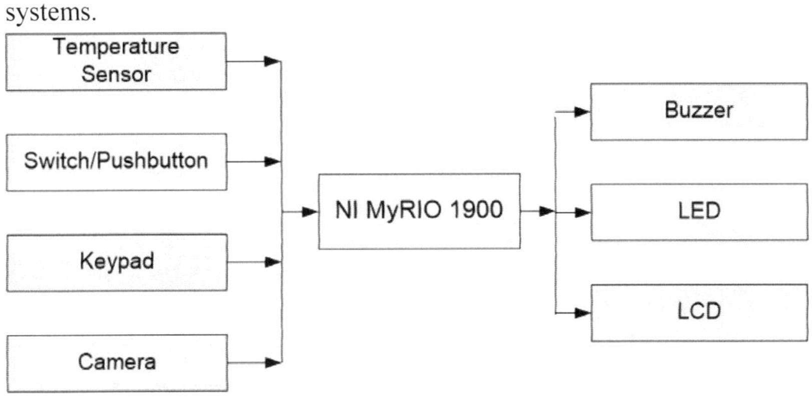

Figure 2. Block diagram of smart home systems prototype

IOP Conf. Series: Materials Science and Engineering **215** (2017) 012034 doi:10.1088/1757-899X/215/1/012034

In the Figure 2, it can be seen several sensors and actuators that are used in the design of smart home systems prototype, as follows.

1. Thermistor, to measure the temperature.
2. LED as light controlled.
3. Keypad, enter the homeowner's password.
4. LCD, to display the password.
5. The USB camera, to take pictures of people who type or input the password.
6. Buzzer, as an alarm.
7. MyRIO, as a controller.
8. Switch, as a modifier condition and value assumptions of the proximity sensor in the gate of a house.

2.2. Software Design of Smart Home System in MyRIO

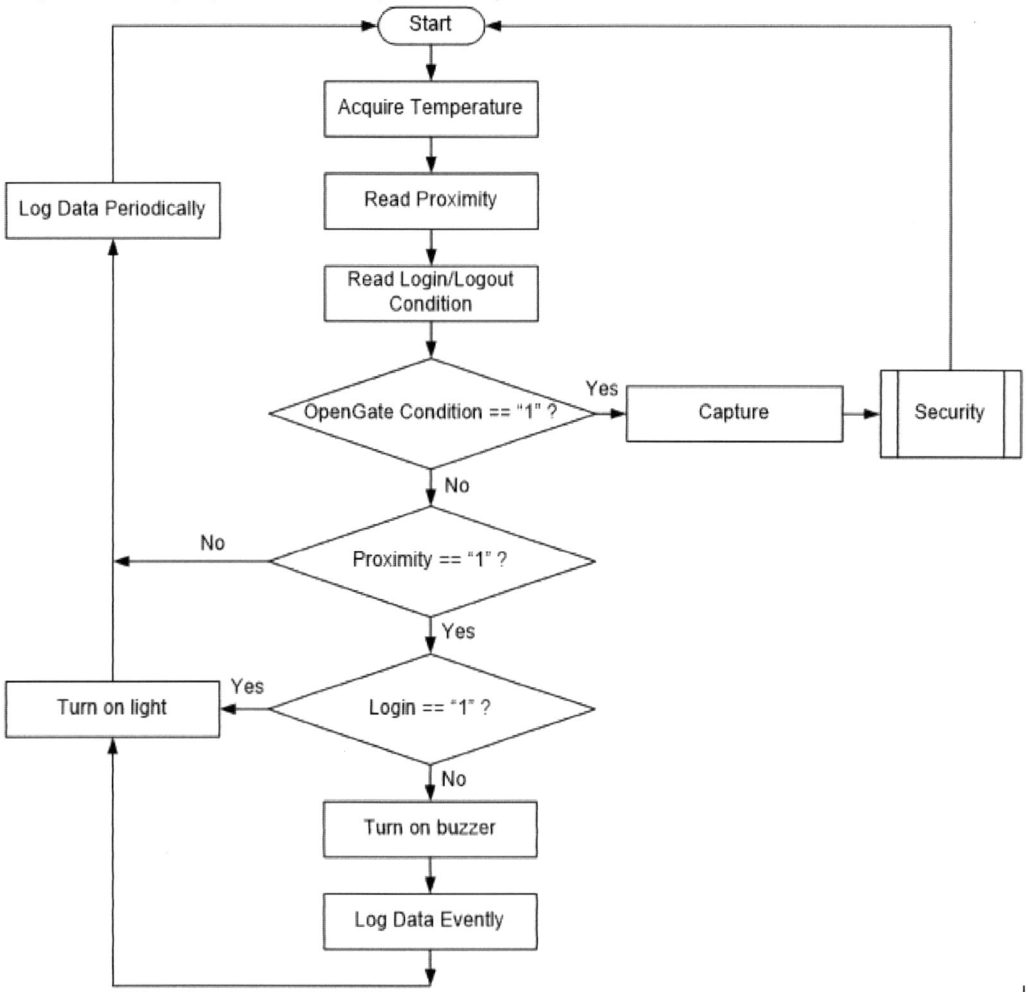

Figure 3. Flowchart of smart home systems prototype

Figure 3 shows the flowchart of main program in MyRIO, while Figure 4 shows a flowchart of security subroutine. The system is continuously read temperature, proximity sensor, login/logout conditions and the state of the gate. If the gate is open or "1", then the camera will take a picture of the person who enter the house. If else, the user is prompted to enter a password on the keypad. If the password is correct, then the gate will open. However, if the password is incorrect, then the gate is not open and the alarm will turn on.

Then if conditions login/logout is "1" and the proximity change the conditions to "1" which means the homeowner is coming, then the light goes on. Meanwhile, if the conditions login/logout is "0" and the proximity change conditions which means that it is not the homeowner who enter the house, then the light is off, the alarm will be lit and the system will send an alert to the user/homeowner via email. All data (temperature, proximity, conditions login/logout, conditions open the gate and the picture of person who open the gate) will be sent to the user's email once a day, which can be set delivery time.

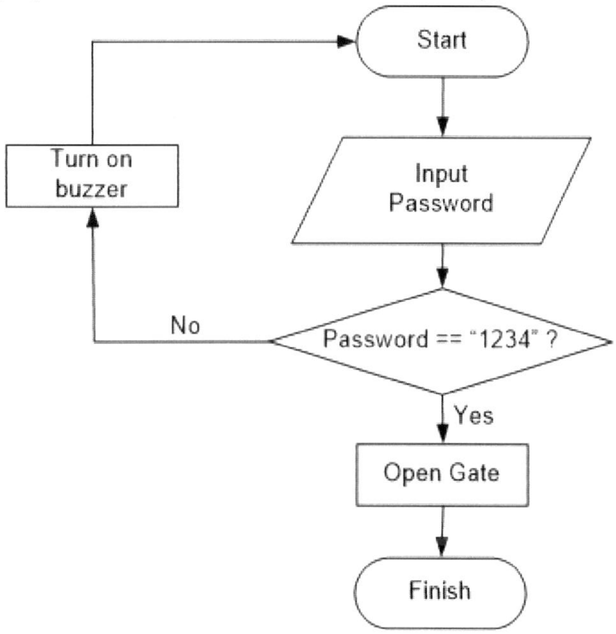

Figure 4. Flowchart of security subroutine

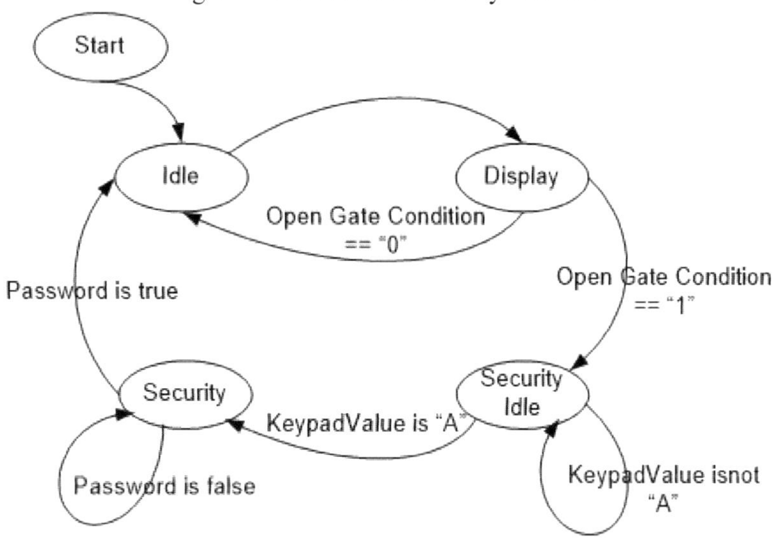

Figure 5. State diagram of smart home systems

Figure 5 shows the state machine diagram of the program. Starting from the *Start* and then jump to *idle*. *Idle* condition is a condition in which MyRIO waiting for input from the user to determine the circumstances which are to be executed next, while *Display* is a subroutin for displaying the sensor and output state. If the value of OpenGateCondition is “1”, then the condition to be executed is a condition *SecurityIdle* in which the system will wait for password entered by user. If the value of OpenGateCondition is “0”, then the conditions remain in the *Idle* state. When the user press the “A” key means the user has finished entering the password then the *Security* conditions is executed. If the password is correct, then the state will move to the *Idle* state. Meanwhile, if the password is incorrect, then the condition will remain on *security* conditions. The front panel for smart home systems prototype is shown in Figure 6.

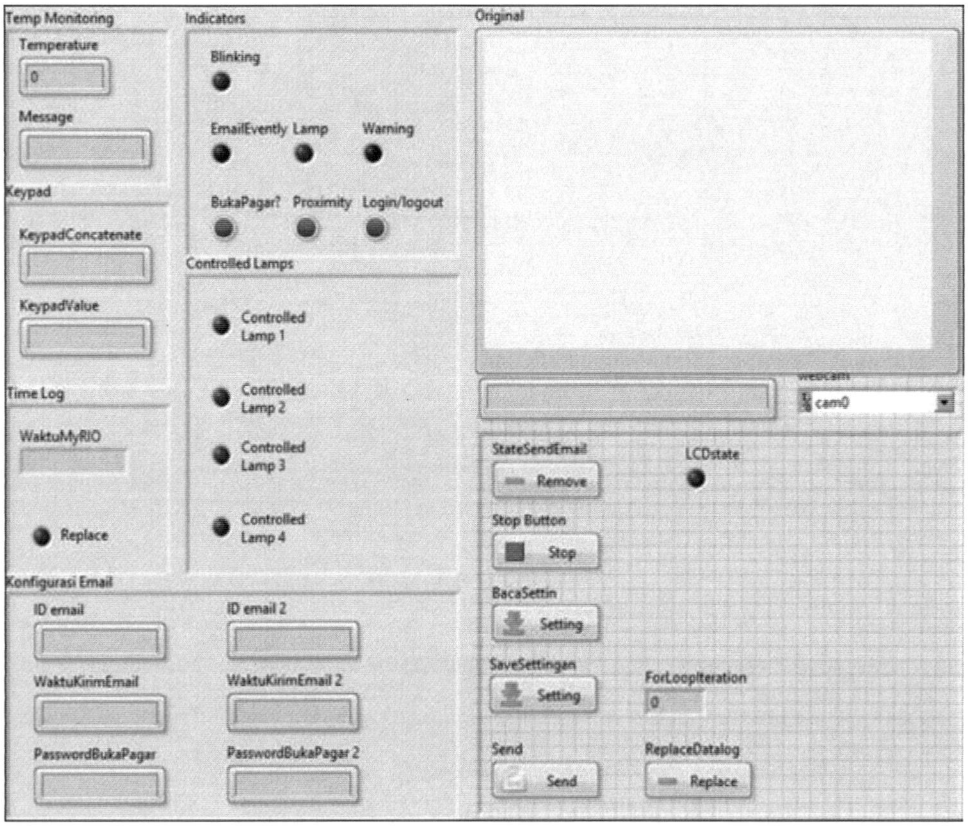

Figure 6. *Smart Home System Front Panel*

2.3. *Software Design of Smart Home Wireless Monitoring System*

Smart home wireless montoring systems is an application that will be embedded in the PC as the wireless monitoring system. This application can be used to monitor, control and change parameter settings on smart home systems. Figure 7 shows the block diagram of it. It can be seen from the figure that the communication between the PC and MyRIO are using wifi network. The protocol used is User Datagram Protocol (UDP). UDP is a protocol that makes it possible to transmit short data packets to one or more recipients. Unlike TCP (transmitter control protocol), UDP does not guarantee the data sent completely accepted by the recipient. So it is very possible that the data packets are missing or not in order. So that, the UDP should be used for short data transmission and not the transmission of information requiring data accuracy. Limit the maximum size of each packet of approximately 65 thousand.

IOP Conf. Series: Materials Science and Engineering **215** (2017) 012034 doi:10.1088/1757-899X/215/1/012034

Figure 7. Block diagram of smart home wireless monitoring systems

The monitoring features are as follows.

1. The condition of the lights. So when users turn on the lights, the lights indicator on the monitoring application will also on.
2. Temperature. The average value of current temperature reading will be displayed on the system monitoring applications in real time.
3. The condition of proximity sensors. Proximity sensors are used to determine whether there is movement of people in the house. If there are changes in the value of the proximity sensor, the indicator will light up on the monitoring application. So that users know that there is someone in the house.
4. The condition of login/logout. If the user wants to leave the house, then the user should tell the system by changing the conditions of switches login/ logout. This condition is also displayed on the monitoring application.
5. The condition of the gate. If a user wants to open the gate, then the user should press the pushbutton to open the gate. This condition will be displayed on the monitoring application. Then the user is required to enter a password through the keypad. If the password is incorrect, then the alarm will sound indicating the house in a state of warning. This condition will also be displayed in the monitoring application.
6. Control light conditions by using monitoring application.
7. Change the settings on the system includes a user's email address, report time and password to open the fence.

Figure 8 shows the state machine diagram of wireless monitoring application. The first state to be executed is *InitSetup*. The next state is the *Startup* state. In this state the user is prompted to press Start on LabVIEW pushbutton to start the application. If the pushbutton is pressed, then the state will be executed next is *Initialize*. But if the pushbutton is not pressed, then the next state is the *Startup* and executed continuously. Then the next state is the *Monitoring*. Then if *configuration* pushbutton is pressed, then the state will be executed next is *Configuration* to display the configuration setting. If the *setting* pushbutton is pressed, then the state will be executed next is state *Configure*. If the *stop* pushbutton is pressed, then the state will be executed next is state *Stop*. Whereas if none pushbutton is pressed, the state executed next is *Manipulated*. Then it backs to the state *Monitoring*. Front panel of the monitoring application is shown by Figure 9.

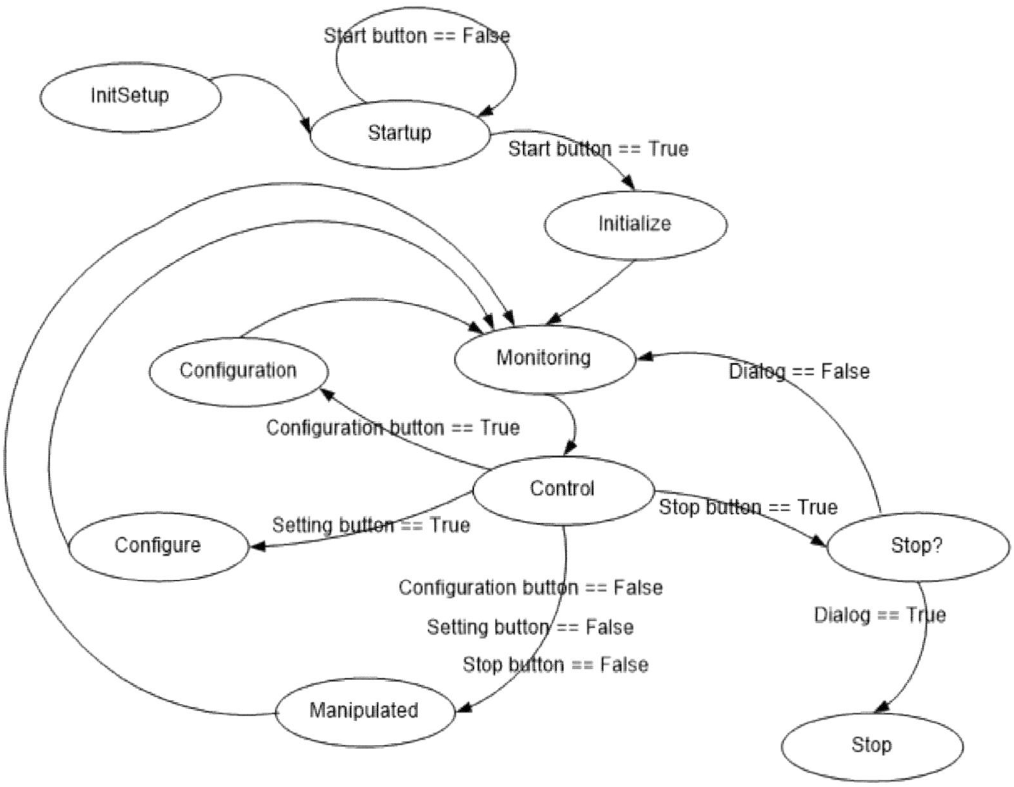

Figure 8. State diagram of wireless monitoring system

Figure 9. Front panel of wireless monitoring system

3. Result and Analysis

Figure 10. Hardware testing

To determine the performance of the systems, both hardware and software were tested. Figure 10 shows the hardware testing. Testing was conducted on the thermistor for temperature measurement, LED as a light-controlled, the keypad to enter a password, LCD to display the passwords, USB camera to take pictures of people who are entering the password as well as Buzzer as alarm. From these tests, all hardware is working properly as designed. Further testing is functional of wireless monitoring system. Figure 11 shows an example of the monitoring system testing when the room temperature is 25,244 °C.

Figure 11. Wireless monitoring systems testing

The next test is checking the email notifications which are sent to homeowner. Figure 12 is an example of the display in the email containing the email notificaton of smart home systems. The email includes two attachments namely images of user entering to the house and data monitoring as shown in Figure 13.

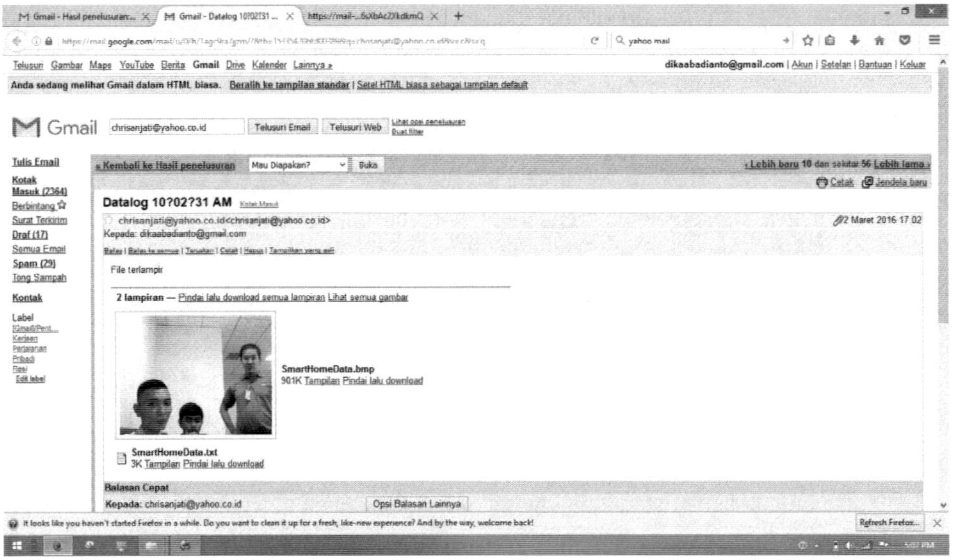

Figure 12. Example of an email notification

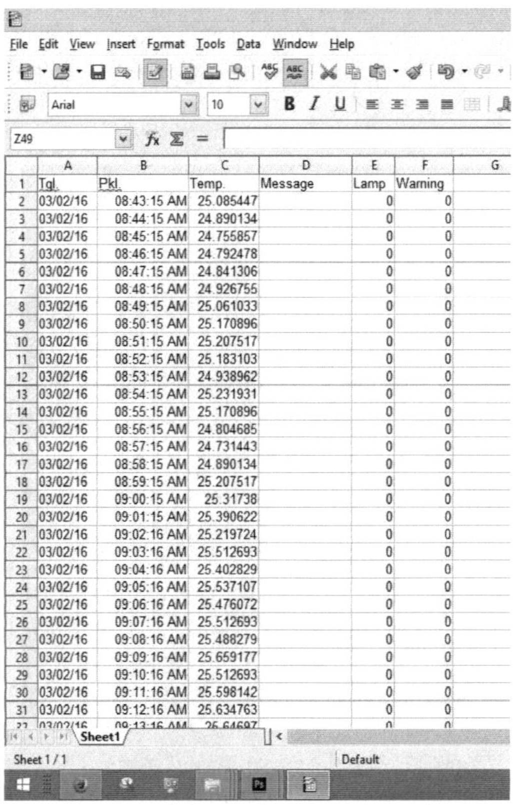

	A	B	C	D	E	F
1	Tgl.	Pkl.	Temp.	Message	Lamp	Warning
2	03/02/16	08:43:15 AM	25.085447		0	0
3	03/02/16	08:44:15 AM	24.890134		0	0
4	03/02/16	08:45:15 AM	24.755857		0	0
5	03/02/16	08:46:15 AM	24.792478		0	0
6	03/02/16	08:47:15 AM	24.841306		0	0
7	03/02/16	08:48:15 AM	24.926755		0	0
8	03/02/16	08:49:15 AM	25.061033		0	0
9	03/02/16	08:50:15 AM	25.170896		0	0
10	03/02/16	08:51:15 AM	25.207517		0	0
11	03/02/16	08:52:15 AM	25.183103		0	0
12	03/02/16	08:53:15 AM	24.938962		0	0
13	03/02/16	08:54:15 AM	25.231931		0	0
14	03/02/16	08:55:15 AM	25.170896		0	0
15	03/02/16	08:56:15 AM	24.804685		0	0
16	03/02/16	08:57:15 AM	24.731443		0	0
17	03/02/16	08:58:15 AM	24.890134		0	0
18	03/02/16	08:59:15 AM	25.207517		0	0
19	03/02/16	09:00:15 AM	25.31738		0	0
20	03/02/16	09:01:15 AM	25.390622		0	0
21	03/02/16	09:02:16 AM	25.219724		0	0
22	03/02/16	09:03:16 AM	25.512693		0	0
23	03/02/16	09:04:16 AM	25.402829		0	0
24	03/02/16	09:05:16 AM	25.537107		0	0
25	03/02/16	09:06:16 AM	25.476072		0	0
26	03/02/16	09:07:16 AM	25.512693		0	0
27	03/02/16	09:08:16 AM	25.488279		0	0
28	03/02/16	09:09:16 AM	25.659177		0	0
29	03/02/16	09:10:16 AM	25.512693		0	0
30	03/02/16	09:11:16 AM	25.598142		0	0
31	03/02/16	09:12:16 AM	25.634763		0	0

Figure 13. Sample of data logging

IOP Conf. Series: Materials Science and Engineering **215** (2017) 012034 doi:10.1088/1757-899X/215/1/012034

4. Conclusion

This study has been designed the prototype of smart home systems using MyRIO 1900 with simulated sensor devices and actuator. The system is equipped with a wireless monitoring system and email notifications as well as data logging. Based on the results and analysis of the design as well as testing, the smart home system using MyRIO generate real-time output responses when given a particular input. MyRIO performance as the controller is excellent and produces a stable system.

References

[1] P. Rajiv, R. Raj, and M. Chandra, "Email based remote access and surveillance system for smart home infrastructure ☒," *Perspect. Sci.*, pp. 3–5, 2016.

[2] Z. Liu, "Hardware Design of Smart Home System Based on zigBee Wireless Sensor Network," *AASRI Procedia*, vol. 8, no. Secs, pp. 75–81, 2014.

[3] J. Puustjärvi and L. Puustjärvi, "The role of smart data in smart home : health monitoring case," *Procedia - Procedia Comput. Sci.*, vol. 69, pp. 143–151, 2015.

[4] A. Hussein, M. Adda, M. Atieh, and W. Fahs, "Smart Home Design for Disabled People based on Neural Networks," *Procedia - Procedia Comput. Sci.*, vol. 37, pp. 117–126, 2014.

[5] G. Mehdi and M. Roshchin, "Electricity consumption constraints for smart-home automation : An overview of models and applications," *Energy Procedia*, vol. 83, pp. 60–68, 2015.

[6] T. Teich, F. Roessler, D. Kretz, and S. Franke, "Design of a Prototype Neural Network for Smart Homes and Energy Efficiency," *Procedia Eng.*, vol. 69, pp. 603–608, 2014.

[7] H.-C. Sun and Y.-C. Huang, "Optimization of Power Scheduling for Energy Management in Smart Homes," *Procedia Eng.*, vol. 38, pp. 1822–1827, 2012.

Minimization of Defective Products in The Department of Press Bridge & Rib Through Six Sigma DMAIC Phases

YA Rochman[1] and A Agustin[2]

[1,2] Department of Industrial Engineering, Faculty of Industrial Technology, Islamic University of Indonesia, Jl. Kaliurang Km 14.5 Yogyakarta, Indonesia

E-mail: gusti@uii.ac.id

Abstract. This study proposes the DMAIC Six Sigma approach of Define, Measure, Analyze, Improve/Implement and Control (DMAIC) to minimizing the number of defective products in the bridge & rib department. There are 5 types of defects were the most dominant are broken rib, broken sound board, strained rib, rib sliding and sound board minori. The imperative objective is to improve the quality through the DMAIC phases. In the define phase, the critical to quality (CTQ) parameters was identified minimization of product defects through the pareto chart and FMEA. In this phase, to identify waste based on the current value stream mapping. In the measure phase, the specified control limits product used to maintain the variations of the product, the calculation of the value of DPMO (Defect Per Million Opportunities) and the calculation of the value of sigma level. In analyze phase, determine the type of defect of the most dominant and identify the causes of defective products. In the improve phase, the existing design was modified through various alternative solutions by conducting brainstorming sessions. In this phase, the solution was identified based on the results of FMEA. Improvements were made to the seven priority causes of disability based on the highest RPN value. In the control phase, focusing on improvements to be made. Proposed improvements include making and define standard operating procedures, improving the quality and eliminate waste defective products.

1. Introduction

Intense competition in the business world today makes the company must constantly improve itself in order to survive and be able to continue its business. Companies should always strive to improve the quality and productivity levels for the use of resources more effectively and efficiently so that it will reduce production costs, produce higher quality products and produce better services.

Musical instrument manufacturing company has several divisions in the manufacture of piano, one of which is the press department of the bridge and rib. This department is responsible for the process of pinning cabinet treble and bass bridge, the pressing process rib on the sound board and the process of scraping residual glue on the sound board after the pressing process. In the course of the work, a product defect is difficult to avoid, product defects that often occur in the form of a broken rib, broke sound board, strained rib, rib sliding, rib defects, bass bridge chipped, rib curved, sound board striped rib passengers, bridge sliding, one press bridge, sliding bore, uki sound board, sound board peeling, trible broke and fiber sound board upside down.

The six sigma tools and techniques have been applied in various manufacturing area, which encourages continuous improvement in achieving less cost, variation and high quality of end products [1]. Six Sigma philosophy is to continuously monitor the process and aims at the elimination / reduction of defects or failures of the manufacturing process [2]. DMAIC methodology is the Six Sigma's unique approach to continuous process and quality improvement [3]. [4] States that the step by step approach or road map using DMAIC methodology is key success factors to Six Sigma. A systematic approach that Six Sigma DMAIC (Define, Measure, Analyze, Improve, and Control) phase is used for the elimination of waste [5]. Six Sigma is a measure of variation about the average in the service industry or the manufacturing process [6]. The primary objective of six sigma is the elimination of waste or non-value added process activities along with continuous improvement [7]. [8] States that Six Sigma is an approach to improve the efficiency and quality, by eliminating defects and errors. [9] Any deviation in the performance of the critical to quality (CTQ) characteristics are called defect. [10] Proposed that integration of statistical quality systems with six sigma results give effective results to identify the most relevant improvement areas. Based on the literature review, six sigma DMAIC used as a systematic approach to eliminate waste and the best way for improving quality/ reducing waste. This research used to determine the average value of the sigma level, the possibility of defect, the most dominant defect and propose action in an effort to minimize defects.

2. Methodology

2.1. Define

Define aims to describe the problems that occur in the company [6]. This stage contains explanations of the production lines at the press bridge and rib through SIPOC diagram (Suppliers, Inputs, Process, Outputs, and Customer) as shown by Figure 1. This stage also contains the current state value stream mapping as shown by Figure 2. This figure used to determine the flow of the production process from beginning to end. Moreover, this stage identifies waste, especially waste of defective products. Figure 1 and Figure 2 obtained from observation at a musical instrument manufacturing company.

Warehouse provides input to the press in the form of pin bridge and treble rib, rib, and solid sound board. machine bridge provides input in the form of treble and bass bridge, as well as provide input in the form of cold press Sound Board Laminating. The production process starts from the pin bass & treble, press rib, press treble & bass and scrape residual glue. The resulting output is a solid and sound board laminate that has been assembled with the rib, treble & bass bridge that will be sent to the sound board assy painting up.

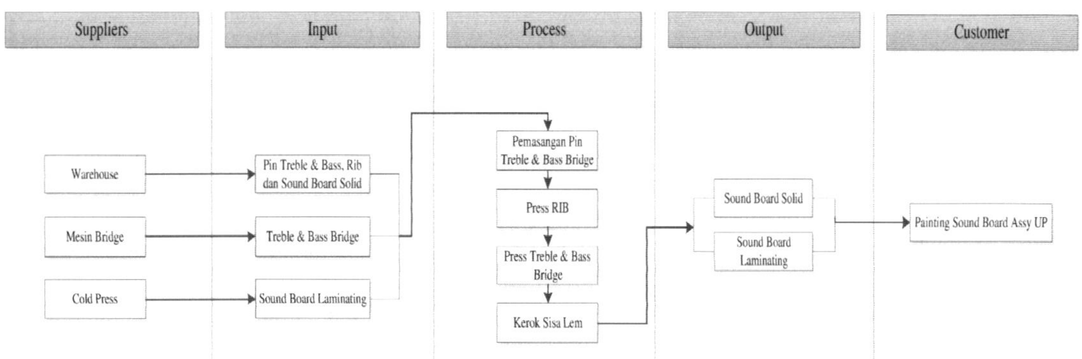

Figure 1. SIPOC Diagram

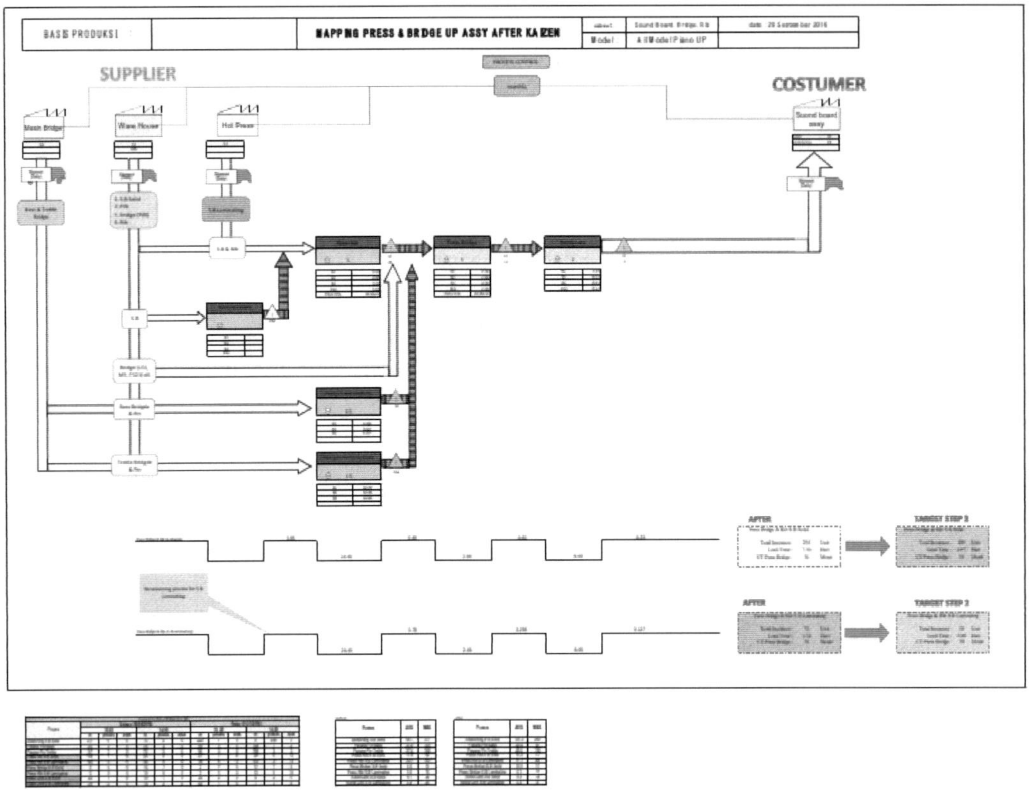

Figure 2. Current State Value Stream Mapping

2.2. *Measure*

In the measure phase, we calculate the value of sigma level at the press bridge and rib. The data needed is total production, the number of defective products, CTQ, DPU, TOP, DPO and Defect per Million opportunities (DPMO). Products control limit are used to maintain product variation (Figure 3 and Figure 4). DPMO used to determine the value of sigma level (Figure 5).

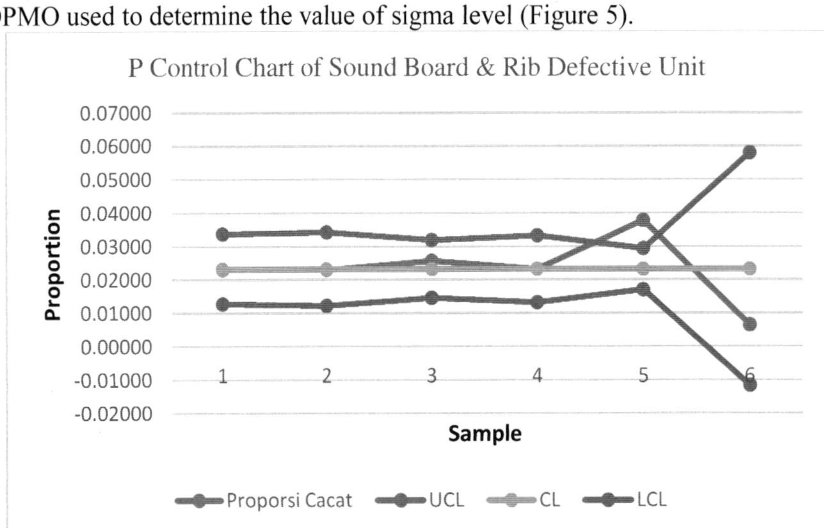

Figure 3. P Control Chart of Sound Board & Rib Defective Unit

Figure 3 shows that the data produced in sample 5 passed the control limits Upper Control Limit (UCL). Data that exceeds the control limits can not be used at a later stage.

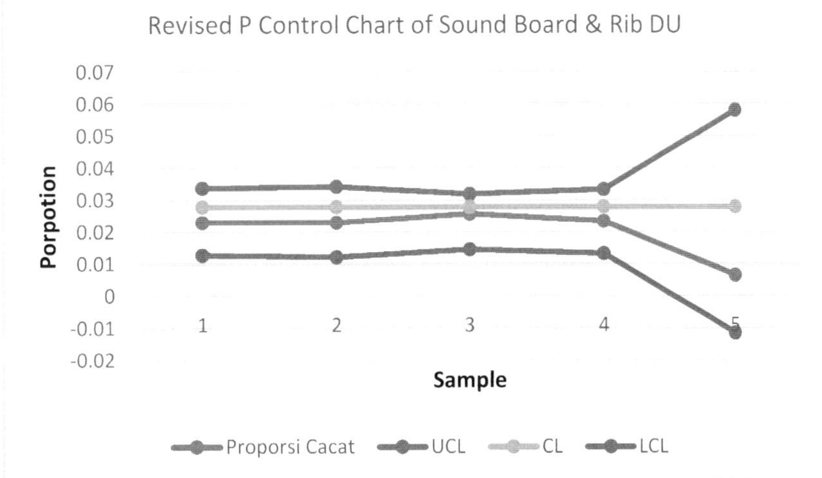

Figure 4. Revised P Control Chart of Sound Board & Rib Defective Unit

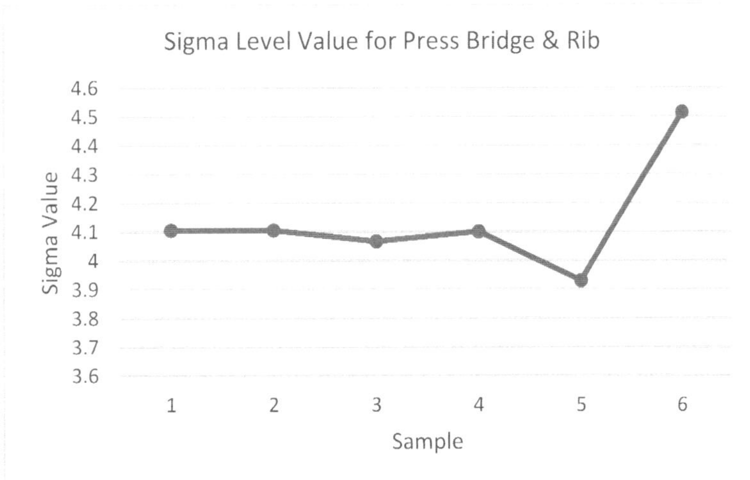

Figure 5 Sigma Level Value for Press Bridge & Rib

2.3. Analyze

This stage determined the most dominant types of defects and identify the causes of defective products. Analyzes were performed using pareto diagrams and fishbone diagrams.

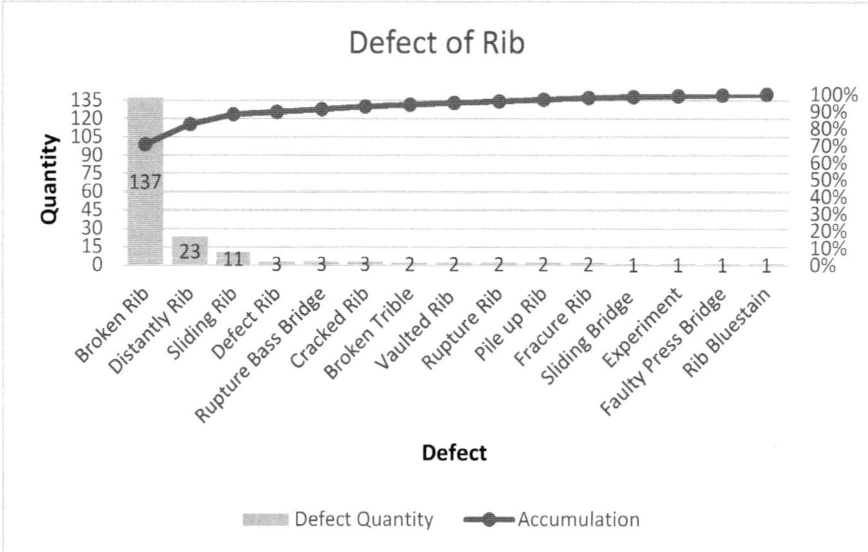

Figure 6 Defect of Rib Graph

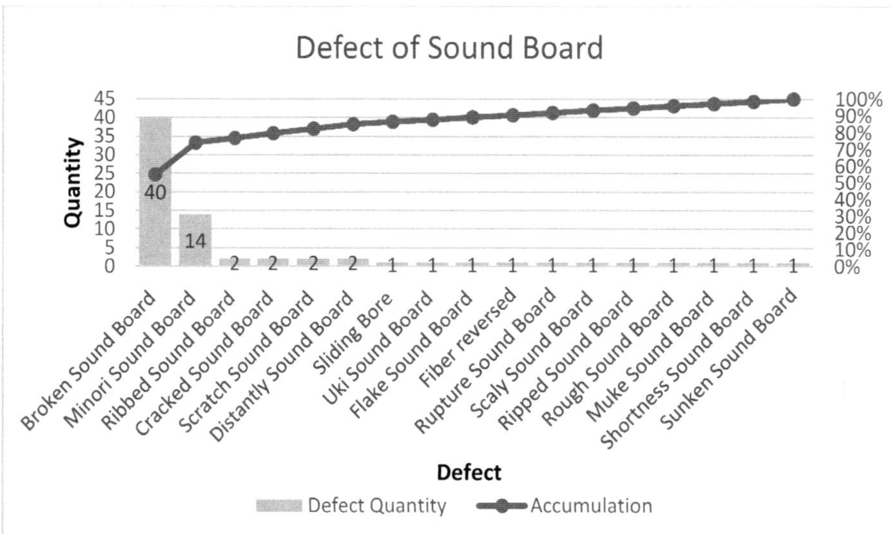

Figure 7 Defect of Sound Board Graph

Analysis of the causes of defects predominant obtained through observation, interviews and brainstorming with the VSM & IE members as well as group head portion press bridge and rib discussing the cause of the defect is rib tableware, sound board tableware, rib cozy, rib slide and sound board minori.

IOP Conf. Series: Materials Science and Engineering **215** (2017) 012035 doi:10.1088/1757-899X/215/1/012035

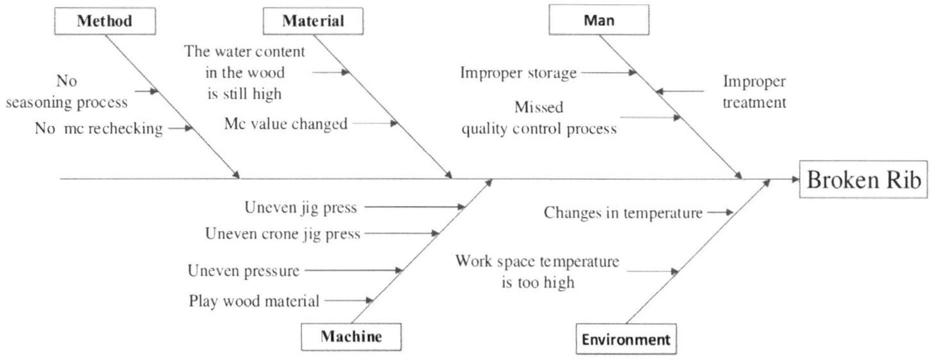

Figure 8 Fishbone diagram for broken rib

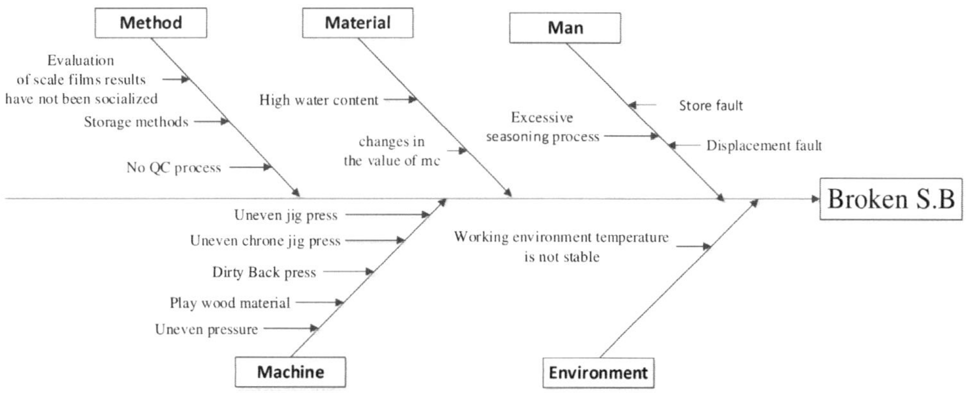

Figure 9 Fishbone diagram for broken sound board

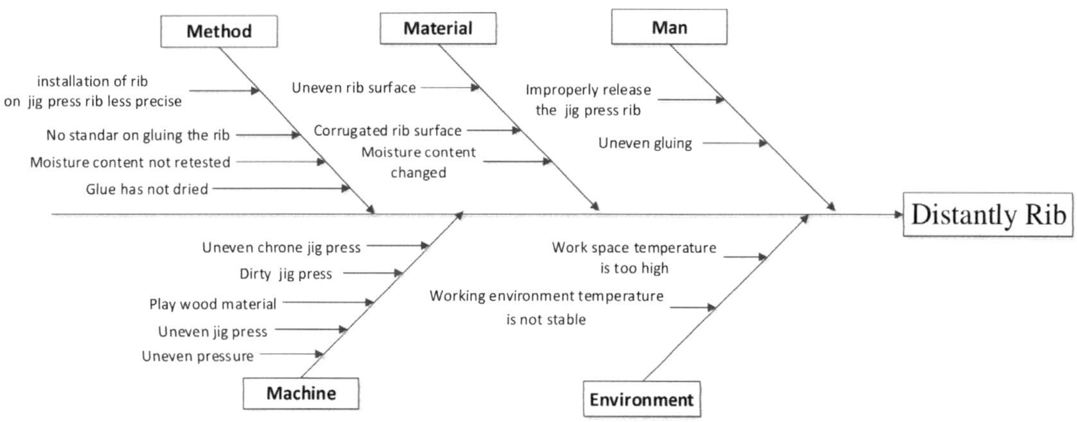

Figure 10 Fishbone diagram for distantly rib

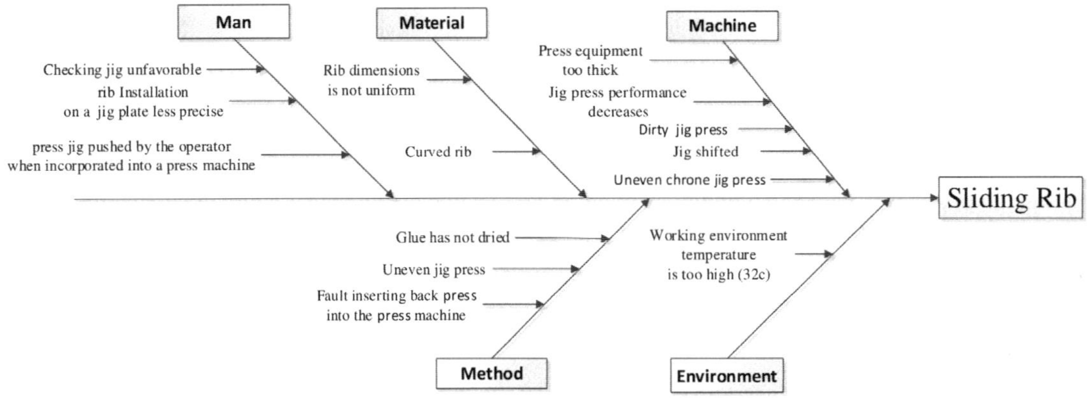

Figure 11 Fishbone diagram for sliding rib

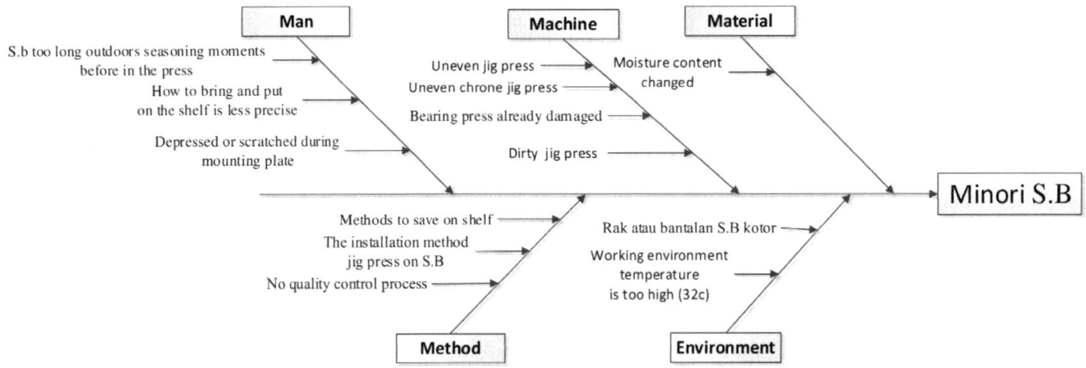

Figure 12 Fishbone diagram for minori sound board

2.4. Improve

At this stage, the calculation of the value of the RPN (risk priority number) in the analysis of FMEA (failure mode and effect analysis) through interviews with members of VSM & IE as well as group head portion press bridge and rib in determining the value of severity, occurrence and detection in determining the value of the RPN for identifies priorities for the cause of the defect.

Table 1 FMEA Analysis

Mode of Failure (Defect)	***Potential Failure***	***Cause of Failure***	***Current Process Control***	***RPN***	***Rank***
Broken Rib	Mc value changed	Work space temperature is too high	Adding AC or blower in the work area	336	1
Broken Sound Board	Mc value changed	Working environment temperature is not stable	Adding AC or blower in the work area	336	1
Distantly Rib	No standard on gluing the rib	No standard operating procedure	Create standard operating procedure	98	1
Sliding Rib	Checking jig unfavorable	Checking is only done at the beginning	The checks carried out 2 times	140	1
Minori Sound Board	Mc value changed	Work space temperature is too high	Adding AC or blower in the work area	336	1

2.5. Control

This phase contains the control that focuses on continuous improvement. Improvements to do is create and define the standard operating procedures for supervising the causes of disabilities in order for defective products can be minimized as well as increasing productivity, improving the quality and eliminate waste defective products on the part of the press bridge and rib. The application of the concept of kaizen (continuous improvement) and 5S are aimed at building a corporate culture to make improvements from small and sustainable.

3. Conclusion

The conclusion of this study are as follows:

1. The sigma level average value is equal to 4,125 and the possibility of defects by 4639 units based on DPMO value calculation.
2. There are 5 types of defects were the most dominant of broken rib with the main causes of EMC (Equilibrium Moisture Content) changes in rib, broken sound board with the main causes of the change in EMC on the sound board, rib cozy with the main causes is no standard in the gluing rib, rib slide the main cause is the lack of checking jig press in a machine press and sound board minori with main causes of EMC changes in the sound board.
3. Based on the results of FMEA with the highest RPN value of each type of defect that occurs then the proposal can be given in an effort to minimize the defects are:
    - To maintain temperature stability, it can be added to the air conditioner or blower the press area and close the emergency door in the area in order to press the external temperature does not mix with the working environment temperature.
    - For the press operator to check the condition of the press jig press machines, to ensure that the press jig installed appropriately.
    - Provide standard sizing in the treble and bass bridge and rib in order to become more flat gluing using tools or using machinery.
    - Creating standard operating procedure (SOP) in the process of quality control in the rib and the sound board as the amount of sampling done, the pressure on the rib and perform quality control for solid sound board.
    - Reviewing the procedures of stacking rib in the warehouse and training to new employees more time to how or techniques work better.

    The future work lies in improving the sigma level to reach 3.2 DPMO by continuously improving on the maintain temperature stability that eliminate the defect.

References

[1] K. Srinivasan, S. Muthu, N. K. Prasad, and G. Satheesh, "Reduction of paint line defects in shock absorber through Six Sigma DMAIC phases," *Procedia Eng.*, vol. 97, pp. 1755–1764, 2014.

[2] R. W. Hoerl, "Six Sigma and the future of the quality profession," *IEEE Eng. Manag. Rev.*, vol. 26, pp. 87–94, 1998.

[3] P. Jirasukprasert, J. Arturo Garza-Reyes, V. Kumar, and M. K. Lim, "A Six Sigma and DMAIC application for the reduction of defects in a rubber gloves manufacturing process," *Int. J. Lean Six Sigma*, vol. 5, no. 1, pp. 2–21, 2014.

[4] J. Antony and R. Banuelas, "Key ingredients for the effective implementation of Six Sigma program," *Meas. Bus. Excell.*, vol. 6, no. 4, pp. 20–27, 2002.

[5] C. Sahay, S. Ghosh, and P. K. Bheemarthi, "Process improvement of brake lever production using DMAIC (+)," in *ASME 2011 International Mechanical Engineering Congress and Exposition*, 2011, pp. 801–826.

[6] T. N. Desai and R. L. Shrivastava, "Six Sigma--a new direction to quality and productivity management," in *Proceedings of the World Congress on Engineering and Computer Science,*

IOP Conf. Series: Materials Science and Engineering **215** (2017) 012035 doi:10.1088/1757-899X/215/1/012035

San Francisco, CA. Elsevier, 2008.

[7] Y. Zare Mehrjerdi, "Six-Sigma: methodology, tools and its future," *Assem. Autom.*, vol. 31, no. 1, pp. 79–88, 2011.

[8] N. Roth and M. Franchetti, "Process improvement for printing operations through the DMAIC Lean Six Sigma approach: a case study from Northwest Ohio, USA," *Int. J. Lean Six Sigma*, vol. 1, no. 2, pp. 119–133, 2010.

[9] G. Eckes, *The Six Sigma revolution: How General Electric and others turned process into profits*. John Wiley & Sons, 2002.

[10] C. Seow, T. Pfeifer, W. Reissiger, and C. Canales, "Integrating six sigma with quality management systems," *TQM Mag.*, vol. 16, no. 4, pp. 241–249, 2004.

A Comparative Study to Predict Student's Performance Using Educational Data Mining Techniques

Annisa Uswatun Khasanah[1)], Harwati [2)]

[1,2)]Department of Industrial Engineering, Faculty of Industrial Technology, Islamic University of Indonesia, Jl. Kaliurang Km 14.5 Yogyakarta, Indonesia
145220102@uii.ac.id[1)], harwati@uii.ac.id[2)]

Abstract. Student's performance prediction is essential to be conducted for a university to prevent student fail. Number of student drop out is one of parameter that can be used to measure student performance and one important point that must be evaluated in Indonesia university accreditation. Data Mining has been widely used to predict student's performance, and data mining that applied in this field usually called as Educational Data Mining. This study conducted Feature Selection to select high influence attributes with student performance in Department of Industrial Engineering Universitas Islam Indonesia. Then, two popular classification algorithm, Bayesian Network and Decision Tree, were implemented and compared to know the best prediction result. The outcome showed that student's attendance and GPA in the first semester were in the top rank from all Feature Selection methods, and Bayesian Network is outperforming Decision Tree since it has higher accuracy rate.
Keywords: student performance, prediction, feature selection, educational data mining

1. Introduction

Data mining is the process of automatically discovering useful information in large data repositories [1]. Data Mining techniques are employed to discover meaningful pattern and rules from large quantities of data that unknown. There are many techniques, methods and rules in Data Mining to extract particular database. Recently, data mining has been widely applied in many applications, such as customer relationship, engineering, medical, marketing, education, etc. There are increasing number of research that used data mining in education, and data mining technique that applied in this filed can be usually called as Education Data Mining. Education Data Mining can be conducted with many methods such as Decision Tree, Neural Network, Naïve Bayes, K-Nearest neighbour, and many other [2].

Education Data Mining has been commonly applied in many universities. Implementing Education Data Mining can help universities to predict their student's performance. Universities today are operating in high competitive environment and have very complex system. It is very important for modern universities to deeply analyse their performance in order to develop strategy and future actions. University management must consider their student's performance as one of important key of the marketing campaign and approach the promising potential of the university [3]. Predicting student's performance can help to predict the weak student and help the university management to make strategy and decision making related to student's performance improvement. The university management can take early step to prevent student fail or student drop out. After knowing the prediction, it is also expected to the weak student that they can improve their performance and achieve better score. Student's performance can be applied based on several factors like personal, psychological, social, and other environment [4][5]. This kind of analysis can help to predict the weak student and help the university management to make strategy and decision making related to student's performance improvement. The university management can take early step to prevent student fail or student drop out. After knowing the prediction, it is also expected to the weak student that they can improve their performance and achieve better score.

Among the major data mining techniques, one of the oldest and most useful technique is classification [6][7]. Tan et al [1] define classification as the task of assigning objects to one of several predefined categories. It uses a set of pre-classified example to build a model that can classify the large record population [8]. Classification methods has been commonly use to predict student's performance [2][3] [5] [6]. Bayesian Network and Decision Tree are two popular classification methods that have been used by many scholars to conduct the student's performance predictions [3][9][10][11][12][13] [14]. Those algorithms are powerful classification algorithm and the rules are easy to be interpret.

This current study will compare Bayesian Network and Decision Tree to whether student will drop out (DO) or not in Department of Industrial Engineering Universitas Islam Indonesia based on several attributes. Feature selection was conducted first before implementing the classification algorithm to select high influence attributes. This study aims to know the best scenario to predict student's performance in Department of Industrial Engineering Universitas Islam Indonesia by conducting feature section and comparing Bayesian Network and Decision Tree. The rest of this paper is organized as follows, Section 2 present the literature review, Section 3 presents the methods research related to this study, Section 4 shows the results and discussion. The concluding is finally made in Section 5.

2. Literature Review

2.1. Bayesian Network

Bayesian Network is as a statistical principle to combine prior knowledge of the classes with new evidence accumulated from the data [1]. The mathematical equation for this method is shown in this equation (1).

$$P(Y|X) = \frac{P(X|Y)P(Y)}{P(X)} \tag{1}$$

X and Y are a pair random attributes. The join probability of X and Y, P(X = x, Y=y), is the probability that variable X will admit a particular value x, while variable Y will admit the value y. A conditional probability is the probability that a random variable will take on particular value given that the outcome for another random variable is known.

There have been many studies that implement Bayesian Network as classification method. Bharadwaj and Pal [5] worked on a study to predict performance of students. The object in this study is 600 students from five different degree colleges in Computer Application course for Bachelor degree, Dr. R. M. L. Awadh University Faizabad, India. This study use Bayesian Network as the classification method and involve 17 attributes. The results show that student's grade in senior secondary exam, living location, teaching media, mother's qualifications, student's other habit, annual income of the family and family status of the student were highly correlated with academic performance of the student.

As well as Bharadwaj and Pal [5], Pandey and Pal [15] also conduct a study using Bayesian Network on category, language, and background. This study concluded that whether new comer students will performer or not. Millian et al [12] also try to use Bayesian Network for student model engineering.

Ramaswami & Rathinasabapathy [16] also used Bayesian Network approach in the field of education to predict student's performance. They conduct an analysis on student's data of the higher secondary schools including 5650 records with 35 attributes, and the class data were varying in five cases. In this study, they implement feature selection before applying the prediction models. The data analysis for this study was conducted using software WEKA. This study used a simple select estimator algorithm, like Hill Climbing, K2, LAGD Hill Climbing, Repeated Hill Climbing, Tabu Search and Network Augmented with Tree, to find the conditional probability tables of the Bayes Network. From the result, it can be concluded that Bayesian Network models with Network Augmented with Tree search algorithm was outperform over others types of search algorithms for all class values types. The result also revealed that the class variable HScGrade highly related with the grade that gained at secondary level, type of transportation to school, medium of instruction, sibling structure (number of brothers and sisters) and economic status of the family.

2.2. Decision Tree

Decision Tree is another classification popular method. Decision tree, as shown in Fig. 1, is as a flow-like tree structure. The tree consists of root node (the first node denoted by oval), internal nodes which can have two or more kinds (the middle node denoted by oval), and leaf nodes (denoted by rectangle). All internal nodes contain splits, which test the value of an expression of the attributes.

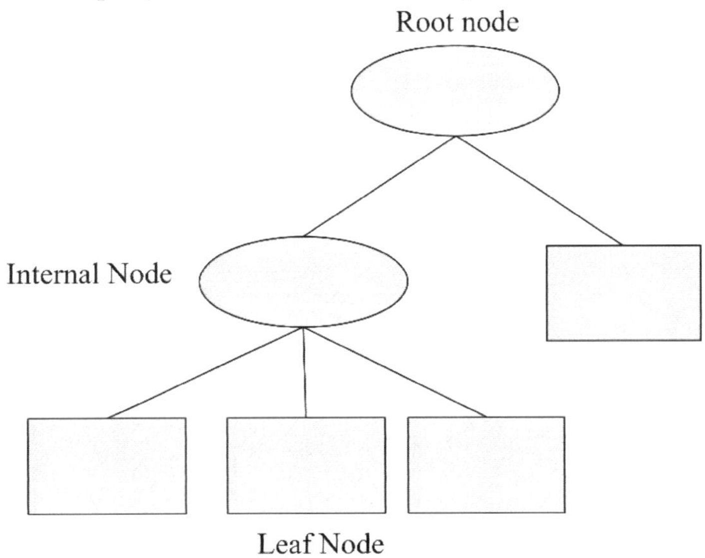

Fig 1. Decision Tree [1]

Yadav & Pal [2] implemented Educational Data Mining Techniques to make a prediction model of engineering student's performance. The CART, ID3, and C4.5 decision tree algorithms were implemented to predict their final exam performance. The result of the decision tree can be used to predict whether the students will be likely to pass, failed or promoted to next year. The dataset in this study was collected for session 2010 from VBS Purvanchal University in Jaunpur (Uttar Pradesh), Institute of Engineering and Technology. Initially data consists of 90 students. The variables used in this study were 13 and for the class attributes final exam results was used. The data was performed in software WEKA. The result showed even though CART, ID3, and C4.5 algorithms gave an acceptable level of accuracy, C4.5 technique outperformed the other with accuracy of 67.7778% compared to other methods.

Yadav et al [12] also implemented decision tree (CART, ID3, and C4.5) to predict academic performance of students in Computer Applications department of course Master of Computer Applications from session 2008 to 2011 in VBS Purvanchal University Jaunpur (Uttar Pradesh), India. Initially size consists of 48 records. The variable that used were marks of previous semester, class test grade, seminar performance, assignment, attendance, lab work, and for the class variable, marks of end semester (first, second, third, fail) were used. The result indicated that a CART was the best with accuracy of 56.25% compared to other techniques, however ID3 algorithm was also showed an acceptable level of accuracy [2][5][14].

As well as previous author, Ahmed & Elaraby [8] also tried to implement Educational Data Mining to conduct student's performance analysis. The decision tree (using ID3 algorithm) method was applied in this study. The dataset was collected from Department of Information System student's database from session 2005 to 2010, and the initial data consists of 1547 records. The variables that included in this study were department of the students, high school degree, midterm marks, lab test grade, seminar performance, assignment, measure of student participate, homework, and the target class is final grade mark. Midterm mark has the biggest gain; then, it was used as the root node. This study concluded that by conducting student performance analysis can help the students to improve performance of the student. In addition, it can identify students who need a special attention to minimize failing ration and take suitable action at right time.

2.3. Bayesian Network compare with Decision Tree

There have been many scholars that try to compare the performance of Bayesian Network and Decision Tree in student's performance prediction. Al-Radaideh et al [9] tried to predict the performance of Technology & Computer Science Faculty, Yarmouk University, Jordan students who took the Programming I course (C++). This paper also tried to compare ID3 algorithm, C4.5 algorithms and Naïve Bayes. There were 12 attributes, included gender, age, department, high school major, high school grade, study type, funding, place of residency, lecturer degree, lecturer gender, lecturer department, number of repetition and the grade of the C++ course (A, B, C, D). WEKA was used to analyse the data. The high school grade had the highest gain ration, then this attribute was considered as the root node of the decision tree. From the result it can be concluded that the classification accuracy for the three different classification algorithms was not so high. It indicated that the samples and attributes that used in this research were not enough to generate a high quality classification model.

Kabakchieva [3] presented the initial results of a research which was conducted in Bulgarian university. The main goal of the research conducted by Kabakchieva [3] was aimed to know if there are any patterns in the collected data which sufficient to predict performance of student at the university based on student's personal characteristics and pre-university characteristics. Nearest Neighbour classifier, Decision tree classifier, and Bayes classifiers were used to analyse the data. The analysis was performed in software WEKA because it is widely implemented for research purposes in the data mining and it is freely available to the public. The initial data included 10330 students and 20 attributes. The final results showed that the decision tree classifier (J48) was outperform other classifier because it has the highest accuracy, then followed by the rule learner (JRip) and the kNN classifier, while the Bayes classifiers are less accurate than the others. Nevertheless, all of the classifiers were showed an overall accuracy below 70 % which means that the error rate was high and the predictions were not able to be trusted.

3. Research Method

This study consists of three steps. The first step is data cleaning, the second step is feature selection and the last step is classification analysis. Data cleaning is important step that must be conducted before analysing the data to identify missing value, noisy and inconsistent data. The second step is Feature Selection. Feature selection algorithm can determine as the combination of a search technique to propose new feature subsets, along with an evaluation measure which scores the different feature subsets [17]. The main objective of feature selection is to select a subset of input attributes by eliminating features, which are irrelevant or give less predictive information [16]. This step is necessary to be applied in order to define the high influencing variables in student's performance prediction. Anuradha and Velmurugan [17] as well as Ramaswami and Bhaskaran [16] used several methods to conduct the feature selection in their study, and the five popular filter feature subset methods with Rank search or Greedy search method are Correlation-based Attribute evaluation, Gain-

Ratio Attribute evaluation, Information-Gain Attribute evaluation, Relief Attribute evaluation, and Symmetrical Uncertainty Attribute. Those methods were also used in this current study.
The last step was conducting the classification analysis using Bayesian Network and Decision Tree. The student's performance parameter that is used in this study is whether the student will be DO or not. The performance parameter to compare both algorithm that used in this study is Accuracy Rate. For the purpose of conducting the experiments, WEKA which is freely open source data mining software was used to implement the algorithm.

4. Result and Discussion

4.1 Data Cleaning

Data in this research were collected from student data base that can be accessed from Universitas Islam Indonesia's information system (UNISYS). The initial data consisted of 178 data of student in academic year 2007 and after cleaning the data, it is only 104 data set available with 12 attributes (including the class attribute, drop out or not) where 13 data (12,5%) classified as drop out, and 91 data (87,5%) classified as not. The attributes are divided into four categories and shown in Table 1.

Table 1. Student data set description

Category	**Attributes**
Personal Information	Gender (GE) Origin (OR)
Family information	Father education (FE) Father occupation (FO) Mother education (ME) Mother occupation (MO)
Pre-university characteristic	Senior high school type (ST) Senior high school department (SD) Senior high school final grade (SF)
University features	first semester Attendance (AT) first semester GPA (GPA) Drop out (DO) or Not (as the Class attribute)

4.2 Feature Selection

Feature Selection revealed the 12 attributes that have high influence with student's performance. The Feature Selection results from the five methods are shown in Table 2.

Table 2. Feature selection result

Feature Selection Method	**Selected Attributes**
Correlation-based Attribute evaluation	MO, **AT, GPA**
Gain-Ratio Attribute evaluation	**GPA, AT**, SD, GE, FO, MO, ME, OR, **ST, FE, SF**
Information-Gain Attribute evaluation	**GPA, AT**, SD, GE, FO, MO, ME, OR, **ST, FE, SF**
Relief Attribute evaluation	**AT, GPA**, FO, SD, **FE**, MO, GE, ME, OR, **ST, SF**
Symmetrical Uncertainty Attribute	**GPA**, AT, SD, GE, FO, MO, ME, OR, **ST, FE, SF**

Correlation-based Attribute evaluation is a Greedy search method that only select the attribute that have high influence with the prediction attribute. Based on Correlation-based Attribute evaluation, only three attributes (MO, AT, GPA) that have high influence with student's performance. The other selection methods are Rank search method that rank the attributes from the highest influencing until the less influencing attributes. Gain-Ratio Attribute evaluation, Information-Gain Attribute evaluation, and Symmetrical Uncertainty Attribute have exactly the same results, while Relief Attribute evaluation have slightly different result. From the whole result it can concluded that two attributes that appear in the first list in almost all selection methods are GPA and AT. It means that first semester GPA and attendance have high influence whether the student will DO or not. For the less high influence attribute the four Rank search method show identical result. ST, FE, SF are in the bottom three of the rank from Gain-Ratio Attribute evaluation, Information-Gain Attribute evaluation, and Symmetrical Uncertainty Attribute, while in Relief Attribute evaluation FE is in the middle rank. Based on this result, the attributes that would be used in further classification analysis are GPA, AT, SD, GE, FO, MO, ME, OR because most of the methods show the same result. This classification analysis would be compare with the analysis that included FE attributes as shown from Relief Attribute evaluation result.

4.3 Classification Analysis

In the Classification analysis three scenarios to compare the performance between Bayesian Network and Decision Tree algorithm to predict whether student DO or Not were conducted. The first scenario used all the attributes, the second scenario compared both algorithms using selected attributes from the Feature Selection result included FE (GPA, AT, SD, GE, FO, MO, ME, OR, FE), and the last scenario only used the most selected attributes (GPA, AT, SD, GE, FO, MO, ME, OR). Table 3 shows the Classification result from each scenarios.

Table 3. Classification results

Scenario	Attributes Used	Classification Algorithm	Accuracy Rate	Number of Incorrectly Classified Instance
1	All	Bayesian Network	95,19%	5
		Decision Tree	93,27%	7
2	GPA, AT, SD, GE, FO, MO, ME, OR, FE	Bayesian Network	97,11%	3
		Decision Tree	94,23%	6
3	GPA, AT, SD, GE, FO, MO, ME, OR	Bayesian Network	**98,08%**	**2**
		Decision Tree	94,23%	6

From the experimental result, it can be seen that within the same algorithm, scenario 1 that include all the attributes has lower accuracy rate than scenario 2 and 3 that only include selected attributes from Feature Selection. Then, it can be proved that Feature Selection can improve the accuracy of prediction model. It also can be seen that scenario 2 has lower accuracy rate that scenario 3. It means that it is better to exclude attribute FE from the analysis. For the classification algorithm comparison, it can be seen that from all scenarios Bayesian Network indicate outperform Decision Tree, and from the whole experiment, prediction with Bayesian Network using selected attributes GPA, AT, SD, GE, FO, MO, ME, OR is the best among the others.

5. Conclusion and Suggestion

Student performance prediction is essential to be conducted for a university to prevent student fail. Number of student drop out is one of parameter that can be used to measure student performance and one important point that must be evaluated in university accreditation in Indonesia. Of course this number must be minimized. Some high influence attributes in student prediction that can be considered by university management to minimized student drop out based on the result of this study are First Semester GPA, First Semester Attendance, Senior High School Department, Gender, Father Occupation, Mother Occupation, Mother Education, and Origin. It also can be proved from this study that feature selection can increase the accuracy rate of the prediction, and Bayesian Network is outperforming Decision Tree since it has higher accuracy rate. Then, the best scenario to predict whether the student will drop out or not in Department of Industrial Engineering Universitas Islam Indonesia is using the high influence selecting attributes from feature selection and by implementing Bayesian Network as the classification algorithm.

For future study, it is better if attribute Gender and Origin can be evaluated based on social aspect to make sure that this attributes are not bias and not containing discrimination aspect.

6. Reference

[1] Tan, P. N., Steinbcah, M, Kumar, V., (2006) *Introduction to Data mining, Introduction to Data mining*. MA: Pearson Education.
[2] Yadav, S. K. and Pal, S., (2012) ,Data Mining : A Prediction for Performance Improvement of Engineering Students using Classification, *World of Computer Science and Information Technology Journal WCSIT*, 2(2), 51–56.
[3] Kabakchieva, D., (2013), Predicting student performance by using data mining methods for classification, *Cybernetics and Information Technologies*, 13(1), 61–72.
[4] Hijazi, S. and Raza-Naqvi, S. M., (2006), Factors Affecting Students ' Performance, 3(1), 1–10.
[5] Bhardwaj, B. K. and Pal, S., (2011). Data Mining : A prediction for performance improvement using classification, *(IJCSIS) International Journal of Computer Science and Information Security,* 9(4).
[6] Romero, C., Ventura, S., Espejo, P. G. and Hervás, C., (2008), Data Mining Algorithms to Classify Students, *EDM*, 8–17.
[7] López, M. and Luna, J. M., (2012), Classification via clustering for predicting final marks based on student participation in forums, *Proceedings of the 5th International Conference on Educational Data Mining*, pp. 4–7.
[8] Ahmed, A. B. E. D. and Elaraby, I. S., (2014), Data Mining: A prediction for Student's Performance Using Classification Method, *World Journal of Computer Application and Technology*, 2(2), 43–47.
[9] Al-Radaideh, Q. A., Al-Shawakfa, E. M. and Al-Najjar, M. I., (2006), Mining Student Data Using Decision Trees, *International Arab Conference on Information Technology*.
[10] García, P., Amandi, A., Schiaffino, S. and Campo, M., (2007), Evaluating Bayesian networks' precision for detecting students' learning styles, *Computers and Education*, 49(3), 794–808.
[11] Dekker, G. W., Pechenizkiy, M. and Vleeshouwers, J. M., (2009), Predicting students drop out: A case study, *EDM'09 - Educational Data Mining 2009: 2nd International Conference on Educational Data Mining*, 41–50.
[12] Millán, E., Loboda, T. and Pérez-De-La-Cruz, J. L., (2010), Bayesian networks for student model engineering, *Computers and Education*. Elsevier Ltd, 55(4), 1663–1683.
[13] Kumar, S. A. and Vijayalakshmi, M. D., (2011), Efficiency of Decision Trees in Predicting Student' Academic Performance, © *Cs & It-Cscp 2011*, 335–343.
[14] Yadav, S., Bharadwaj, B. and Pal, S., (2012), Data mining applications: A comparative study for predicting student's performance, *International Journal of Innovative Technology & Creative Engineering*, 1(12), 13–19.
[15] Pandey, U.K. and Pal, S., (2011), Data Mining: A prediction of performer or underperformer using classification, *International Journal of Computer Science and Information Technologies*, 2 (2) , 686-690
[16] Ramaswami, M. and Bhaskaran, R., (2009), A Study on Feature Selection Techniques in Educational Data Mining, 1(1), 7–11.
[17] Anuradha, C. and Velmurugan, T.,(2016), Feature Selection Techniques To Analyse Student Acadamic Performance Using Naïve, *The 3rd International Conference on Small & Medium Business*, 345–350.

IOP Conf. Series: Materials Science and Engineering **215** (2017) 012037 doi:10.1088/1757-899X/215/1/012037

Measuring the Success of E-Cargo Implementation at One of Indonesian Airlines using DeLone and McLean Model

Ruth Monika[1], **Ford Lumban Gaol**[2]

[1] Master of Information System of Bina Nusantara University, Indonesia
[2] Doctor of Computer Science of Bina Nusantara University, Indonesia

E-mail: [1] ruthmonika.sg@gmail.com , [2] fgaol@binus.edu

Abstract. This research is aimed at measuring the success of e-cargo implementation in Indonesia. There are three analyzed districts: Tanjung Pandan (TJQ), Palembang (PLM), and Jakarta (CGK). The research uses information system's success model proposed by Delone and McLean. The research will analyze the factors that measure the success of using DeLone & McLean model of success information systems. Data from 53 respondents were collected through a survey, then the author created a path with the Partial Least Squares (PLS) path analysis using the software SmartPLS (v.3.2.6). The results of this study proved that the information quality variable did not significantly affect use variable, the information quality variable had no significant effect on the user satisfaction variable and the service quality variable had no significant effect on the use variable, while the other variables were tested significant in measuring the success of the use of information systems with R-square value that has a moderate effect of user satisfaction variable by 0.577, the use variable by 0.396, and a net benefit variable by 0.615.

1. Introduction

1.1. Background

Nowadays, information system has integrated and eventually becomes an integral part of company's running. Information system is used to support any operational process at the airline. Thus, almost all of working operational process of the airline is now depended on the information technology. In creating an integrated information system, it must be ensured that the information system already created and implemented benefits the company so that the company's operational processes work faster, more accurate. In addition, information systems can support the achievement of the company's strategy, so that said company can be competitive in the business competition.

In 2010, the airline had expanded its market reach by providing air cargo service, even though still done manually. Subsequently, the airline managed to utilize its information technology by expanding its limit from merely on e-commerce (reservation system or B2C) to air cargo by establishing e-cargo system. The implementation of e-cargo is triggered by the need for efficiency, effectiveness and economic reasons in cargo operations. But in fact, the implementation of e-cargo in the form of replacement from manual operation to automated operation encountered its own obstacles. One reason is because employees as users (end users) are less able to adapt to functioning e-cargo since they have long been accustomed to manage the system manually. To solve it, there needs to do the introduction of workflow for e-cargo systems, user acceptance test (UAT), training to the concerned division, then it is ready to perform the migration by way of conversion of pilots and when the pilot conversion is successful, then it has to deploy live implementation by conversion phases (phase-in conversion) [1] to 43 districts of the airline's routes in Indonesia.

Implementation stages of the pilot conversion and phase-in conversion of system e-cargo have been carried out in the Airport Ticketing Office (ATO) of Tanjung Pandan with a three-letter code TJQ, Palembang with three letter code PLM, Jakarta with a three letter code CGK which are some of cargo routes from districts owned by the airline. Therefore, there must be a measure of the success of this e-cargo system whether it is already in line or not with the needs of companies. To measure the success of an information system in a company, there are companies using measurement in financial terms, which is return of investment. The researchers [2] also have made a success measurement model information system that emphasizes the need for better and more consistent success metrics. One way of measuring the success of information systems model is DeLone and Mclean information system success model [3]. In this study [3], there are six variables: the system quality, information quality, service quality, use, user satisfaction and net benefit. DeLone and McLean Model is considered to measure the complexity of information systems because each of the variable is not independent but all depend on each other (interdependent) and interrelationship. The mutual connectedness between variable and this system means that the ultimate goal of net benefit will not increase if the five other variables are not perceived by the user, meaning that the five other variables must be developed to achieve improved net benefit perceived by the users.

1.2 Conversion System

Conversion system is a stage that is used to operate the new system in order to replace the old system or the process of changing from the old to the new system [4]. The level of difficulty and complexity of the conversion from the old system to the new system depends on a number of factors. If the new system is a software package wrapped (canned) that will run on the new computer, then the conversion will be relatively easier but if the conversion utilizing the new customized software, new databases, computers and new control software, the new network and a drastic change in the procedure, the conversion becomes rather difficult and challenging, such factors on activity data collection and new generated data implemented, and developing interface mechanism of the old system. This interface allows the old system operates with the new input data. Then the activities of the new database access, storage, and calling are implemented and another segment of the system being implemented. Hence, it has the advantages of speed changes in a particular company that can be minimized, and the sources of the data processing can be acquired gradually over a period of extensive time. Moreover, it also provide a challenge because it has the disadvantage that the purpose of the cost that must be established to develop a temporary interface with older systems, limited applied power, and setbacks in the spirit of the organization since people feel that they never be able to complete the system.

1. Direct Conversion
   The conversion is done by ending the old system and replace it with a new system [4].
2. Parallel Conversion
   In this conversion, the new system and the old system are equally run. After a certain period, if the new system has been acceptable to replace the old system, the old system is immediately terminated [4].
3. Phase-In Conversion
   Conversion is done by replacing a part of the old system with a new system [4].
4. Pilot Conversion
   This approach is done by implementing a new system only in specific locations that are treated as a pioneer [4].

1.3 DeLone and McLean Model

DeLone and McLean [3] defined six different variables of IS success: system quality, information quality, users, user satisfaction, the net benefit as in the Figure 1.

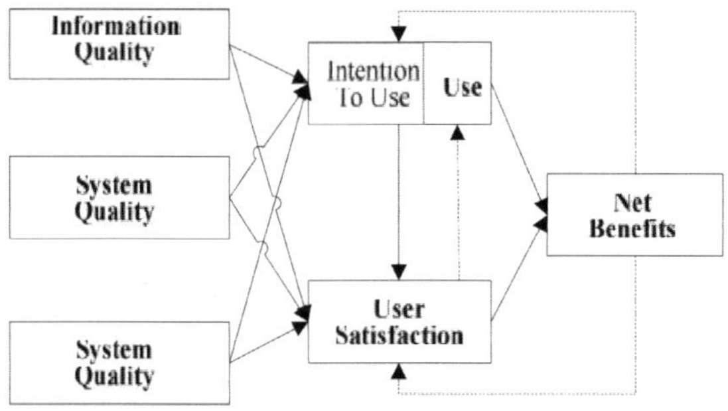

Figure 1. DeLone and McLean Success Model
Source: DeLone and McLean Model [3]

1.4 Research Questions

Seen from the success' perspective of information system with DeLone and McLean model as the basis, the determiner of the condition of the system and the behavior of the users of e-cargo can be identified their inter-variable relationships from six variables: system quality, information quality, service quality, use, user satisfaction, and net benefit. Thus, the problem of the research is:

1. Migration of manual cargo into e-cargo using a pilot conversion method and phase-in conversion. What does the result from the quality system, the information quality, service quality on the e-cargo affect the use and user satisfaction?
2. If the user behavior is the use and user satisfaction, do the use and user satisfaction affect the net benefit?

1.5 Research Goals

To measure the factors that affect the successful implementation of the airline e-cargo using six variables on DeLone and McLean model on information systems success.

2. Research Method

2.1. Data Collection Method

This research was conducted at the cargo division in the districts of Tanjung Pandan (TJQ), Palembang (PLM), and Jakarta (CGK). The process of distributing the questionnaire for data collection was conducted from September 3, 2016 to January 20, 2017. The population of the research is the users of e-cargo at the analyzed airline, particularly those of cargo division. The data collection of the research was conducted by circulating questionnaire.

2.2. Research Variables

2.2.1 Independent Variables (Exogenous Variabel)

The independent variables in this study are the quality system, the quality of information and the service quality. Further explanation of these three variables will be described below.

1. System quality
   This variable assesses whether the characteristics possessed by the system has been in accordance with the users. This variable is measured using the items of navigation, design, usability, functionality, responsiveness, and availability.
2. Information quality
   Items that need to be taken into account in calculating the information quality are understandable, accuracy, usability, attractiveness, reliability, completeness and timeliness.
3. Service quality
   Variable measurement of this variable uses the items of responsiveness, perceptive, empathy, assurance and training.

IOP Conf. Series: Materials Science and Engineering **215** (2017) 012037 doi:10.1088/1757-899X/215/1/012037

2.2.2 Intervening Variables

Intervening variables are ones that arise when the independent variables affect the dependent variable. Intervening variables in the IS Succes model [3] are as follows.

1. Use

   This research uses 'use variable; as an intervening variable. The objective of the use is the attainment use of e-information system capabilities cargo for cargo staff who use it while the measurement of variables using daily use, and the level of use of the features.

2. User satisfaction

   According to [3], user satisfaction is the users' response to the use of information system output. User satisfaction plays an important role to determine the responses of users of information systems toward applied information system. User satisfaction will increase if there is a match between the expected to the output of the information. Measurement of this variable applies adequacy, efficiency, effectiveness, and satisfaction.

2.2.3 Dependent Variables (Endogenous Variables)

1. Net benefit

   Net benefit is a measure of the contribution of information systems towards users of the system. Measurement of this variable applies performance, achievement, productivity, effectiveness, simplicity, usefulness or benefits.

2.3. Framework of thinking

Framework of thinking is a chart that illustrates the flow of thought in doing this research can be seen in the figure 2.

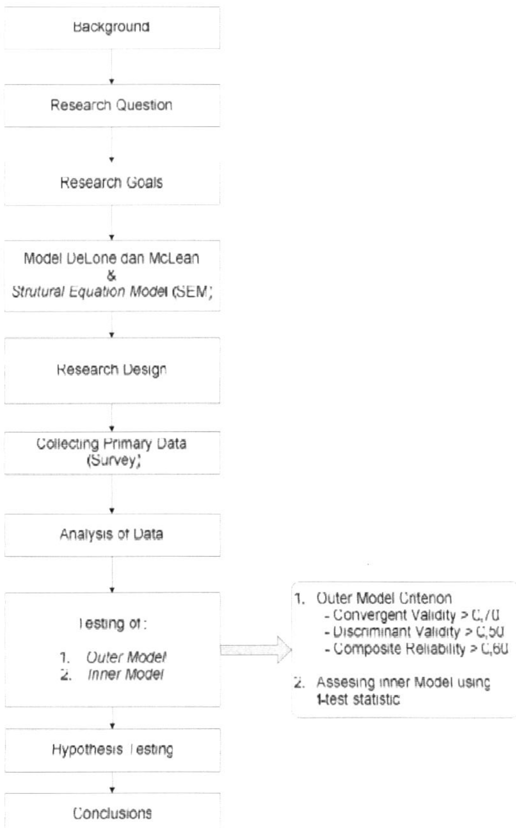

Figure 2. Framework of Thinking

2.4. Theoretical Framework and Hypothesis

2.4.1 Theoretical Framework

Theoretical framework model is based on the theory from DeLone and McLean model on information systems success [3] The relationship between these variables can be seen in figure 3.

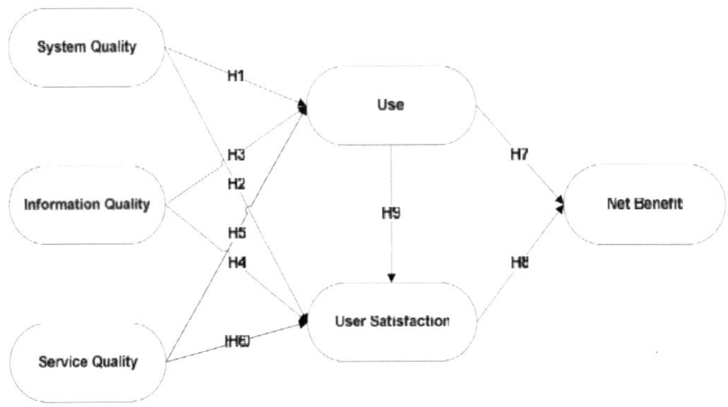

Figure 3. Theoretical Framework
Source: DeLone and McLean Model [3]

2.4.2 Hypothesis

From the DeLone and McLean model above, a proposed initial hypothesis is obtained and detailed as follows:

H1 : System quality (KS) affects the use (PE)
H2: System quality (KS) affects the use satisfaction (KP)
H3: Information quality (KI) affects the use (PE)
H4: Information quality (KI) affects the user satisfaction (KP)
H5 : Service quality (KL) affects the use (PE)
H6 : Service quality (KL) affects the user satisfaction (KP)
H7 : Use (PE) affects net benefits (MB)
H8: User satisfaction (KP) affect the net benefits (MB)
H9: Use (PE) affects user satisfaction (KP)

2.4.3 Conceptual Framework

The conceptual framework used in this study is explained in Figure 4.

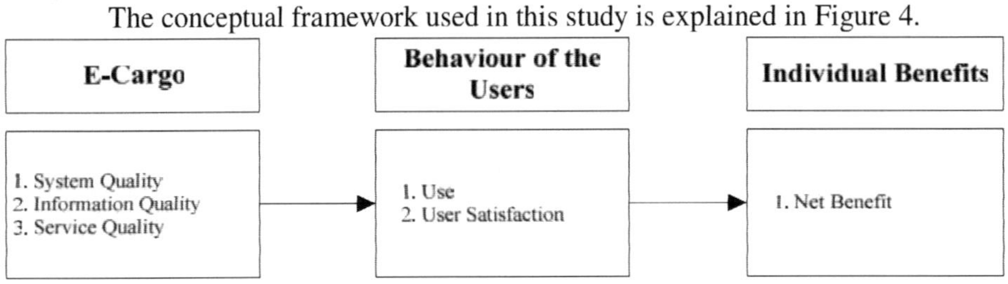

Figure 4. Conceptual Framework

The purpose of the model in figure 4 is to measure the successful implementation of e-cargo, and the first thing to do is to assess the condition of the e-cargo system. E-cargo condition assessment is determined by variables such as the quality of the system, information quality, and quality of service that it generates [5]. The existence of the e-cargo will affect the behavior of the user, and to study the behavior of users, it is examined from the use variable and user satisfaction variable, so that user behavior will affect the net benefit variable.

2.4.4 Data Analysis Technique

Analysis of the data used in this study is a measurement model test (validity and reliability) followed by Structural Equation Model (SEM) using software SmartPLS (v.3.2.6), which is testing the relationship between the variables and hypothesis testing. Structural equation model used in this study can be seen in table 1.

Table 1. Classification Variables and Indicators

No	Laten Variables Name	Symbol	Indicators
1	System Quality	KS1	Navigation
		KS2	Design
		KS3	Usability
		KS4	Functionality
		KS5	Responsiveness
		KS6	Availability
2	Information Quality	KI1	Understandable
		KI2	Accuracy
		KI3	Attractiveness
		KI4	Reliability
		KI5	Completeness
		KI6	Timeliness
3	Service Quality	KL1	Responsiveness
		KL2	Perceptive
		KL3	Empathy
		KL4	Assurance
		KL5	Training
4	Use	PE1	Daily use
		PE2	Level of use
5	User Satisfaction	KP1	Adequacy
		KP2	Efficiency
		KP3	Effectiveness
		KP4	Satisfaction
6	Net Benefit	MB1	Performance
		MB2	Achievement
		MB3	Productivity
		MB4	Effectiveness
		MB5	Simplicity
		MB6	Usefulness/benefits

3. Result and Discussion

3.1. The Description of Research Result

3.1.1 The Identity of Research Result

The identity of these respondents were observed by gender, age, education, districts they lived in, the duration of using the e-cargo.

- Gender

  Respondents were most likely to be male, as many as 41 people (77%), and the rest of the respondent is sex women, as many as 12 people (23%). This indicates that most of the subjects of this study are male.

- Age

  Most respondents are still in the productive age; as many as 37 people or 70% are in the range of age 21-30 years, as many as 14 people or 26% are the range of age 31-40 years, and followed by the age range of 41-50 years as many as 2 or 4 %.

• Education
Based on past levels of education, the average respondent, which is the 92% or 49 people, have had a bachelor's degree (S1), whereas for diploma (D3), there is only a small portion that is 8% or 4 people from the total respondents.
• Districts
For the district where the distribution of the questionnaire is done, the average respondents, at most, 91% or 48 people are in the district of Jakarta (CGK), then 5% or 3 people are in the district of Tanjung Pandan (TJQ), while for the district of Palembang (PLM), at least 4% or 2 of total respondents scored the lowest.
• The Duration of Using the E-Cargo
Most of the respondents who had known e-cargo are seen from the data respondents who had known e-cargo less than 1 year are 53 people or 100%, which is caused by the newly implemented e-cargo, run for only seven months in the airline.

3.2. Descriptive Statistics
In Table 2. constructs on descriptive statistics indicate that the system quality variable has the empirical range of 6 to 10 with an average score (mean) of 8.15 and a standard deviation of 1.16. The average score (mean) on the system quality when higher than 0.15 compared to the median value, 8, indicates that the user feels that the system has good quality. The value of standard deviation indicates a deviation above the average value of the respondents' answers to questions about the system quality by 1.16.

Table 2. Constructs At Descriptive Statistics

Constructs	N	Min.	Max.	Mean	Median	Standard Deviation
System Quality (KS)	53	6	10	8.15	8	1.16
Information Quality (KI)	53	9	15	12.04	12	1.98
Service Quality (KL)	53	9	20	15.11	16	2.75
Use (PE)	53	6	10	8.15	8	1.26
User Satisfaction (KP)	53	12	20	16.49	16	2.68
Net Benefit (MB)	53	14	25	21.04	20	2.92

Source: Processed Primary Data, 2017

Information quality has the empirical range of 9 to 15 with an average value (mean) of 12.04 and a standard deviation of 1.98. The average value (mean) of the information quality that is 0.04 higher than the median value, 12, indicates that the users feel that the information quality has been good. The value of the standard deviation indicates a deviation above the average value of the respondents' answers to questions about the information quality by 1.98.

The service quality has the empirical range of 9 to 20 with the average value (mean) of 15.11 and a standard deviation of 2.75. Respondents considered the service quality provided is inadequate. This can be seen in Table 2. that the difference between the median and the average value (mean) is 0.89 with the average value (mean) of 15.11 compared to the median magnitude which reached 16.

Use has a range of 6 to 10 with an average value (mean) of 8.15 and a standard deviation of 1.26, which indicates the deviation from the average (mean) respondents. The median value, 8, which approaches the average value (mean) of 8.15 indicates that the user considers the use of the system strongly supports the acquisition of the benefits of the used system.

User satisfaction has a data range of 12 to 20 with an average value (mean) of 16.49 and a standard deviation of 2.68. The 2.68 value of standard deviation illustrates the deviation from the average value (the mean). The average value of user satisfaction is quite high, 16.49 higher than the median value, 16, which shows the user satisfaction of the e-cargo.

The net benefit has the empirical range of 14 to 25 with an average value (mean) of 21.04 and the 2.92 standard deviation. The net benefits also has attained a positive opinion, it is seen from the 1.04 difference between the average value (mean) and the median. The 20 median value would be lower than the value of the 21.04 average value (mean).

3.3. The Evaluation of Measurement Model or Outer Model

3.3.1 Validity Costruct

a. *Convergent Validity*

Convergent validity test is done by looking at the score of average variance extracted (AVE) and loading factor. A factor or indicator is valid when the loading score value is > 0.7 and can be tolerated up to 0.5 on the theory that is being developed, at the moment.

Table 3. Cross Loading (Invalid Indicators)

	User Satisfaction	Information Quality	Service Quality	System Quality	Net Benefit	Use
KI1	0.475	0.731	0.510	0.509	0.485	0.467
KI2	0.497	0.567	0.410	0.559	0.336	0.219
KI3	0.400	0.717	0.602	0.457	0.534	0.496
KI4	0.533	0.787	0.645	0.599	0.674	0.519
KI5	0.375	0.595	0.726	0.337	0.414	0.320
KI6	0.566	0.597	0.281	0.640	0.570	0.431
KL1	0.528	0.627	0.782	0.427	0.566	0.408
KL2	0.478	0.617	0.807	0.469	0.535	0.380
KL3	0.414	0.543	0.817	0.381	0.594	0.368
KL4	0.569	0.752	0.850	0.536	0.671	0.554
KL5	0.296	0.437	0.635	0.521	0.625	0.438
KP1	0.873	0.708	0.577	0.642	0.651	0.485
KP2	0.795	0.549	0.476	0.497	0.575	0.502
KP3	0.786	0.454	0.400	0.560	0.544	0.411
KP4	0.893	0.654	0.535	0.586	0.682	0.541
KS1	0.372	0.689	0.508	0.610	0.587	0.432
KS2	0.462	0.532	0.406	0.723	0.609	0.198
KS3	0.371	0.506	0.410	0.661	0.499	0.347
KS4	0.345	0.530	0.448	0.619	0.489	0.476
KS5	0.512	0.418	0.238	0.719	0.472	0.285
KS6	0.610	0.469	0.388	0.668	0.462	0.418
MB1	0.552	0.585	0.492	0.519	0.734	0.730
MB2	0.639	0.627	0.612	0.687	0.783	0.368
MB3	0.502	0.530	0.488	0.561	0.764	0.469
MB4	0.371	0.506	0.551	0.554	0.737	0.407
MB5	0.464	0.574	0.689	0.468	0.694	0.498
MB6	0.726	0.637	0.651	0.703	0.825	0.491
PE1	0.511	0.549	0.533	0.453	0.618	0.925
PE2	0.566	0.613	0.502	0.565	0.618	0.933

Source : Data Processing using SmartPLS (v.3.2.6)

Based on table 3. there are nine indicators, which indicate it should be removed because it is invalid, it is because the nine indicators have no sufficient cross loading value i.e., KS1, KS3, KS4 and KS6 each only has valuse of loading factor 0.610, 0.661, 0.619 and 0.668 from the the construct of the quality system (KS). Furthermore, KI2, KI5 and KI6 from the constructs of the information quality (KI), which only has factor loading values of 0.567, 0.595, and 0.597, then KL5 from the construct of service quality (KL) whose loading factor is 0.635. Lastly, one indicator of net benefits is the MB6 indicator which is one of six indicators of net benefits (MB). It only has a loading factor of 0.694 thus these indicators should also be removed.

b. *Discriminant Validity*

The research's validity test, later, tested the discriminant validity of the measurement model (outer model) with reflexive indicators assessed by cross loading measurements with the constructs. If the correlation between the constructs and the measurement item is higher than the other constructs, then it indicates that the latent constructs predict that the size of said block is better than other block sizes. This study shows that the entire constructs (variables) has a better correlation with the items themselves (indicators) than the other constructs items.

Table 4. Cross Loading (Valid Indicators)

	User Satisfaction	Information Quality	Service Quality	System Quality	Net Benefit	Use
KI1	0.477	0.804	0.558	0.293	0.465	0.468
KI3	0.398	0.743	0.597	0.265	0.497	0.496
KI4	0.531	0.834	0.629	0.366	0.635	0.519
KL1	0.529	0.628	0.819	0.328	0.525	0.408
KL2	0.477	0.575	0.803	0.166	0.464	0.380
KL3	0.412	0.480	0.799	0.140	0.576	0.367
KL4	0.569	0.750	0.881	0.335	0.655	0.554
KP1	0.873	0.614	0.587	0.563	0.654	0.485
KP2	0.791	0.456	0.457	0.353	0.559	0.502
KP3	0.789	0.316	0.422	0.450	0.558	0.412
KP4	0.893	0.568	0.554	0.439	0.685	0.541
KS2	0.463	0.456	0.362	0.886	0.610	0.198
KS5	0.513	0.265	0.200	0.917	0.499	0.284
MB1	0.552	0.631	0.482	0.354	0.723	0.730
MB2	0.641	0.574	0.570	0.613	0.789	0.368
MB3	0.501	0.393	0.421	0.429	0.809	0.469
MB4	0.371	0.458	0.492	0.420	0.748	0.408
MB5	0.726	0.534	0.644	0.553	0.834	0.491
PE1	0.510	0.557	0.505	0.232	0.580	0.924
PE2	0.565	0.598	0.475	0.270	0.619	0.934

Source: Data Processing using SmartPLS (v.3.2.6)

The validity test can be seen in Table 4. for instance, item KI1 has a 0.804 loading value on the construct of information quality (KI) which is higher than the construct of user satisfaction (KP), which is 0.477, the construct of service quality (KL), which is 0.558, the construct of quality systems (KS), which is 0.293, the construct of the net benefit (MB), which 0.465, and the construct of the use (PE), which is 0.468.

This proves that the item KI1 predicted that the size of the block KI is better than the other constructs. Something similar happened to other constructs that have a correlation value between the items with the construct itself higher than the other constructs.

Table 5. Latent Correlation Variable

	User Satisfaction (KP)	Information Quality (KI)	Service Quality (KL)	System Quality (KS)	Net Benefit (MB)	Use (PE)
(KP)	1.000					
(KI)	0.593	1.000				
(KL)	0.608	0.749	1.000			
(KS)	0.543	0.391	0.305	1.000		
(MB)	0.736	0.675	0.676	0.610	1.000	
(PE)	0.580	0.622	0.527	0.271	0.646	1.000

Source: Data Processing using SmartPLS (v.3.2.6)

Table 6. AVE and AVE Roots

	Average Variance Extracted (AVE)	Akar Average Variance Extracted (AVE)
User Satisfaction (KP)	0.702	0.837
Information Quality (KI)	0.631	0.794
Service Quality (KL)	0.683	0.826
System Quality (KS)	0.813	0.901
Net Benefit (MB)	0.611	0.781
Use (PE)	0.863	0.928

Source: Data Processing using SmartPLS (v.3.2.6)

The assessment of the validity of a construct also can be done by looking at the value of Average Variance Extracted (AVE). It is done by comparing the root of AVE with the correlation between the constructs. Based on the research results, it can be concluded that the root value AVE is higher than the correlation between one another construct, meaning that the whole construct in the model estimated meets the criteria of discriminant validity. The service quality (KL) variable has a root value AVE of 0.826. This value is higher than the correlation between the variables of user satisfaction (KP) with a value of 0.608 and the information quality variable (KI) with a value of 0749. The whole other variables also showed that the value of AVE root is higher than that of the correlations, such as the net benefit variable (MB), which has its AVE roots of 0.781. Meanwhile, the value of correlation with the other variable, information quality (KI) variable, only reached 0.675. From the two variables, it can be concluded that the value of AVE root is higher than the value of inter-constructs, which means the model used in the study has proven its validity. The comparison between the value of AVE roots with the correlation between constructs can be seen in Table 5. and Table 6.

3.3.2 Costructs' Reliability

Constructs' reliability of measurement model (outer model) with reflexive indicators can be measured by looking at the composite reliability value of the block indicators that measure the construct. A construct is said to be reliable if the composite value reliability is > 0.7 with a received minimum score of 0.5 [6]. The value of the composite reliability can be seen in appendix table 7.

Table 7. Reliability and Validity Constructs

	Cronbach's Alpha	rho_A	Composite Reliability	Average Variance Extracted (AVE)
User Satisfaction (KP)	0.858	0.869	0.904	0.702
Information Quality (KI)	0.707	0.714	0.837	0.631
Service Quality (KL)	0.845	0.862	0.896	0.683
System Quality (KS)	0.771	0.784	0.897	0.813
Net Benefit (MB)	0.842	0.852	0.887	0.611
Use (PE)	0.841	0.844	0.926	0.863

Source: Data Processing using SmartPLS (v.3.2.6)

The results showed that the whole construct has a relatively high composite reliability value, overall is > 0.7. The lowest composite reliability value is on the information quality (KI) variable of 0.837 value while the highest composite reliability value is on the use (PE) variable with a value of 0.926. Constructs' reliability test used Cronbach Alpha too. The results of data processing to see Cronbach Alpha used SmartPLS which can be seen in Table 8.

The suggested Cronbach Alpha value is more than 0.6. The results showed all variables used in this study had a Cronbach Alpha values above 0.6. User satisfaction (KP) variable has the highest value of Cronbach Alpha which is 0.858 while the variable that has the lowest reliability value is information quality (KI) variable with a value of 0.707.

3.4. The Evaluation of Sructural Model or Inner Model

Testing of the structural model or inner model aims to determine the causality relationship between latent variables, significance values, and R-squared values of the research model. The structural model can be determined based on the R-squared value of the endogenous construct that can be explained by the exogenous construct. The dependent construct employs a t-test as well as significance of the structural path parameter coefficient. Evaluation of the structural model using PLS begins with examining the R-squared value of each latent dependent variable. The R-squared values estimated using SmartPLS are presented in table 8.

Table 8. Structural Model R-Square

	R-Squared
KI	
KL	
KP	0.577
KS	
MB	0.615
PE	0.396

Source: Data Processing using SmartPLS (v.3.2.6)

According to [6], an R-squared value by 0.67 indicates a substantial effect, an R-squared value greater than 0.33 indicates a moderate effect, and an R-squared value greater than 0.19 indicates a little or insignificant effect. Findings of this research reveal that use (PE) by 0.396 belongs to the moderate category, meaning that information quality (KI), service quality (KL), and system quality (KS) have an effect on use (PE) by 39.6%, as for the remaining effect by 60.4%, it is generated by variables other than those studied. The r-squared value of the variable net benefit (MB) by 0.615 belongs to the moderate category, meaning that information quality (KI), service quality (KL), and system quality (KS) have an effect on net benefit (MB) by 61.5% and the remaining effect by 38.5% is generated by variables other than those studied. The R-squared value of the variable User Satisfaction (KP) by 0.577 belongs to the moderate category, meaning that information quality (KI), service quality (KL), and system quality (KS) have an effect on user satisfaction (KP) by 57.7% and the remaining effect by 42.3% is generated by variables other than those studied.

3.4.1 Hypothesis Testing

The marks in testing the hypothesis must be consistent with the theory that was hypothesized, it can be assessed from the t-test through bootstrapping procedure. If the value of t-test is higher than 1.96 (significant at 5%), then it is a significant relationship. The value of t-test in the output path coefficients can be seen in Table 9.

Table 9. Path Coefficient

	Hypo thesis	Original Sample (O)	Sample Mean (M)	Standard Deviation (STDEV)	T Statistics (|O/STDEV|)	Signifikansi (α = 5%)
KS → PE	H1	0.031	0.031	0.136	0.227	Not significant, have a positive relationship
KS → KP	H2	0.352	0.343	0.100	3.514	Significant, have a positive relationship
KI → PE	H3	0.506	0.533	0.168	3.014	Significant, have a positive relationship
KI → KP	H4	0.030	0.022	0.163	0.184	Not significant, have a positive relationship
KL → PE	H5	0.138	0.120	0.160	0.862	Not significant, have a positive relationship
KL → KP	H6	0.322	0.334	0.160	2.015	Significant, have a positive relationship
PE → MB	H7	0.331	0.340	0.133	2.493	Significant, have a positive relationship
KP → MB	H8	0.544	0.539	0.117	4.650	Significant, have a positive relationship
PE→ KP	H9	0.296	0.304	0.130	2.277	Significant, have a positive relationship

Source: Data Processing using SmartPLS (v.3.2.6)

4. Conclusion

This research aims to measure the factors that affect the successful implementation of e-cargo using six variables of model of information system success created by DeLone and McLean [3]. Furthermore, it has been identified from a total of 9 hypothesis, there are six hypotheses that are empirically proven to achieve success.

From the results of the data with SmartPLS, there are two variables that will greatly affect the success of the implementation of e-cargo; quality system variable with a 3.514 value of t-statistic with the determinants of success that the system can display the route (from and to), date, price, the availability of lots of baggage on air cargo, the system can read the label cargo receipt (for the check-in process incoming and outgoing cargo), OR data (office receipt) financial statement on sales of cargo can be inspected and reported in realtime. Then the highest is user satisfaction variable with the value of t-statistic of 4.650 and the determinants of success is the satisfaction of users as they can easily make a reservation for the cargo, no need to write the manual for a cargo label, and when entering the goods into cargo to the aircraft, by means of a barcode scanning on the label, it does not need to be checked manually anymore because all these activities are already recorded realtime on the system, so users find the labor more quickly and efficiently.

Satisfaction in using the system had a significant effect on the net benefits, because satisfaction is a feeling of significant use by one's feelings of like or dislike (user) of the e-cargo. Thus, it can be interpreted that the quality of the e-cargo system has been so successful that the user has been satisfied and it has the benefit in an individual level as they use the system. Thus, the general model of DeLone & Mclean on information system success is a framework that can measure the successful implementation of e-cargo in Tanjung Pandan (TJQ), Palembang (PLM), and Jakarta (CGK).

Based on the research that has been done, some suggestions that can be submitted for future works are as follows.

- Information is in conformity with the application of existing theory, but the science will continue to evolve so it is necessary to develop management activities for the maintenance of the information system for the long term.
- Hopefully, the results of this study can be beneficial for the company. For the three hypotheses are not significant and have positive relationships that should continually be improved by providing training on a regular basis, providing knowledge sharing for the division of cargo in order to determine what updates are the latest in e-cargo, so that the result of the output of e-cargo can provide real benefits for the users adn eventually ended up at users' satisfaction. Last, the company also gain benefit by the increasing profit and revenue on cargo selling.

References

[1] Ullianjari, A. (2015). Proses Konversi Sistem Informasi Pada Perusahaan. Bogor: MB-IPB Program Pasca Sarjana Manajemen dan Bisnis .

[2] Seddon, Peter B. (1997). A Respecification and Extension of the DeLone and McLean Model of IS Success. Information Systems Research, 8(3):240-253.

[3] DeLone, W. H., & McLean, E. R. (2003). The Delone and McLean Model of Information Systems Success: A Ten-Year Update. Journal of Management System, 9

[4] Sundari, R. (2005). Teaching Materials & Files Implementasi Sistem Informasi. Retrieved from Universitas Gunadarma, teaching materials and files: http://staffsite.gunadarma.ac.id, diakses 8 maret 2017

[5] Suchban, N., Siti, E., & Kertahadi. (2012). Mengukur Efektivitas Sistem Informasi dan Mengetahui Kesuksesan Portal Akademik (Siam) On Line (Studi Kasus Terhadap Pengguna di Program Pendidikan Vokasi Universitas Brawijaya). ejournalfia, 4-6.

[6] Sofyan Yamin, H. K. (2014). Generasi Baru Mengelola Data Penelitian Dengan Partial Least Square Path Modeling: Aplilasi dengan Software XLSTAT, SmartPLS, dan Visual PLS. Jakarta: Salemba Infotek.

IOP Conf. Series: Materials Science and Engineering **215** (2017) 012038 doi:10.1088/1757-899X/215/1/012038

DEVELOPING AND EVALUATING PROTOTYPE OF WASTE VOLUME MONITORING USING INTERNET OF THINGS

Mohamad Fathhan Arief[1] and Ford Lumban Gaol[2]

[1]Master of Information Technology
Bina Nusantara University, Indonesia

[2]Doctor of Computer Science
Bina Nusantara University, Indonesia

Email :

[1]fathhan.arief@gmail.com
[2]fgaol@binus.edu

Abstract. *In Indonesia, especially Jakarta have a lot of garbage strewn that can be an eyesore and also cause pollution that can carry diseases. Garbage strewn can cause many things, one of her dues is bins are overflowing due to the full so it can not accommodate the waste dumped from other people. Thus, the author created a new method for waste disposal more systematic. In creating new method requires a technology to supports, then the author makes a prototype for waste volume monitoring. By using the internet of things prototype of waste volume monitoring may give notification to the sanitary agency that waste in the trash bin needs to be disposal. In this study, conducted the design and manufactured of prototype waste volume monitoring using LinkItONE board based by Arduino and an ultrasonic sensor for appliance senses. Once the prototype is completed, evaluation in order to determine whether the prototype will function properly. The result showed that the expected function of a prototype waste volume monitoring can work well.*

1. Introduction

Waste strewn on the street can be an eyesore and also cause pollution, especially pollution which can carry diseases. all those problems existed in Jakarta. Therefore, the author had the idea to devise a method of transporting waste more systematic. When creating new method requires a technology that can provide the ability to give notification to the sanitary agency that waste in the trash bin needs to be disposed, so waste would not be overflow from the trash bin. Given this capabilities, the sanitary agency can inform decision more efficiency such as how many human resources and truck should be sent to the location, and when the sanitary agency could deploy for disposing the waste in that location without disturbing the activities of other people that live in that area.

When constructing the waste volume monitoring prototype the author required a platform like the internet of things. Internet of things (IoT) is the latest communication model, where objects and tools of daily life will be equipped with technology to communicate, and the appropriate protocol that will enable them to communicate with each other and the user, which become an integral part of the internet [1]. In the recent paper of Harmon said [2], that IoT can describe the architecture of the integrated system between sensors, software, networks, and the appropriate interfaces that will provide real-time awareness, process, and knowledge to gather intelligence to be able to make good decisions. IoT will encourage the development of applications that has great potential in a wide range of data generated by those objects to simplify the use of the old and new services to citizen, companies and public administration, with allows easy access and interaction with a wide range of devices.

The application of IoT can be classified into various types based on network availability, size, coverage, and repeatability [3].

- Smart Home / Smart Building
- Healthcare
- Smart Business
- Utilities
- Mobile
- Environmental monitoring

The focus of this prototype is the internet of things for environmental monitoring and utilities. This waste volume monitoring prototype have a capability of sensing and distributing data so they can give an information to the sanitary agency, so they can make a better decision to manage their resources.

In order to make this new system for waste disposal, the author need an IoT platform for its base, In this paper the author used Sensor-Cloud Platform [4], this platform aims to manage physical sensor by connecting them to the cloud - or can be called cloud computing as a computing paradigm where many systems are connected each other, to provide scalable infrastructure of application and file storage [5]. And provide services such as automation data retrieval which will be submitted to the cloud to be processed and can be part of a resource.

this paper will explain of developing and evaluating waste volume monitoring prototype. what kind of devices and application this prototype used, and then the author will evaluate the function to determine whether the prototype will work properly as expected.

2. Monitoring of Waste Volume

In order to conserve the environment, important features were needed to make the environment has a degree of autonomy, adjusting to changes in the environment, and communicate with humans in a natural way. Automation that can affect the amount of interaction required by the population, as well as reducing the consumption of utilities and other potential losses [3][6].

In such an environment, waste disposal is important in order to society to live hygienically. To safeguard public health and the environment from potential harm because of the delay in disposing the waste/garbage, environmental management must be developed and implemented. Waste management is an important requirement for ecologically sustainable development in many countries [7].

With this kind of technology- waste volume monitoring prototype, will give a lot of advantages Like Bashir, A. has written, such as [8]:

1. Intelligent planning. with the waste volume monitoring technology, sanitary agency can schedule the departure for transporting waste [9].
2. Business efficiency increases. When sanitary agency can schedule their work, the use of resources can be used appropriately [10].
3. Increase the quality of life (reducing pollution due to the waste was strewn). With the reduction of pollution, the environment would be more comfortable than ever.

That point shows that the new method of managing waste in an efficient way may reduce the inappropriate use of human resource, time and cost [8][10]. And also by comparing to the conventional static scheduling of transporting waste, this system scheduling are said to allow cost reduction, by reducing the number of trucks, and manual labor cost [10].

To create the prototype, the author used one sensor, which is sensor ultrasonic, a LinkItONE board, and ubidots application as a server to collect data, analyze and sending the notification.

2.1. Ultrasonic Sensor (HC-SR04)

Just like the authors of [10][11] did, in this paper the author using an ultrasonic sensor. The workings of this sensor is automatically send a signal ultrasonic periodically and will detect whether there will be a signal ultrasonic that will be reflected back to the ultrasonic sensor, if the sensor receives the reflection, the author could mentioned the time signal sent by the sensor until the sensor detects the reflected signal is called duration / time. After the authors obtain data duration / time, the author could find the distance between waste and ultrasonic sensor using this formula:

$$s = t \times v$$

which,

t = duration / time (second)
v = velocity (m/s)
s = distance (m)

in this case the author need to use the speed of sound which is 344 m/s, because sensor ultrasonic using ultrasonic sound when release a signal. After that the author need to divide by 2 because the sound had travel 2 times, one from sensor to garbage/waste, and one more from garbage/waste to sensor. And the formula becomes like this:

$$s = (t \times 344 \text{ m/s}) / 2$$

2.2. LinkItONE Board

According to wiki.seeed.cc [12] LinkIt One is a PCB board made by collaboration of Seeed Studio and MediaTek to made a prototype of devices embedded with internet of things. The specifications of LinkIt One PCB board are listed on table below.

Table 1 LinkItONE Specification

Specification	Value
Dimensions	3.3 x 2.1 inches
RAM	4 MB
Digital Output	3.3v
Analog Input	5v
DC Current per I/O pin	1mA

2.3. Ubidots

According to www.ubidots.com [13] Ubidots is an engineering services firm, specializing in hardware and software development for internet of things projects. In ubidots, the data from sensor will be analyzed and presented as a line diagram. Ubidots also provide the feature to send SMS and e-mail.

3. Metodology

By using the prototype, the author constructs a new process for new waste disposal method, there are 5 step for the new method [14][15].

Step 1 : Using the ultrasonic sensor to obtain data duration of the round trip of signal ultrasonic to the trash and back to the sensor again. This phase will be carried out periodically, so the available data will continue to generate real-time data.

Step 2 : after that, the data of duration receive by the PCB, then the PCB will do the calculation process to obtain the distance between the sensor and waste/garbage.

Step 3 : Distance data will be received by the ubidots to be managed and processed so it can be seen and read by users with ease.

Step 4 : Then Ubidots will process the distance data to be measured, if the distance data exceed or equals to the predetermined value of the limit, it will proceed to the next step, but if it turns out not reach out to the predetermined value, then it will go back to the first step.

Step 5 : Ubidots will generate notification alert if waste bin already reached the point of full and the garbage/waste need to be disposed.

3.1. Data Resources
In this study the author conducted trials using waste volume monitoring prototype that the authors created. The data consist of time, distance, and the sent of notification alerts. 3 sessions will be conducted trials which have different condition, ie:

1. At the time when waste bin is empty
2. At the time when waste bin will be filled with trash periodically until half full.
3. At the time when waste bin will be filled with trash periodically until reach the limit value that have been set.

Where, the first session will take 30 data based on roscoe said that in the experimental study recommended 30 samples or more [16]. As for the trial of the 2nd and 3rd only done 10 times due simply to see whether the function of waste volume monitoring are work properly.

3.2. Evaluation
When conducting this evaluation the author has a target whether the waste volume monitoring technology would function properly. therefore in this study the author conducted three trials, in which trial has a different assessment requirements.

In the first trial whether the monitoring waste volume prototype can give the exact distance data as expected, by calculating the accuracy of the results using this formula:

$$Accuracy = \frac{b}{a} * 100\%$$

Where,

- b is number of correct data
- a is amount of data from the trial

The second trial is whether this prototype can give actual distance data in real time. And for the 3rd trial the author wanted to know whether the function of notification alert will be sent successfully in accordance with a predetermined distance.

4. Result
In this study the author use a container that has a length of 30 cm, width 27 cm, and 36.5 cm high with volume of 29565 cm^3. In the first trial the author get a result like in the table below:

Table 2 Result of the second trial

No.	Time	Distance of trash with sensor (in cm)	Notification alert
1	8:00	36.50	-
2	8:05	36.50	-
3	8:10	36.51	-
4	8:15	36.50	-
5	8:20	36.47	-
6	8:25	36.50	-
7	8:30	36.50	-

8	8:35	36.50	-
9	8:40	36.50	-
10	8:45	36.52	-
11	8:50	36.50	-
12	8:55	36.50	-
13	9:00	36.50	-
14	9:05	36.52	-
15	9:10	36.50	-
16	9:15	36.50	-
17	9:20	36.50	-
18	9:25	36.50	-
19	9:30	36.50	-
20	9:35	36.50	-
21	9:40	36.49	-
22	9:45	36.50	-
23	9:50	36.50	-
24	9:55	36.50	-
25	10:00	36.50	-
26	10:05	36.50	-
27	10:10	36.42	-
28	10:15	36.50	-
29	10:20	36.50	-
30	10:25	36.50	-

From the table above can be seen that, this prototype get the numbers 36.50 cm for 24 times (result with red font) from 30 result data, With the height of the container used is 36.50 cm. with the calculation of accuracy that have been discussed in the evaluation, this prototype have 80% of accuracy.

In the 2nd trial the author get result like the table below:

Table 3 Result of the second trial

No.	Time	Distance of trash with sensor (in cm)	Notification alert
1	11:00	36.47	-
2	11:05	33.38	-
3	11:10	30.47	-
4	11:15	28.52	-
5	11:20	26.49	-
6	11:25	25.58	-
7	11:30	24.43	-
8	11:35	24.43	-
9	11:40	22.52	-
10	11:45	23.42	-

From the table above can be seen that distance from sensor and trash are going smaller from 36.47 cm to 23.42 cm. prove that the prototype can provide data in real time without error.

And for the last trial the result are like this:

table 3 Result of the third trial

No.	Time	Distance of trash with sensor (in cm)	Notification alert
1	12:00	22.51	-
2	12:05	20.37	-

3	12:10	17.45	-
4	12:15	15.57	-
5	12:20	14.49	-
6	12:25	13.49	-
7	12:30	12.51	-
8	12:35	10.43	-
9	12:40	8.58	✓
10	12:45	8.43	✓

Can be seen that at the result of number 9 and 10 the notification has been sent. Prove that the main function of this prototype is work properly.

5. Conclusion and Suggestion

Based on the research that has been done, the prototype of monitoring the waste volume can function well up to receives notifications alert when a container filled up, Can be seen from the table in chapter 4. So with this prototype sanitary agency can use a new method of waste disposal.

Although the prototype of monitoring the waste volume had given a good results, there remains some suggestions for development. As it takes its expert systems in order to make a better system and prevent errors that might occur later. It also calls for additional technologies such as GPS to detect trash anywhere and temperature sensor for more actual data of distance calculation.

References

[1] Zanella, A., & Bui, N. (2014). Internet of Things for Smart Cities. IEEE INTERNET OF THINGS JOURNAL, Vol. 1, No. 1, pp. 22 - 32

[2] Harmon, R.R., Enrique, G., & Bhide, S. (2015). *Internet of Things. 2015 Proceedings of PICMET '15: Management of the Technology Age, pp.* 485 -494

[3] Hallaj Asghar, M., Mohammadzadeh, N., & Negi, A (2015). Principle Application and Vision in Internet of Things (IoT)

[4] Petrolo, R., Loscri, V., & Mitton, N. (2014). *Towards a Smart City based on Cloud of Things.* Inria Lille – Nord Europe, France.

[5] Torry Harris. CLOUD COMPUTING – An Overview. http://www.thbs.com/downloads/Cloud-Computing-Overview.pdf

[6] Sajal, K., & Cook, J. (2010). *Designing Smart Environments: A Paradigm Based on Learning and Prediction.* Department of Computer Science and Engineering University of Texas at Arlington.

[7] Glouche, Y., & Couderc, P. (2013). *A Smart Waste Management with Self-Describing objects.* The Second International Conference on Smart System, Device and Technologies, No.1, pp. 63 – 70

[8] Bashir, A. (2013). *Concept, Design and Implementation of Automatic Waste Management System.* International Journal on Recent and Innovation Trends in Computing and Communication, Vol. 1, Issue. 7, pp. 604 – 609

[9] Navghane S.S., Killedar M.S. (2016). IoT Based Smart Garbage and Waste Collection Bin. IJARECE, Vol 5, Issue 5

[10] Yusof, N.M., Jidin, A.Z., & Rahim, M.I. (2016). Smart Garbage Monitoring System for Waste Management. MATEC Web of Conferences 97, 01098 (2017).

[11] Monika, K.A., Nikitha Rao, Prapulla, S.B., & Shobha, G. (2016). Smart Dustbin-An Efficient Garbage Monitoring System. IJESC Volume 6 Issue No. 6

[12] http://wiki.seeed.cc/LinkIt_ONE/

[13] http://www.ubidots.com/about

[14] Parkash, Prabu. V. (2016). IoT Based Waste Management for Smart City. IJIRCCE, Vol. 4 Issue 2

[15] Vikrant Bhor. (2015). Smart Garbage Management System. IJERT, Vol. 4 Issue 03

IOP Conf. Series: Materials Science and Engineering **215** (2017) 012038 doi:10.1088/1757-899X/215/1/012038

[16] Roscoe. (1975). http://sola.siu.ac.th/sola/public/FacSites/GE1701/reading/Sampling.pdf

IOP Conf. Series: Materials Science and Engineering **215** (2017) 012039 doi:10.1088/1757-899X/215/1/012039

Failure Mode and Effect Analysis (FMEA) Applications to Identify Iron Sand Reject and Losses in Cement Industry : A Case Study

V N Helia[1], W N Wijaya[2]

[1,2]Industrial Engineering Department, Faculty of Industrial Technology, Universitas Islam Indonesia, Indonesia

Abstract. One of the main raw materials required in the manufacture of cement is iron sand. Data from the Procurement Department on XYZ Company shows that the number of defective iron sand (reject) fluctuates every month. Iron sand is an important raw material in the cement production process, so that the amount of iron sand reject and losses got financial and non-financial impact. This study aims to determine the most dominant activity as the cause of rejection and losses of iron sands and suggest improvements that can be made by using the approach of FMEA (Failure Mode and Effect Analysis). Data collection techniques in this study was using the method of observation, interviews, and focus group discussion (FGD) as well as the assessment of the experts to identify it. Results from this study is there are four points of the most dominant cause of the defect of iron sand (mining activities, acceptance, examination and delivery). Recommendation for overcoming these problem is presented (vendor improvement).

1. Introduction

The main objective of the company is getting profit by reducing or even eliminating the waste. So that the process of improvement (continuous improvement) must be done to improve the quality of the product. Not only the number of failed products to be reduced, but also customer satisfaction will be increased so as to win the market. One of the raw materials that focused in the fulfilment of quality is the quality of the iron sand raw material. The role and function of iron sand as one of the key raw material in the cement manufacturing industry makes the fulfilment of the quality and quantity of iron sand as a requirement that must be met. Procurement Department's data from XYZ company shows the amount of iron sand reject and losses were fluctuating every month. Iron sand is also very important for the cement manufacturing process compared with other raw materials so that the presence of iron sand reject and losses would impact on financial and non-financial. Although the factors that lead to low productivity of the manufacturing industry quite a lot, but the quality and quantity of raw materials had the largest percentage compared to other factors. It is in line with the statement that the quality will always directly proportional to productivity [7]. Company utilizing limestone, silica sand, clay, iron sand as the main raw material of cement production before the addition of gypsum and trass.

Company get iron sand from supplier in Cilacap. The use of iron sand a day is 250-400 tons, and if counted in months, the use of iron sand per month can reach 10000-15000 ton. Therefore, iron sand is the main ingredient that is very important in the cement manufacturing process. Based on this background we conducted research aiming to identify the dominant cause of failure in achieving quality and quantity of iron sand by using FMEA method (Failure Mode and Effect Analysis). FMEA is one of the most powerful methods available for measuring the reliability of products or process. It provides an easy tool to determine which risk has the greatest concern and therefore an action is needed to prevent a problem before it arises and can considerably decrease the loss to the industry in term of both money and time [1, 12]. It is the correct method to identify the cause of the problem and prevent the reoccurence of the problem in a system such as previous research in Semen Padang Company [9] and in Ceramic Tiles Manufacturing Plant [17]. Other research discuss about Pareto Analysis and FMEA to identify the failures stemming from forklift break, pilling, machine failure and wet wedge mark in XYZ business firm [5]. The final output of this study was able to determine the most dominant activity as the cause of rejection and losses of iron sand. Then, propose improvement suggestions to reduce iron sand reject and losses in XYZ company.

2. Literature Review

2.1. Quality

In everyday life we often hear people talk about quality issues. Quality is a prime concern for any organization. A quality product must fulfill the customer need and meet the specification [16]. The concept of quality itself is often regarded as a measure of the relative goodness of products and services consisting of design quality and quality compliance. The design quality is a function of product specifications, while the quality of conformity is a measure of how much a product meets the requirements or specifications of a predetermined quality. However, this aspect is not the only aspect of quality. TQM (Total Quality Management) is a much broader concept that not only emphasizes the aspects of the results but also the human qualities and the quality of the process [4].

2.2. Quality Control

To be able to always maintain good quality consistently, required an activity called quality control. It defines quality control in general as a system that is used to preserve or maintain the desired level of quality in a product or service [2]. Quality control also has an idea of the use of techniques and activities in order to achieve, maintain, and improve the quality of a product or service. Quality control can be divided into two [11]: 1) On-line Quality Control is a quality control during the production process is running, such as diagnosing and adjustment process, process control, and inspection results of the process, 2) Off-line Quality Control. It is the efforts that aim to optimize the design of products and processes as business support of On-Line Quality Control. Off-Line Quality Control is done before and after the process. According to [6], quality control consists of three aspects, known as the trilogy concept of quality, namely Quality Planning, Quality Control, and Quality Improvement.

2.3. FMEA

FMEA is a structured procedure to identify and prevent as much as possible modes of failure (failure mode). FMEA is used to identify the sources and root causes of quality problems [2]. A failure mode is what is included in the defect / failure in design, conditions beyond the limits established specifications, or changes in the product that causes disruption of the function of the product. FMEA can be done by [10]:

a. Identifying and evaluating the potential failure of a product and its effects.
b. Identify actions that could eliminate or reduce the chance of potential failure occurs.
c. Recording process (document the process)

The purpose can be achieved by the company with the implementation of FMEA:

a. To identify the mode of failure and the severity of the effect

IOP Conf. Series: Materials Science and Engineering **215** (2017) 012039 doi:10.1088/1757-899X/215/1/012039

b. To identify critical characteristics and significant characteristic
c. To sort order potential design and deficiency process
d. To help focus engineer in reducing attention to the products and processes,
e. and help to prevent problems

2.4. *Implementation Steps using FMEA* [3]
The basic steps in the process of Failure Mode and Effect Analysis (FMEA):

1. Identify the functions of the business process being studied.
2. Identify potential failure modes in the business process being studied.
3. Identifying potential effects of failure on the business process being studied
4. Identify the causes of the failure of the business process being studied.
5. Identify modes of detection on the business process being studied.
6. Determine the rating of the severity, occurrence, detection and Risk Priority Number (RPN) on business processes that are being studied.
7. Proposed improvements.

Measurement of the value of severity, occurrence and detection are as follows [8, 13]:

a. Severity value
Severity is the first step to analyse risks, which calculates how much impact or intensity of incident affect the final outcome of the process because a failure mode in one activity can lead to a failure mode in another activities. The impact on the rating scale ranging from 1 to 10 in the table 1. These numbers will conduct an expert to prioritize the failure modes. A severity rating of 1 means there is no danger or no effect on the quality. Otherwise, a severity rating of 9 or 10 is made for those effects that could cause bad effect or injury to the user. If it is happened, the actions to eliminate the failure mode such as change the design are considered.

Table 1. Severity ratings

Classification	Scale	Example
No result/ none	1	No effect on the quality
Very minor	2	Iron sand quality characteristics is not disturbed
Minor	3	Consequently little to the quality of iron sand
Very low	4	Iron sand quality small impaired
Low	5	Failure resulted in some dissatisfaction on the quality of iron sand
Moderate	6	Failure cause inconvenience
High	7	Iron sand quality unsatisfactory
Very high	8	Iron sand quality is very unsatisfactory
Hazardous with warning	9	Potential cause bad effect on iron sand
Hazardous without warning	10	Failure mode effect is fatal to the quality of iron sand

b. Occurrence value
If predetermined rating on the severity, the next step is to determine the rating of the value of occurrence. Occurrence is a possibility that the cause of the failure will occur and result in the failure during the period of business processes running. Determining the value of occurrence can be seen by the table 2. If the occurance is high that means more than 4 or more than 1 when the severity rating is 9 or 10, the actions needed. This step is also means the detailed development section of the FMEA process. The occurance rating 1 means no failure. Then, the occurance rating from 2 to 10 shows elevated level of the occurrence of failure.

Table 2. Occurrence ratings

Classification	Scale	Criteria
Never	1	History showed no failure
Rarely	2	Possible failure is very rare
Very little	3	The possibility of failure is very little
Bit of all	4	Iron sand quality suffered minor annoyances
Low	5	Some of the possible failure
Medium	6	The possibility of failure occurs
Quite high	7	The possibility of failure is high enough
High	8	The high number of failures
Very high	9	The extremely high number of the possibility of failures
Almost certainly	10	Failure is almost certainly

c. Detection Value
Having obtained the value of occurrence, it is time to determine the value of detection. Detection function for preventing potential business processes are being researched and reduce the failure level on business process. Rate determination detection can be seen in the table below.

Table 3. Detection ratings

Classification	Scale	Criteria
Almost certainly	1	Controls definitely detect
Very High	2	Control almost detect
High	3	Controls possessed greater chances to detect
Mod. High	4	Controls may detect high enough
Moderate	5	Controls may detect moderate
Low	6	Controls may detect low
Very low	7	Controls possessed very low opportunities to detect
Remote	8	Control possessed very little opportunities to detect
Very remote	9	Controls may not detect
Impossible	10	Controls certainly not detect

After getting the value of severity, occurrence, and detection, it will obtain the value of Risk Priority Number (RPN), by multiplying the value of severity, occurrence and detection (RPN = S x O x D). Then, do the sorting based on the value of Risk Priority Number (RPN) highest to the lowest. After that, the activities that have great value Risk Priority Number (RPN) and has an important role in a business process activities studied, performed the proposed improvements to reduce the failure rate of the business processes.

2.5. Analysis of Variance (ANOVA)

According [15] analysis of variance (ANOVA) is a method to describe the total diversity of data into components that measure the various sources of diversity. In application, ANOVA was used to test an average of more than two samples of whether there are differences in average significant / not between the groups. It is can be classified into several criteria [18], as follows:

- Classification in one direction (One-Way ANOVA)
  One way ANOVA classification is based on the observation ANOVA criteria or the factors that cause variation.

- Classification in two directions (Two-Way ANOVA)
  Two-way ANOVA classification is based on the observation of two factors that cause variation.
- Classification in many directions (MANOVA)
  Multi-way ANOVA is based on observations of many criteria.

3. Methodology
The final output of this study was able to determine the most dominant activity as the cause of rejection and losses of iron sand. Research stage are as follow to solve this problem (figure 1). Data collection methods that used in this research are observation, interview and group discussion forum. Observations carried out to observe the process of iron sand supply chain activities, orderings business process in the company and the needs of quality iron sand that must be met by supplier's specifications iron sand quality. Its spesification obtained from Quality Control Department. Following up with interviews were conducted to identify the cause of the failure to reject and loses in iron sand as well as to get an assessment of the four respondents who have been determined. Then, group discussion forum used to determine whether the results of expert assessment of the dominant cause occurs reject and quantity of iron sand loses no element subjectivities.

Figure 1. Research stage

4. Result and Discussion

4.1. The Benefits of Iron Sand
At XYZ Company, iron sand is used as a corrective raw material. Iron sand that content of Fe_2O_3, SiO_2, MgO and concrete 80-100 mesh size has the potential to be used as a substitute for cement in the production of high-performance concrete. Iron sand has a tendency to heat up in direct sunlight, causing the temperature high enough to cause minor burns. According to the results of research [14] demonstrate the benefits and usefulness of iron sands are:

1. The use of iron sand is 80% of the total weight of the sand provides maximum compressive strength between the levels of iron sand is 42.65 MPa and compressive strength can increase by 28.41% compared normal concrete.
2. The use of iron ore by 80% of the total weight of the sand increases tensile strength divided by 4.84% compared to normal concrete. In this iron sand increases the compressive strength and tensile strength divided by 80%, this is possible because in addition to the nature of the filler is also the chemical properties of iron sand containing SiO_2 so as to help the performance of cement as a binder.
3. Iron sand forming the colour of cement

4.2. Identification of the Ordering Business Process for Iron Sand

In the figure 2 can be seen a description of activities undertaken in the ordering supply chain to the use of iron sand. Some of the activities carried out by the company and suppliers. Through a number of these activities can be identified type of failure or potential failure modes that can occur. The description of the types of failures and possible causes other potential failure of the supply chain ordering iron sand that is on the next point.

IOP Conf. Series: Materials Science and Engineering **215** (2017) 012039 doi:10.1088/1757-899X/215/1/012039

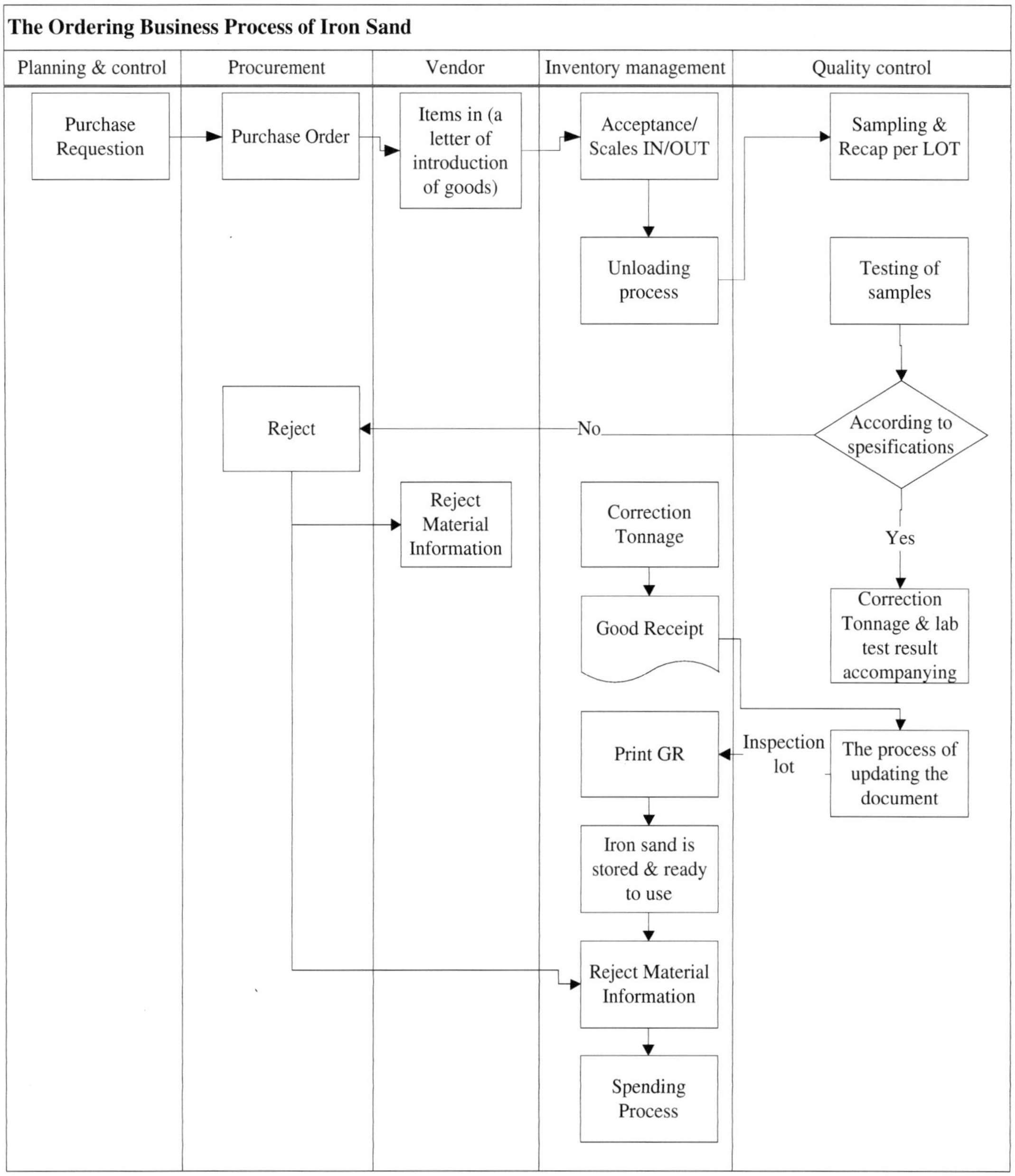

Figure 2. Identification of the Ordering Business Process for Iron Sand

IOP Conf. Series: Materials Science and Engineering **215** (2017) 012039 doi:10.1088/1757-899X/215/1/012039

4.3. Identification of failure modes

Here is the identification mode failure of iron sand reject and losses.

Table 4. Identification of Potential Failures

Activity	Potential Failure Mode	Symbol	Causes	Effect	Control
Mining	unfavorable quality map	A1	human error	reduced iron sand quality because one determines the location of mining iron sand which has the highest Fe content.	data accuracy
	iron sand contaminated soil	A2	mining techniques less than perfect and unfavorable separation systems (magnetic separator)	reduction in the tonnage of iron sand and low of Fe content in iron sand	determine the specifications, iron sand sampling test,the effectiveness of the penalty clause of the purchase contract iron sand to suppliers
Shipping	iron sand splattered on the streets	A3	damaged roads, the rear cover truck is less than perfect	reduced iron sand tonnage	perform a physical examination on a truck used to send iron sand
	iron sand becomes wet	A4	rainy weather/ humid during the delivery process	affect the tonnage of iron sand when weighed	covered with tarpaulins
	truck number is not optimal	A5	for the inability of suppliers to provide truck	delay the fulfillment of supplies	determine the standard of the minimum number of trucks will be used, delivery deadline
	Wrong unloading site instruction	A6	human error	double handling/cost incurred is high	to establish coordination with the unloading iron sand workers and the inventory workers
Acceptance	sample test results iron sand from the supplier and the actual iron sand in a test by the company as a whole is not equal	A7	fraud by supplier that carries a good sample but actual-his iron sand that was sent has a low Fe quality	receiving iron sand with a lower quality and made losses to financial	quality checks iron sand before unloading iron sand into the truck
Checking	take a sample error	A8	lack of tools and knowledge about the correct sampling	misidentified quality iron sand	modification tools that can represent the whole population sample of iron sand and do trainning workers who has the integrity to sampling activities
Storage	bad drainage stockpile	A9	flow of water does not drain	the water content in iron sand increased/ stagnant water and difficult to dry	drainage maintenance periodically for stockpile
	limited storage area	A10	storage techniques (pileup) iron sand less precise	companies into limited capacity	engineering storage (pileup) improved
	sloppy in the process of loading unloading/storage	A11	lack of knowledge of handling iron sand	not optimal warehouse	storage techniques (pileup) improved
	iron sand spilled when loading	A12	over carrying capacity	the road becomes dirty, splattered iron sand	loading techniques improved

IOP Conf. Series: Materials Science and Engineering **215** (2017) 012039 doi:10.1088/1757-899X/215/1/012039

From the identification of the failure modes of the next process is weighting conducted by the expert. In this study researchers determine the four experts who come from Procurement Department at XYZ Company. Fourth experts will assess the severity, occurrence and detection on failure mode that has been made above. Then, the experts will assess the Risk Priority Number (RPN). It is calculated value to determine the cause of the most dominant in the high issues the iron sand reject and losses.

4.4. Risk Priority Number (RPN) calculations

Simbol	*Severity*				*Occurance*				*Detection*				RPN				Rata rata RPN (μ)
	1	2	3	4	1	2	3	4	1	2	3	4	RPN1	RPN2	RPN3	RPN4	
A1	9	8	7	7	7	8	5	8	5	5	4	2	315	320	140	112	221,75
A2	8	8	8	9	7	8	8	9	6	5	5	2	336	320	320	162	284,5
A3	4	2	3	5	3	1	2	2	4	4	5	2	48	8	30	20	26,5
A4	7	8	6	8	6	8	7	7	4	4	6	2	168	256	252	112	197
A5	6	6	5	8	4	6	6	5	5	5	5	2	120	180	150	80	132,5
A6	4	5	4	5	3	1	3	2	4	4	5	2	48	20	60	20	37
A7	8	8	8	10	8	8	6	9	3	4	4	2	192	256	192	180	205
A8	7	8	6	9	6	8	8	8	5	4	4	2	210	256	192	144	200,5
A9	6	6	6	7	4	5	5	4	5	5	4	2	120	150	120	56	111,5
A10	5	6	7	7	4	5	5	5	6	5	5	3	120	150	175	105	137,5
A11	4	6	5	6	4	5	4	6	5	5	5	2	80	150	100	72	100,5
A12	5	6	5	7	5	5	4	6	6	5	5	2	150	150	100	84	121

Figure 3. Calculation of RPN Value

Risk Priority Number (RPN) value assessment carried out by interview and group decision forum. The results of the assessment based on the table above can be seen the results of 12 different types of Risk Priority Number (RPN) value and potential failure of the most dominant can be seen in Table 5.

Table 5. Sequence RPN

No.	Activities	Failure Mode	RPN
1	Mining	quality map poor	284.5
2		iron sand contaminated soil	221.8
3	Acceptance	sample test results iron sand from suppliers and actual iron sand in a test by the company as a whole does not equal	205
4	Examination	mistake taking samples	200.5
5	Delivery	iron sand getting wet	197

4.5. ANOVA

ANOVA is a statistical technique that is applied to compare the average of two samples or more, ANOVA analysis was performed to compare the value of the RPN of 12 failure modes. To prove whether the average RPN twelve failure modes are different, then the test hypothesis is:

- $H_0 = RPN_1 = RPN_2 = RPN_3 = RPN_4$
- H_1 = average value of RPN different for at least two failure modes

One-way ANOVA uses SPSS software performed in this research. In this statistical test some assumptions must be met, namely:

a. Samples derived from a group of independent
b. data each failure mode normally distributed
c. among the groups should be homogeneous Variants

The test output results of statistical tests One Way ANOVA using SPSS software in figure 4.

ANOVA

Failure Mode

	Sum of Squares	df	Mean Square	F	Sig.
Between Groups	68085,167	3	22695,056	3,573	,021
Within Groups	279513,500	44	6352,580		
Total	347598,667	47			

Figure 4. ANOVA Output

After the fourth variant proved similarly, new ANOVA test done to test whether the four samples had an average the same or different. ANOVA test showed the value or significance probability is 0.021. This means significance = 0.021 <0.05 then H_0 is rejected. Can be deduced by using ANOVA showed that there was no value Risk Priority Number (RPN) of twelve identified failure modes of equal value, so that the predominant failure mode for the failure of iron sand reject and quantity losses in XYZ can be directly determined by that failure mode with Risk Priority Number (RPN) average value.

5. Conclusions and Recommendations

The activities were dominant as a cause of rejection and quantity losses of iron sand in XYZ company is on mining activities, the quality map is not good, and iron sand contaminated soil, then test results of samples of iron sand from suppliers and iron sand actually tested by the company as a whole is not the same, the activity of examination / inspection, such as errors in sampling and at the time of delivery, namely iron into the wet sand. Proposed improvements to reduce this failure rate is company need to re-review of the contract clause with the vendor. A review of the contract clause can be done by increasing the penalty sanctions in violation of the contract clause and accelerate the due date making status iron sand reject.

Based on the conclusion, recommendations are given by the author. Firstly, for company, the advice that can be given is to assess the scale of priorities vendor based on the consistency of the quality of iron sand, the fastest time of reservation and within the nearby mine. As well as upgrading the effectiveness of sanctions penalties for vendors who violate the rules of a contractual agreement. Company also need to provide training for workers in charge of the sampling. For further research, the advice given to the development of the research is to determine the fundamental causes of failure and the risk of failure suggested for using house of risk to quantify how big a risk that will result from failure. Another recommendation is by using quality methods such as Fault Tree Analysis (FTA) and the methods of other qualities associated with finding the fundamental causes of product failure. In analysing the causes of the failure of the product to the next process is to continue the improvements that have been proposed to expand research on the value engineering analysis, which is an advanced technique in analysing the causes of product failure.

REFERENCES

[1] Ambekar S, Edlabadkar A and Shrouty V 2013 A Review : Implementation of Failure Mode and Effect Analysis *Int. J. Of Eng. And Innovative Tech. (IJEIT)* vol.1 issue 8

[2] Foster S 2010 *Managing Quality (Integrating The Supply Chain, 4th Edition)* (New Jersey: Pearson Education, Inc)

[3] Gasperz V 2002 *Pedoman Implementasi Program Six Sigma* (Jakarta: Gramedia Pustaka Utama)
[4] Gasperz V 2005 *Total Quality Management (TQM)* (Jakarta: Gramedia Pustaka Utama)
[5] Gorener A, Toker K 2013 Quality Improvement in Manufacturing Processes to Defective Products using Pareto Analysis and FMEA *Beykent University J. Of Social Sciences-BUJSS* Vol. 6 No.2 ISSN : 1307-5063
[6] Juran J M and Godfrey A B 2005 *Juran's Quality Handbook Fifth Edition* (United States of America : McGraw-Hill)
[7] Liker J K 2004 *The Toyota Way :14 management principles from the world's greatest manufacturer* (New York : McGraw –Hill)
[8] McDermott R E, Mikulak JE and Beauregard MR 1996 *The Basics of FMEA* (New York: Productivity Press)
[9] Nandhiroh S, Rahmattullah 2015 Analisis FMEA untuk Identifikasi Terjadinya Batu Bara Reject dan Losses *Simp. Nasional RAPI XIV-2015* ISSN 1412-9612 pp 159-166
[10] Narayanagounder 2009 A New Approach for Prioritization of Failure Modes in Design FMEA using ANOVA *J World Academy of Sc., Eng. of Tech.* vol 25 pp 95-98
[11] Nasution 2001 *Manajemen Mutu Terpadu* (*Total Quality Management*) (Jakarta: Ghalia Indonesia)
[12] Parsana T S and Patel M T 2014 A Case Study : A Process FMEA Tool to Enhance Quality and Efficiency of Manufacturing Industry *Bonfring Int. J. Of Industrial Eng. and Mgt Sc.* Vol. 4 no.3 pp 145-152
[13] Prajapati D R 2012 Implementation of Failure Mode and Effect Analysis : A Literature Review *Int. J. Of Management, IT and Engineering IJMIE* Vol.2 Issue 7 ISSN : 2249-0558 pp 264-292
[14] Rasyid K 2000 Desain Eksperimen Taguchi Untuk Memaksimumkan Kuat Tekan Semen *Skripsi Institut Teknologi Surabaya* Surabaya
[15] Singgih S 2008 *Panduan Lengkap Menguasai SPSS 16* (Jakarta: PT. Alex Media Komputindo)
[16] Tripathi S 2014 A Survey on Quality Perspective and Software Quality Models *IOSR J. Of Comp. Eng. (IOSR-JCE)* e-ISSN : 2278-0661, p-ISSN: 2278-8727 vol.16 issue 2 ver. VI pp 63-72
[17] Tsarouhas P H, Arampatzaki D 2010 Application of Failure Modes and Effect Analysis (FMEA) of a Ceramic Tiles Manufacturing Plant *Proceedings of 1st Olympus International Conference on Supply Chains*
[18] Wallpole R E 1992 *Pengantar Statistika* (Jakarta: PT Gramedia)

Acknowledgments

Thanks to Industrial Engineering Department of Universitas Islam Indonesia for the support and to cement company from Indonesia as respondent so that the primary and secondary data obtained.

Fabrication of punch and die of micro-blanking tool

Aida Mahmudah [1], Gandjar Kiswanto [2] and Dedi Priadi [3]

[1] Manufacturing Design Engineering Department, Politeknik Manufaktur Bandung, Bandung, Indonesia
[2] Mechanical Engineering Department, Universitas Indonesia, Depok, Indonesia
[3] Metallurgy and Materials Engineering Department, Universitas Indonesia, Depok, Indonesia

Corresponding author: gandjar_kiswanto@eng.ui.ac.id

Abstract. The difficulties of tool fabrication in micro level provide an enormous challenge due to the dimension limitation. Reducing the difficulty and tool cost fabrication should always be considered by determining the simplicity of tool geometry, precision tolerance, and an appropriate material selection. This paper study about the manufacturing of micro-blanking punch-die. The mechanical material removal method was chosen to fabricate punch-die since this is a single piece production and intended for small-scale production of micro parts. The manufacturing process consists of punch-die design and fabrication, tool assembly, and clearance measurement. The stress analysis is also conducted to ensure the strength of the punch-die. A micro-blanking experiment was held to observe the tool performance. The result showed that micro-blanking tool can be manufactured successfully and used well to conduct the micro-blanking experiment.

Keywords: Micro-blanking tool, punch-die, tool fabrication

1. Introduction

To date, the growing needs of micro-parts still indicate an increase. Therefore, the manufacturing process of micro-parts still required to meet the demand. Micro-sheet metal forming is one of the popular micro manufacturing processes to produce the micro-parts. In this technology, the tool is an aspect that holds a very important role since it directly affects the geometry and the quality of the micro-part. The difficulties of manufacturing process is an enormous challenge since the size of the tool is very small. A simple geometry of tool, a required tighter tolerances, a material selection, and a simple assembly must be determined very carefully in order to reduce the difficulty level and the cost of the tool component fabrication.

The most likely manufacturing method to produce the micro-tool components are additive, subtractive, and mechanical material removal. In this study, the main components of micro-blanking tool is fabricated by mechanical material removal method. This is suitable to be applied on a single tool piece production and intended for small-scale production of micro parts [1]. In the tool components machining process, the entire machining system requirements must be met well, such as machine tool itself, environment, sub-system, and reliable data management.

Some previous researches have been held in micro-tool manufacturing, such as Yi et al. [2] fabricated a punch piercing for 15μm hole using micro-EDM and Chern et al. [3] who introduced a technique to produce a variation of clearance of micro-blanking tool by controlling the spark gap in micro-EDM. Further, Jie Xu et al.[4] developed a micro-forming system for micro-punching process in high production volume. Meanwhile, in bending process, Fujimoto et al. [5] investigated a surface

IOP Conf. Series: Materials Science and Engineering **215** (2017) 012040 doi:10.1088/1757-899X/215/1/012040

treatment of the punch with an ion beam irradiation and diamond-like carbon (DLC) coating. Another die manufacturing was also explained by Gau et al. [6] in micro-deep drawing tool.

This paper study about the manufacturing process of the main components of the micro-blanking tool, i.e. the punch and die. The study consists of tool design process, stress analysis, tool manufacturing, tool measurement, and micro-blanking experiment. The results showed that the manufacturing processes have been successfully performed and produce a reliable tool to be used in micro-blanking experiment.

2. Micro-blanking

Blanking process is a cutting process of the sheet material (strip material/foil material) by applying shear forces on the material being cut. The cuts would occur if the shear stress on the material beyond the limits of the shear strength. Therefore, this process is also called shearing process. Other similar shearing process is the piercing. The difference between both of them is shown in Fig. 1.

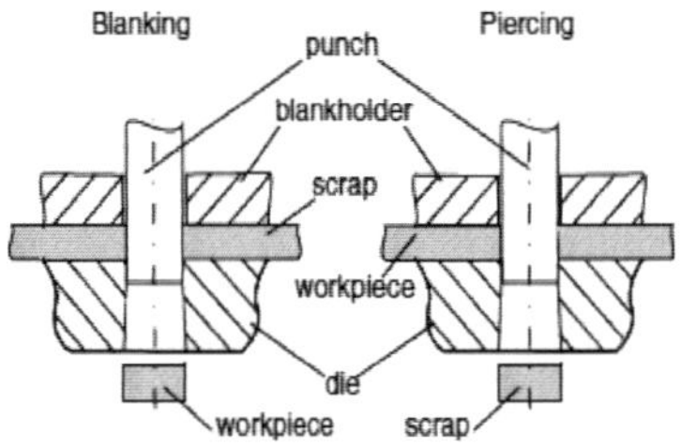

Figure 1. Shearing process: blanking and piercing [1]

In general, the micro tool components are divided into two groups, namely the main components and support components, as shown in Fig. 2. The manufacturing of main components, i.e. punch and die, will be discussed further in this study. A gap between punch and die is called clearance, which are designed to facilitate the occurrence of fracture on the material to be cut. In micro level, the behavior of micro-blanking deformation is influenced by the ratio of clearance and the grain size, c/d [7].

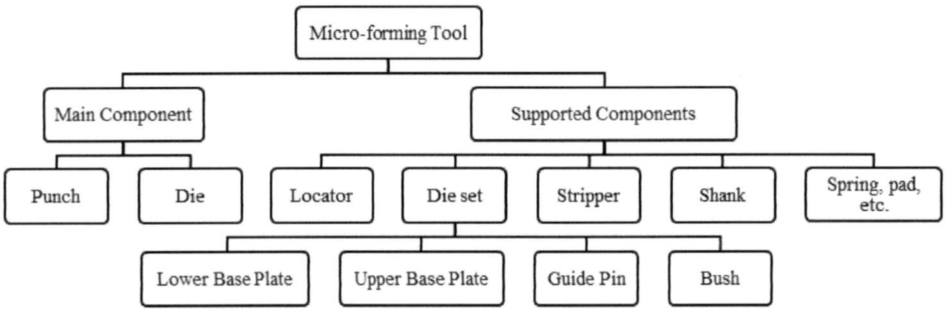

Figure 2. Micro-forming tool components

Material that can be used for the main component and has a good machinability is from the tool steel group, while the other components can use the material from the machinery steel group. In this research, the SKD11 (AISI D2) was used for the punch and die. Whereas, the S45C (AISI 1045) was chosen from machinery steel for the rest non-standard components of micro blanking tool. The mechanical properties for SKD11 and S45C are shown in table 1.

Table 1. Mechanical properties of S45C dan SKD11

		Machinery Steel S45C (AISI 1045)	Cold Work ToolSteel SKD11 (AISI D2)
Ultimate tensile strength	MPa	569	
Yield Strength	MPa	343	1650
Elongation	%	20	
Modulus of Elasticity	GPa	205	209.9
Poissons Ratio		0.29	
Machinability	%	55	
Shear Modulus	GPa	80	

For blanking process, the necessary force to shear the sheet material, i.e. $F_{blanking}$ is estimated with the following equation [8]

$$F_{blanking} = L \cdot T \cdot \tau_m = 0.7 \cdot L \cdot T \cdot UTS \quad (1)$$

where:
L = the total length sheared (perimeter) [mm]
T = thickness of the material [mm]
UTS = the ultimate tensile strength of the material [MPa]

3. Punch-die design and fabrication

This sub-section discuss about punch-die design and fabrication, tool assembly, and clearance measurement. The stress analysis is also conducted to ensure the strength of the punch-die.

3.1. Design of punch - die

It was determined that in one blanking stroke, four specimen materials should be produced. So, the blanking lay-out was designed (Fig. 3b.) based on the blanking part geometry (Fig. 3a). Thus, the geometry of raw material for blanking process was obtained (Fig. 3c). Clearance between the punch and die was set at a maximum of 0.05mm / side.

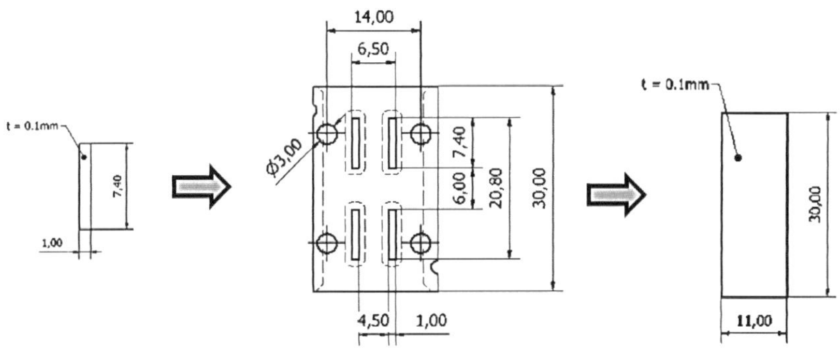

(a) (b) (c)

Figure 3. Initial design (a). Blanking part geometry, (b). Blanking lay-out, (c). Raw material dimension

Hereafter, the punch and die was designed based on the blanking lay-out. The detailed drawing of the punch and the die are showed in Fig. 4 and 5. For ease of machining and assembly processes, the pin locator was implemented in the construction (Fig. 5c). The locator pins was set 2mm above the die surface to facilitate the raw material placing.

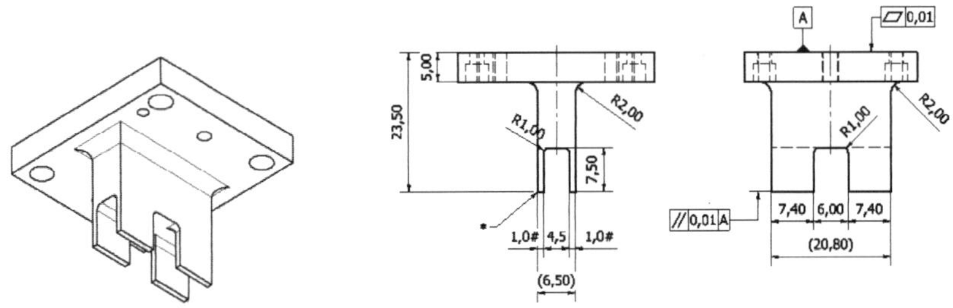

(a) (b)

Figure 4. Micro blanking Punch

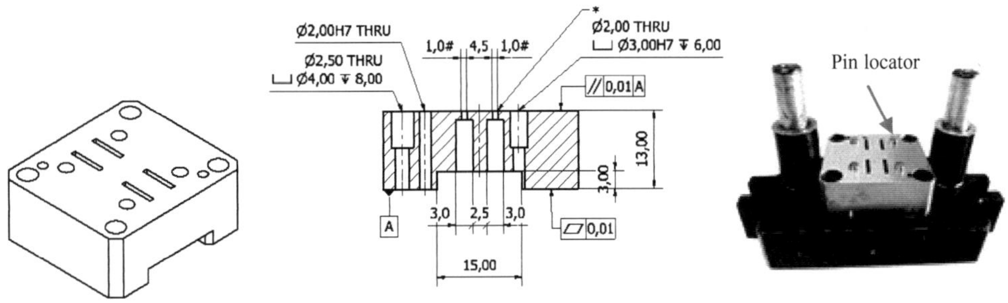

(a) (b) (c)

Figure 5. Micro blanking Die

For stripper components, the urethane was selected since the construction becomes simple and does not damage the material foil (Figure 6c). The stripper serves as the holder of sheet material before it is cut and keep the scrap material does not stick to the punch when it is raised. The punch-stripper assembly is showed in Figure .6c

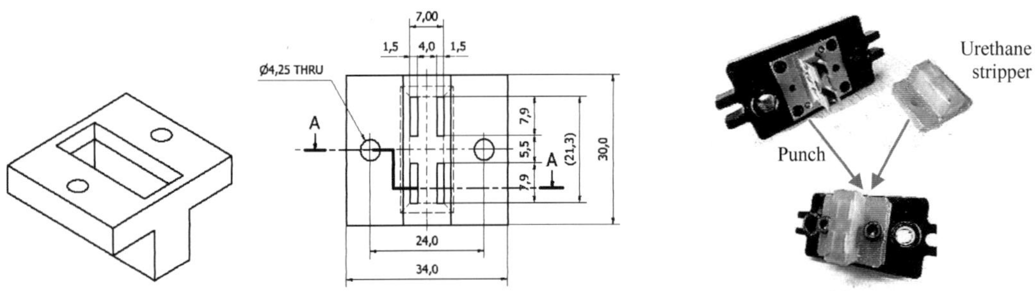

(a) (b) (c)

Figure 6. Urethane Stripper

The compressive force urethane ($F_{urethane}$) is calculated by the following equation [9]

$$F_{urethane} = \frac{E \cdot \Delta H \cdot S}{H} \tag{2}$$

where,

E = the elastic modulus for urethane [kgf/cm^2]
S = Load area [mm^2]
ΔH = the change in height [mm]
H = the initial height (free height) [mm]

The elastic modulus for urethane is obtained from the graphic in figure 7. Thus, with E = 108 kgf/cm^2, S = 253mm^2, ΔH = 2.5mm, H = 12mm, thus the $F_{urethane}$ = 557.36 N

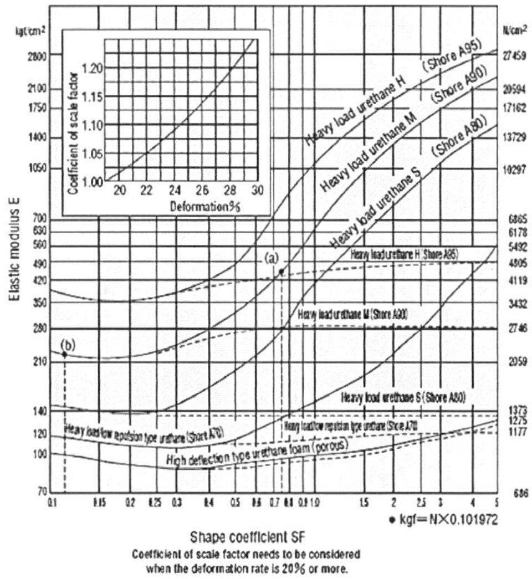

Figure 7. Elastic modulus (E) for urethane [9][a]

[a] SF = load surface/size surface

3.2. Stress Analysis of punch - die

The major load that applied at punch is came from the calculated blanking force based on the Eq. 1. The sheet of material that will be used in the blanking experiment are aluminum, brass, copper, and steel with 0.1mm thickness. The mechanical properties of those materials (Table 2) are obtained from tensile test using Shimadzu Servopulser 20T tensile machine.

Table 2. Mechanical properties of the sheet material

		Aluminum	*Brass*	*Copper*	*Steel*
Tensile Strength	[MPa]	127	379	657	1338
Yield Strength	[MPa]	58	192	-	1206
Elongation	[%]	15.8	23.13	2.11	5.33

So, the maximum blanking force is estimated reached with steel sheet as a material specimen (tensile strength = 1338 MPa), i.e.:

$$F_{blanking} = 0.7 \cdot 16.8 \cdot 0.1 \cdot 1338 = 1573.5N$$

Based on the free body diagram of the punch (Fig. 8), the stress analysis are conducted (Fig. 9). The analysis showed that the maximum stress are 97.18 N, and the maximum displacement is 0.016mm.

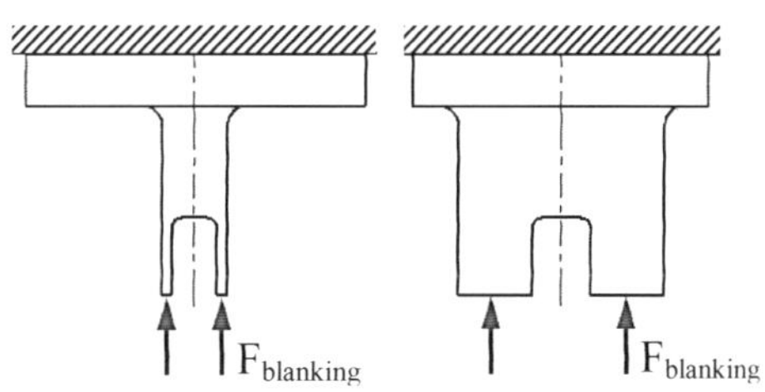

Figure 8. The free body diagram for punch

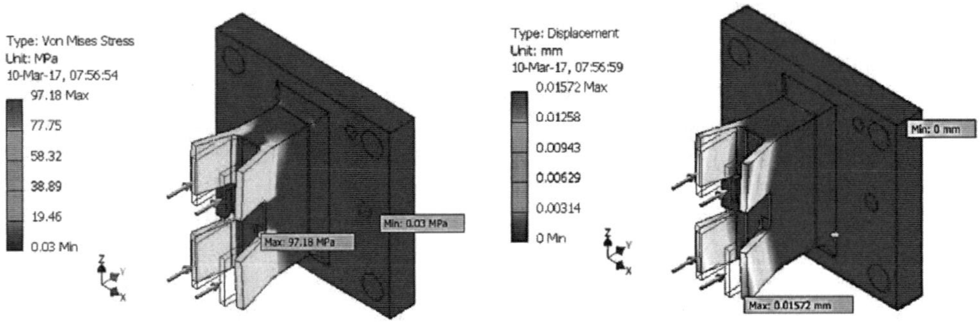

(a) (b)

Figure 9. Stress analysis for punch (a) Stress distribution (b) displacement distribution

The major load that applied at die is came from the calculated urethane force, $F_{urethane}$ = 557.36 N. Based on the free body diagram of the die (Fig. 10), the stress analysis are conducted (Fig. 11). The analysis showed that the maximum stress are 13.69 N, and the maximum displacement is 4.16 x 10^{-4}.

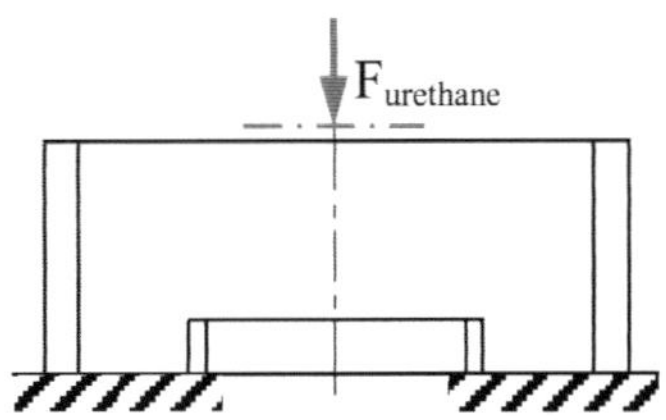

Figure 10. The free body diagram for die

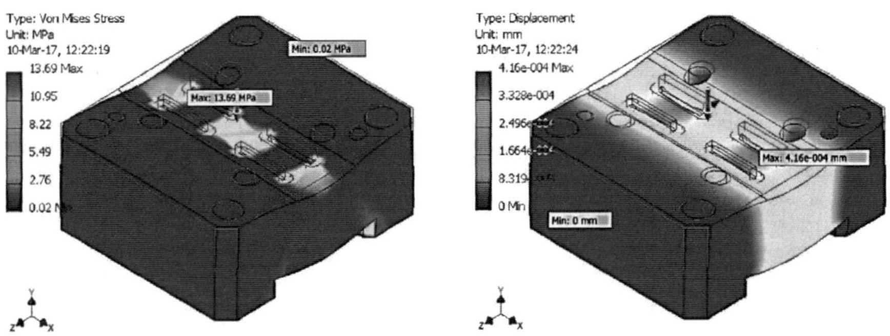

(a) (b)

Figure 11. Stress analysis for die (a) Stress distribution (b) displacement distribution

3.3. Fabrication of punch-die

The technology used in the punch-die fabrication are conventional milling and drilling, CNC milling, grinding, hardening, wire EDM, and grinding for finishing. The fabrication flow-chart for punch and die are showed in Fig. 12a and 12b, respectively.

For CNC machine, it was used Vertical Machining Centre (VMC) Vista 1000 with a speed of 4500rpm and feeding tool 1500mm/minutes. To make the hole in the die, the WEDM Makino with a gap 0.01mm and 0.005mm precision was used. The diameter of the wire was 0.15mm.

For micro-blanking tool, the final dimension of the main components are obtained from grinding process. It can also achieve a surface roughness value until 0.2μm. To achieve 63HRc hardness, punch and die were hardened at temperatures 850-1030°C, and quenching temperature was 150°C. After all components have been fabricated, the assembly process was conducted according to the guidance. The assembled micro blanking too is showed in Fig13.

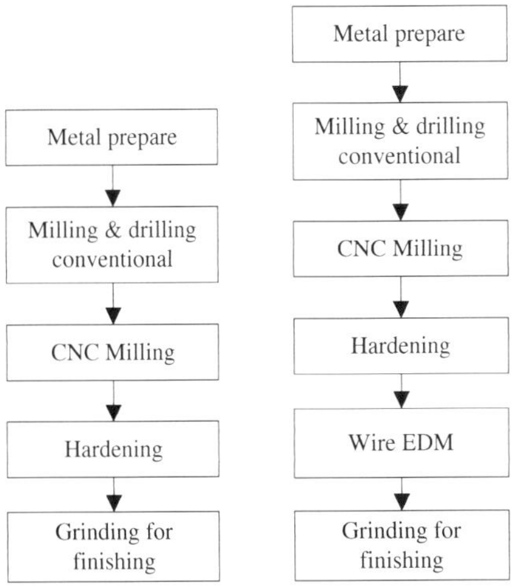

(a) (b)

Figure 12. Fabrication flow chart for (a). punch, and (b). die

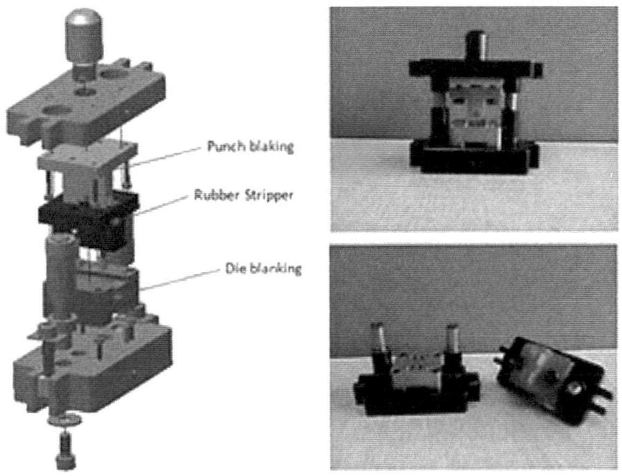

Figure 13. Assembly of micro-blanking tool

3.4. Clearance Measurement

Punch and die clearance measurements was conduct by using vision-based measuring machine TESA - VISIO 300 and CMM machines Crysta Plus M443. The determination of datum and element to be measured should be definite previously. On this measurement, the point "O" on the guide rode (Fig. 14) is determined as a datum. The measurement results are shown in Table 3, and it was obtained the clearance per side for the length direction is 0.045-0.06mm and 0.03-0.035mm for the width direction.

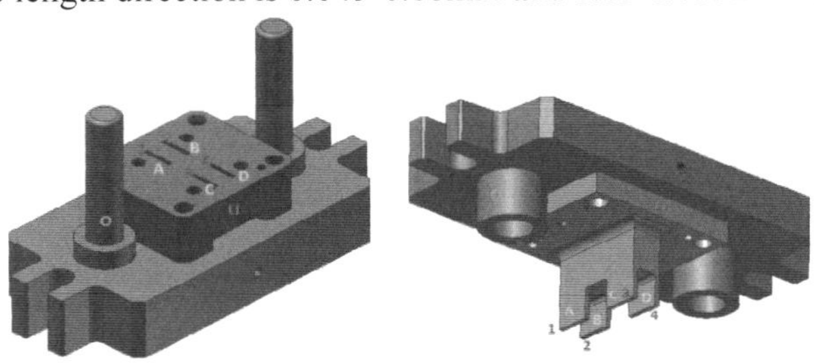

Figure 14. Measurement geometry set-up

Table 3. Punch-die measurement result

No.	Die		Punch		Punch-die clearance	
	Length	Width	Length	Width	Length	Width
A	7.43	1.03	7.31	0.96	0.12	0.07
B	7.42	1.03	7.32	0.96	0.10	0.07
C	7.42	1.03	7.33	0.96	0.09	0.07
D	7.42	1.02	7.32	0.96	0.10	0.06

4. Experiment of micro-blanking process

The 5kN μ-Forming machine was used to conduct a micro-sheet metal forming process. The micro-blanking tool was mounted in the machine and executed some blanking processes (Fig. 15). The blanking process was held to a copper foil with a variation of punch velocity, which is 0.5 mm/s and 5 mm/s. No lubrication was implemented in the process.

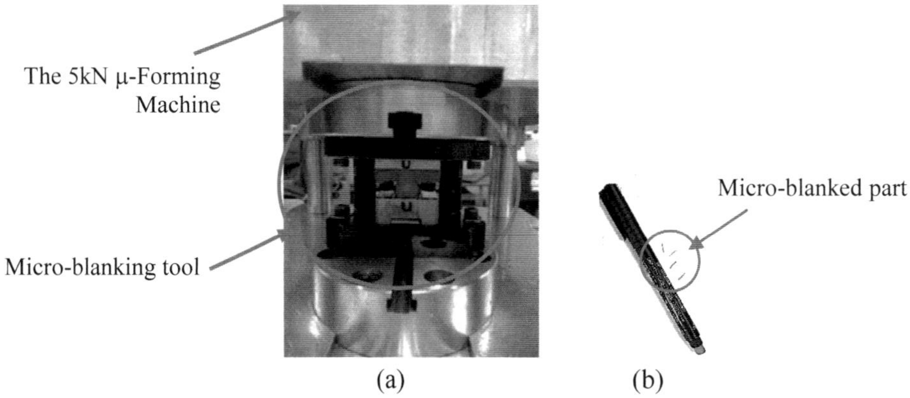

Figure 15. Micro-blanking process: (a). the mounted tool and (b). the blanked-part

The blanked-parts quality was figured out by observing the shear edge on length and width orientation of material. Since the clearance on length orientation was bigger than the other one, the shear edge observation was conducted by using Dyno-Lite digital microscope PRO2 AD-4113ZT. The shear edge of the copper foil in length orientation is shown in Fig. 16.

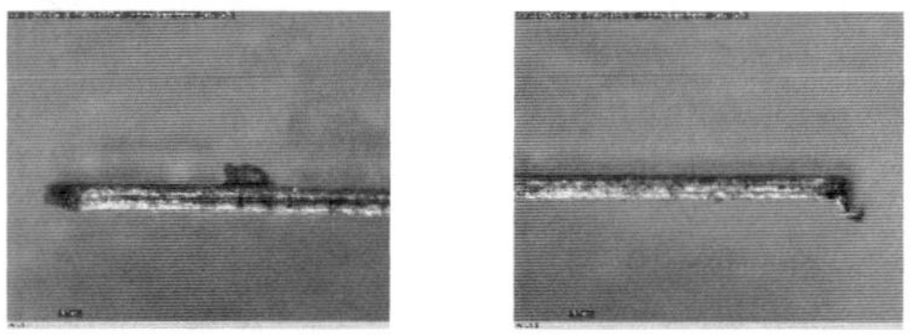

(a) (b)

Figure 16. Shear edge observation of copper foil in length orientation with (a). 0.5mm/s and (b) 5mm/s punch velocity [10]

However, the shear edge on width orientation was observed by using scanning electron microscope Jeol JSM-6510LA. The shear edge of the copper foil in wide orientation is shown in Fig. 17. A proportion of shear edge is occurred in micro blanking part same as in macro range, that are zone of rollover (R_z), shear (S_z), fracture (F_z), and burr (B_z).

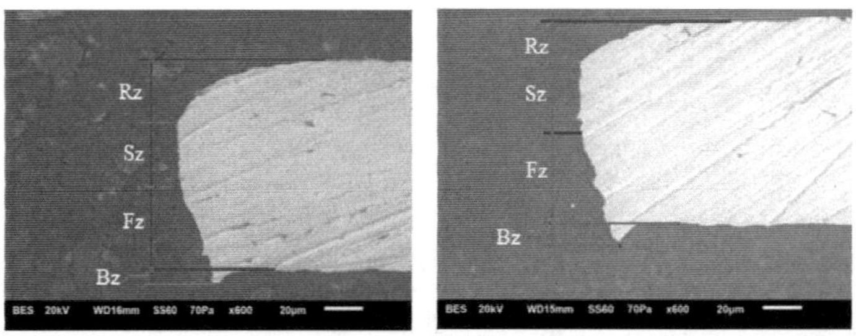

(a) (b)

Figure 17. Shear edge proportion of copper foil in width orientation with (a). 0.5mm/s and (b) 5mm/s punch velocity

5. Conclusions

The micro-blanking tool can be manufactured successfully and used well to conduct the micro-blanking experiment. Nevertheless, since the displacement exceeds 10% of the material thickness, it should to be controlled periodically, especially when the steel foil is used as a blanking material. A more tight clearance can be obtained if the machining method and parameters are chosen appropriately. For further works, another smaller punch-die manufacturing should be investigated. A tool wear analysis and investigation should be conduct to figure out the punch-die performance.

A urethane material can be used as a stripper, although it was stickier with the punch because it has a rough roughness and a light mass. Therefore, urethane cannot be returned to its original position by itself. However, because the simplicity of construction, the design and fabrication of urethane in micro tool should be considered to be developed.

References

[1] F. Vollertsen, *Micro Metal Forming*. Germany: Springer, 2013.
[2] S. M. Yi, B. Y. Joo, M. S. Park, C. N. Chu, and S.-I. Oh, "Mechanical punching of 15 μm size hole," *Microsystem Technologies,* vol. 12, pp. 877-882, 2006.
[3] G.-L. Chern and S.-D. Wang, "Punching of noncircular micro-holes and development of micro-forming," *Precision engineering,* vol. 31, pp. 210-217, 2007.
[4] J. Xu, B. Guo, D. Shan, C. Wang, J. Li, Y. Liu, *et al.*, "Development of a micro-forming system for micro-punching process of micro-hole arrays in brass foil," *Journal of Materials Processing Technology,* vol. 212, pp. 2238-2246, 11// 2012.
[5] K. Fujimoto, M. Yang, M. Hotta, H. Koyama, S. Nakano, K. Morikawa, *et al.*, "Fabrication of dies in micro-scale for micro-sheet metal forming," *Journal of Materials Processing Technology,* vol. 177, pp. 639-643, 7/3/ 2006.
[6] J.-T. Gau, S. Teegala, K.-M. Huang, T.-J. Hsiao, and B.-T. Lin, "Using micro deep drawing with ironing stages to form stainless steel 304 micro cups," *Journal of Manufacturing Processes,* vol. 15, pp. 298-305, 4// 2013.
[7] J. Xu, B. Guo, C. Wang, and D. Shan, "Blanking clearance and grain size effects on micro deformation behavior and fracture in micro-blanking of brass foil," *International Journal of Machine Tools and Manufacture,* vol. 60, pp. 27-34, 9// 2012.
[8] P. D. VUKOTA BOLJANOVIC, *SHEET METAL FORMING PROCESSES AND DIE DESIGN*. 200 Madison Avenue New York, New York 10016: Industrial Press Inc, 2004.
[9] M. Coorporation, "MiSUMi Standard Component for Press and Die," M. Coorporation, Ed., ed, 2012, pp. 1121-1130.
[10] A. M. Gandjar Kiswanto, Sugeng Supriadi, and Ken-ichi Manabe, "The Influence of Die Clearance and Punch Velocity in Micro-Blanking Process," in *The 14th International Conference on QIR (Quality in Research)*, Lombok, Indonesia, 2015, pp. 231-235.

The Improvement of the Closed Bounded Volume (CBV) Evaluation Methods to Compute a Feasible Rough Machining Area Based on Faceted Models

Himawan Hadi Sutrisno[1], Gandjar Kiswanto [2,*], Jos Istiyanto [2]
[1] Mechanical Engineering Dept., Universitas Negeri Jakarta, Indonesia
[2] Dept. of Mechanical Engineering, Universitas Indonesia, Depok, Indonesia

corresponding author: gandjar_kiswanto@eng.ui.ac.id

Abstract. The rough machining is aimed at shaping a workpiece towards to its final form. This process takes up a big proportion of the machining time due to the removal of the bulk material which may affect the total machining time. In certain models, the rough machining has limitations especially on certain surfaces such as turbine blade and impeller. CBV evaluation is one of the concepts which is used to detect of areas admissible in the process of machining. While in the previous research, CBV area detection used a pair of normal vectors, in this research, the writer simplified the process to detect CBV area with a slicing line for each point cloud formed. The simulation resulted in three steps used for this method and they are: 1. Triangulation from CAD design models, 2. Development of CC point from the point cloud, 3. The slicing line method which is used to evaluate each point cloud position (under CBV and outer CBV). The result of this evaluation method can be used as a tool for orientation set-up on each CC point position of feasible areas in rough machining.

1. Introduction

Along with the development in the manufacturing industry, milling machine models have gone through a transformation with the purpose of fulfilling the consumers' needs. Although this transformation has significantly improved the complex models, the 5-axis milling still has its limitation. Also, there hasn't been any improvement of the machining method especially for rough machining. Until now, the rough machining commonly adopts 3-axis machining methods. Certain models which contain the CBV need specific handling. This type of machining depends on machine operator's experience so it is likely for error to occur in the product.

At the same time, the developers are still trying to improve the model to solve other difficulties in the manufacturing. To improve productivity, a new development method is needed for the machine to reduce the total machining time. This paper introduced a more advanced method called 5-axis milling for improving the rough machining efficiency.

2. Related work

In order to reduce errors during handling made by the operators, a few researchers have attempted to optimize the rough machining methods[1-5], the tool orientation methods[6-8], the prevention of collision [9, 10], integrated CAM system[11], and the improvement of tool sequence selection[12]

There have been a few studies conducted discussing the CBV evaluations. Woo[13] introduced the concept of visibility cone which can be mapped onto a unit sphere. This visibility method provides an alternative way for detection if there is an obstacle on the workpiece surface during the machining process. Suthunyatanakit, *et al.* [14]evaluated CBV by putting an obstacle on a light ray on the surface machining. If the light ray is blocked by the obstacle, then the area will be focused to calculate inaccessibility indication where it is not visible by the simple method.

Balasubramaniam *et. al.* [10] proposed that graphic engine was used for visibility computation for generating collision free in 5-axis machining process especially for finishing. This concept uses a graphic hardware to show the machine limitation, the tool tilt and the accessibility tool for surface machining. Li, *et al.* [15] used the concept of visibility to describe the reachability of a light ray to the object surface.

In previous research, development CBV grouping method for complex faceted models has been presented by Kiswanto and Panuju[16] by pairing two triangles of normal vectors in one particular field and the two triangles were grouped in a bucket. The two triangles of normal vectors were grouped according to the machining characteristic surface which are CBV area and OBV area. The OBV area is usually called free surface without any obstacles for other surfaces. It needs a normal vector visualization which occupies the workpiece surfaces. In one workpiece model, the many triangles can be used to simplify the calculation process and for this purpose, bucketing method is required.

3. Improvement of CBV evaluation methods to compute feasible rough machining

In this study, the database for the calculation is the faceted models (triangulation), or also known the triangular mesh, polyhedral models, as well as tessellated models. This model has been used by many researchers due to its advantages in data processing and time computing. Since the CAD design is represented by a triangle, the shape of the triangulation result is an approximation. The accuracy of triangulation depends on the number of triangle resolution. Higher resolution results in a better shape visualization.

To get visualization needed, the computation uses Matlab software with STL file as the first data. The STL file contains information related with the workpiece model, the features and the dimension of the workpiece which should be computed. The Figure 1 below is the CAD models while figure 2 illustrates STL file formed. It can be seen in figure 2, the information needed in STL file is the triangle positions in an indexed list as calculation objects.

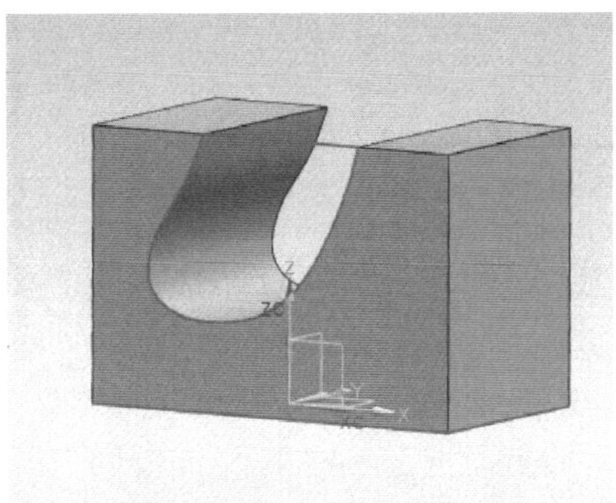

Figure 1. CAD models

```
solid
  facet normal -1.0000000E-00 +1.5789838E-16 +1.7368822E-15
    outer loop
      vertex   -5.0000000E+01 -1.2500000E+02 +0.0000000E+00
      vertex   -5.0000000E+01 -1.2500000E+02 +9.0000000E+01
      vertex   -5.0000000E+01 -3.5000000E+01 +9.0000000E+01
    endloop
  endfacet
  facet normal -1.0000000E-00 +1.7368822E-15 +1.5789838E-16
    outer loop
      vertex   -5.0000000E+01 -3.5000000E+01 +9.0000000E+01
      vertex   -5.0000000E+01 -3.5000000E+01 +0.0000000E+00
      vertex   -5.0000000E+01 -1.2500000E+02 +0.0000000E+00
    endloop
  endfacet
  facet normal +0.0000000E+00 +0.0000000E+00 -1.0000000E-00
    outer loop
      vertex   +1.4000000E+02 -1.2500000E+02 +0.0000000E+00
      vertex   -5.0000000E+01 -1.2500000E+02 +0.0000000E+00
      vertex   -5.0000000E+01 -3.5000000E+01 +0.0000000E+00
    endloop
  endfacet
  facet normal +0.0000000E+00 +0.0000000E+00 -1.0000000E+00
    outer loop
      vertex   -5.0000000E+01 -3.5000000E+01 +0.0000000E+00
      vertex   +1.4000000E+02 -3.5000000E+01 +0.0000000E+00
      vertex   +1.4000000E+02 -1.2500000E+02 +0.0000000E+00
    endloop
  endfacet
  facet normal +0.0000000E+00 -1.0000000E+00 +0.0000000E+00
    outer loop
      vertex   -5.0000000E+01 -1.2500000E+02 +9.0000000E+01
      vertex   +3.8759826E+00 -1.2500000E+02 +6.6545677E+01
      vertex   +4.9288276E+00 -1.2500000E+02 +6.8940863E+01
    endloop
  endfacet
  facet normal +0.0000000E+00 -1.0000000E+00 +0.0000000E+00
    outer loop
      vertex   +5.6132195E+01 -1.2500000E+02 +2.5406290E+01
      vertex   +5.3258520E+01 -1.2500000E+02 +2.5743900E+01
      vertex   +5.0592502E+01 -1.2500000E+02 +2.5678484E+01
    endloop
  endfacet
```

Figure 2. STL file form

Referring to the explanations above, the principles of this method are :

3.1. STL reader and positioning to coordinate system

The STL files (ASCII format) contains the following data:

```
solid
     Facet normal---- ---- ----
       Outer loop---- ----- -----
          Vertex ---- ----- -----
          Vertex ---- ----- -----
          Vertex ---- ----- -----
      Ensloop
     Endfacet
--------
-----
End
```

The content of STL file above are elaborated below:

1. **Solid, End Solid**, marks faceted models representation
2. **Facet normal**, provides information on the position of normal vector on a triangle
3. **Outer loop, end loop** are the looping of the coordinates of the vertices
4. **Vertex** is the point position of the triangle on the coordinate system.

The following is the data processing model used to manage the STL file which contains the data facet and vertices that form the triangle. The function of STL reader can be seen in psedocode below:

```
folder = 'folder name';
filename = 'file name';

stlpath = strcat(folder, '/', filename, '.txt');
triangles_csv = strcat(folder, '/', filename, '_t.csv');
vertices_csv = strcat(folder, '/', filename, '_v.csv');

if exist(triangles_csv) && exist(vertices_csv)
 disp(['Read from file...',triangles_csv,' and ', vertices_csv, '.']);
 T = csvread(triangles_csv);
 V = csvread(vertices_csv);
else
  [T, V] = stlreader(stlpath);
 csvwrite(triangles_csv, T);
 csvwrite(vertices_csv, V);
end
```

The Matlab software is used to compute the above function to get information and position of all triangles at coordinated positions. In addition to the STL read import function, here are two STL reads with *m* files available on the Matlab file exchange.

3.2 Generating of point clouds

By triangulating the workpiece, the length, width and height of the workpiece are calculated from triangle position against a three-dimensional plane. Location of the furthest triangle on X axis identifies the maximum value of x and the same thing shall apply to Y axis and Z axis. The workpiece dimension will be counted from X max to X min, for Y axis, from Y max to Y min, for Z axis, Z max to Z min

In the point cloud creation process, which was explained in figure 3, the distances between point cloud towards the flat plane as well as vertical plane are determined by the density value. Initial formation of point cloud is carried out along a flat plane for example XY plane from the minimum value to the maximum value. The point cloud here is a virtual point construction to simplify calculation and it is used as initial reference at the work piece surface. As explained before, every point cloud on XY place at elevation z = 0 is also made on each altitude level (Z level) with the value between levels of density equal to density of point cloud. The making of point cloud on every Z level is a layering step on the feeding process, considering the depth of cut on the machining process.

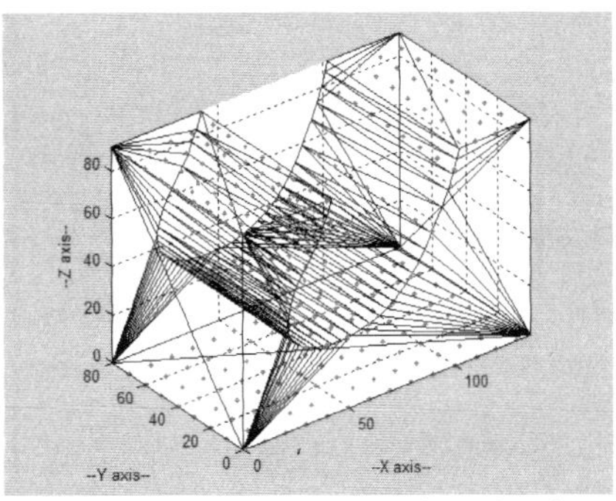

Figure 3. generated point cloud

3.3 Slicing line method

By using ray triangle intersection method, by Moller & Trumbore[17], this process produces less memories in computation. This step is needed to compute the intersection of ray with a triangle and then to compute the barycentric coordinates of that intersection. Slicing method been applied in research[18]. Look at figure 4 below:

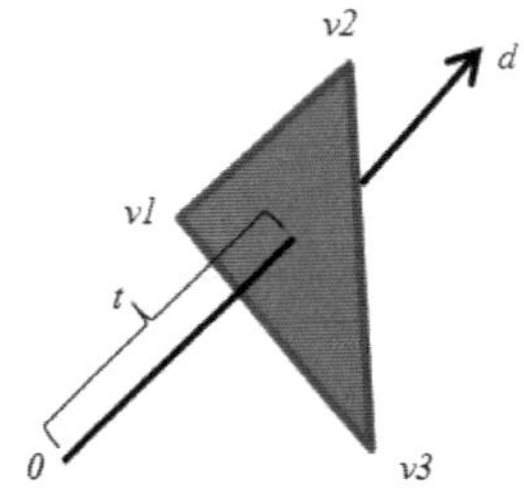

Figure 4. slicing line with ray triangle intersection method[17]

A ray with origin *o* and direction *d* intersects a triangle defined by its vertices, v_o, v_1, and v_2 and the ray intersects with a point at triangle areas by the equation of plane:

$$d = ax + by + cz \quad (1)$$

When *a, b, c* are coefficient forms, a vector of the plane is $n = [a\ b\ c]^T$ so the equations are

$$n\,.x = d \quad (2)$$

And

$$x = [x\ y\ z]T \quad (3)$$

To calculate the intersection of the ray and the triangle, the equation above becomes:

$$f(t) = o + t.d \quad (4)$$

and

$$n\,.f(t) = d \quad (5)$$

$$n\,.[o + t\,.d] = d \quad (6)$$

$$t = \frac{d - n.\ o}{n\,.d} \quad (7)$$

The function and simulation result (figure 5) are shown below:

```
function [flag, u, v, t] = rayTriangleIntersection (o, d, p0, p1, p2)
 Input:
    o : origin.
    d : direction.
    p0, p1, p2: vertices of the triangle.
 Output:
    flag: (0) Reject, (1) Intersect.
    u,v: barycentric coordinates.
    t: distance from the ray origin.
    epsilon = 0.00001;

    e1 = p1-p0;
    e2 = p2-p0;
    q  = cross(d,e2);
    a  = dot(e1,q); % determinant of the matrix M
     if (a>-epsilon && a<epsilon)
        [flag, u, v, t] = deal(0,0,0,0);
        return;
    end;
```

IOP Conf. Series: Materials Science and Engineering **215** (2017) 012041 doi:10.1088/1757-899X/215/1/012041

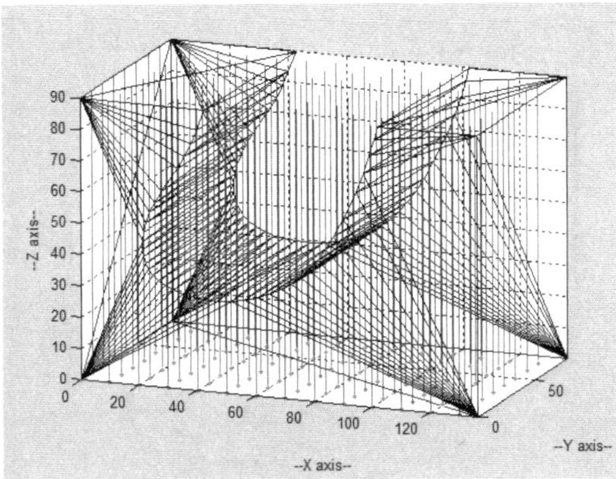

Figure 5. Slicing line method

3.4 Slicing line that cutting thru a triangle

This subchapter will explain how to get the intersection of the slicing line with a model (triangles). As described in the previous subchapter, the virtual point cloud was computed by utilizing of a ray triangle intersection method. The distance of point cloud was formed by the density value. This means the position of the point cloud as the object computation has been determined as the origin of the ray. The ray direction was determined by the direction of Z axis as well as by adding the value of *(k)* in the *(i, j, k)* vectors. Pointed on the equation above, each slicing line will calculate of *(t)* value where the value of *(t)* indicates intersection with triangle. Next, each slicing line will look like a triangle intersection by iterations and will be stored in the data index. The steps to get the triangle to intersect with the slicing line as following:

```
%% cutting_sl:  find all SL that cutting thru a triangle.
%% This returns sub elements of points cloud 'sub_points'.
function sub_points = cutting_sl(tri_vertices, points_cloud)
    mins         = min(tri_vertices);
    maxs         = max(tri_vertices);
    density      = points_cloud(1,2,1) - points_cloud(1,1,1);
    min_x        = points_cloud(1,1,1);
    from_x       = mins(1);
    sl_x_from    = ceil( (from_x + density - min_x) / density );
    to_x         = maxs(1);
    sl_x_to      = floor( (to_x + density - min_x) / density );
    min_y        = points_cloud(1,1,2);
    from_y       = mins(2);
    to_y         = maxs(2);
    sl_y_from    = ceil( (from_y + density - min_y) / density );
    sl_y_to      = floor( (to_y + density - min_y) / density );
    sub_points   = points_cloud( sl_y_from:sl_y_to, sl_x_from:sl_x_to, : );
end
```

3.5 CBV determination

According with the previous description, for each triangle intersected by the slicing line will be stored at the database and will be evaluated for CBV area grouping. The grouping method is illustrated in figure 6 below. The basic concept is to classify the location of each point cloud based on position at the models. The point cloud which is on CBV area is located between two solid models of a slicing line whereas the point cloud as OBV area is located outside of CBV area and beyond solid models. The point cloud at CBV area can be seen in figure 7 below.

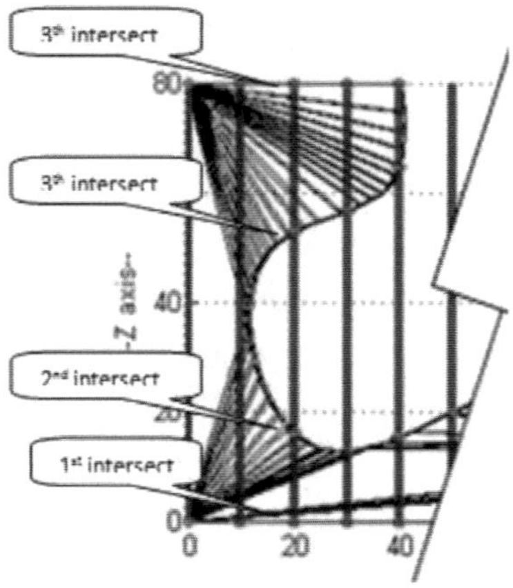

Figure 6. the number of triangle intersection

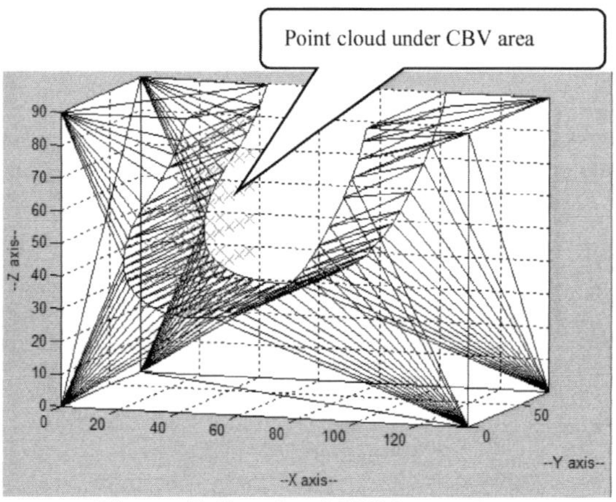

Figure 7. Result of CBV grouping

4. Result And Discussions

In this study, the CC points under the CBV area were determined by using virtual points in the form of point clouds. The distances among the point clouds were used as the base value, or what is known as the density value. Each point cloud became the reference to determine the grouping: CBV or OBV area. Meanwhile, in the previous study, the CBV and OBV are was determined before the point clouds were made and therefore, there were steps previously done to determine the machining area which was simplified in this study.

Table 1. the difference between the previous and the new study.

	Steps for determination	Previous method	New method
1	STL reader	*yes*	*yes*
2	Plot normal vector	*yes*	*no*
3	Calculate of average point	*yes*	*no*
4	Bucketing method	*yes*	*no*

5	Generating of point cloud	*yes*	*yes*
6	Paired normal vector evaluation	*yes*	*no*
7	Map matrix	*yes*	*no*
8	CBV grouping	*yes*	*yes*
9	Double CBV detection	*yes*	*no*

5. Conclusion

Compared to the previous concept, this method is simpler and more suitable for rough machining because the points used as computation objects are virtual points with certain distances so that the result data is still rough. The previous method is more suitable for finishing process because main point of object computation uses triangles at surface curves.

References

[1] B. L. G. K. J. P. Kruth, "Development of a Five axis Milling Tool Path Generation Algorithm based on faceted model," 2003.
[2] S. C. Park and Y. C. Chung, "Tool-path generation from measured data," *Computer-Aided Design,* vol. 35, p. 9, 2003.
[3] H.-T. Young, L.-C. Chuang, K. Gersehwiller, and S. Kamps, "A five axis Rough Machining Approach for a Centrifugal Impeller," *The International Journal of Advanced Manufacturing Technology,* p. 16, 2003.
[4] H. T. Young, L. C. Chuang, K. Gerschwiler, and S. Kamps, "A five-axis rough machining approach for a centrifugal impeller," *The International Journal of Advanced Manufacturing Technology,* vol. 23, pp. 233-239, 2004.
[5] M. Balasubramaniam, P. Laxmiprasad, S. Sarma, and Z. Shaikh, "Generating 5-axis NC roughing path directly from a tesselated representation," *Computer-Aided Design,* vol. 32, p. 17, 2000.
[6] Y.-S. Lee, "Admissible tool orientation control of gouging avoidance for 5 axis complex surfce machining," *computer aided design,* vol. 29, 1997.
[7] K. C. Cha-Soo Jun, Yuan-Shin Lee, "Optimizing tool orientation for 5-axis machining by configuration space search method," *computer aided design,* vol. 35, p. 18, 2003.
[8] G. Kiswanto, H. H. Sutrisno, and J. Istiyanto, "Development of Initial Tool Orientation Method At Close Bounded Area for 5-Axis Roughing Based On Faceted Models," *ICMM,* vol. 3, 2016.
[9] G. Kiswanto, B. Lauwers, and J.-P. Kruth, "Gouging elimination through tool lifting in tool path generation for five-axis milling based on faceted models," *Int. J Adv Manuf Technol,* vol. 32, p. 21, 2007.
[10] M. Balasubramaniam, S. E. Sarma, and K. Marciniak, "Collision-free fininshing toolpath from visibility data," *Computer-Aided Design,* vol. 35, p. 16, 2003.
[11] J. P. K. B. Lauwers, P. Dejonghe, R. Vreys, "Efficient NC programming of Multi axis Milling Machine Through the intregation of tool path Generation and NC simulation," *computer aided design,* 2000.
[12] A. Krimpenis and G.-C. Vosniakos, "Optimisation of Multiple tool CNC Rough Machining of a Hemisphere as a Genetic Algorithm Paradigm Application," *Int. J Adv Manuf Technol,* vol. 20, p. 8, 2002.
[13] W. TC, "Visibility maps and spherical algorithms," *Computer Aided Design,* vol. 26, p. 10, 1994.

[14] K. Suthunyatanakit, E. L. J. Bohez, and K. Annanon, "A new global accessibility algorithm for a polyhedral model with convex polygonal facets," *Computer-Aided Design,* vol. 41, p. 14, 2009.
[15] Y. Li and M. C. Frank, "Computing non-visibility of convex polygonal facets on the surface of a polyhedral CAD model," *Computer-Aided Design,* vol. 39, pp. 732-744, 2007.
[16] G. Kiswanto. and A. Y. T. Panuju, "Development of closed bounded volume(CBV) grouping method of complek faceted model through CBV Boundaries identification.," *IEEE,* vol. 3, p. 5, september 2010.
[17] T. Moller and B. Trumbore, "Fast MinimumStorage RayTriangle Intersection," 1997.
[18] S. C. Park, "Sculptured surface machining using triangular mesh slicing," *Computer-Aided Design,* vol. 36, 2004.

A qualitative and quantitative analysis of vegetable pricing in supermarket

Suci Miranda

Industrial Engineering Department, Faculty of Industrial Technology, Universitas Islam Indonesia, Indonesia

suci.miranda@uii.ac.id

Abstract. The purpose of this study is to analyze the variables affecting the determination of the sale price of vegetable which is constant over time in a supermarket qualitatively and quantitavely. It focuses on the non-organic vegetable with a fixed selling price over time such as spinach, beet, and parsley. In qualitative analysis, the sale price determination is influenced by the vegetable characteristics: (1) vegetable segmentation (low to high daily consumed); (2) vegetable age (how long it can last related to freshness); which both characteristic relates to the inventory management and ultimately to the sale price in supermarket. While quantitatively, the vegetables are divided into two categories: the leaf vegetable group that the leaves are eaten as a vegetable with the aging product (a) = 0 and the shelf life (t) = 0, and the non-leafy vegetable group with the aging group (a) = a+1 and the shelf life (t) = t+1. The vegetable age (a) = 0 means they only last for one day when they are ordered then they have to terminate. Whereas a+1 is that they have a longer life for more than a day such as beet, white radish, and string beans. The shelf life refers to how long it will be placed in a shelf in supermarket in line with the vegetable age. According to the cost plus pricing method using full price costing approach, production costs, non-production costs, and markup are adjusted differently for each category. There is a holding cost added to the sale price of the non-leafy vegetable, yet it is assumed a 0 holding cost for the leafy vegetable category. The amount of expected margin of each category is correlated to the vegetable characteristics.

Keywords: perishable product, vegetable pricing, selling price, vegetable characteristics, cost plus pricing

1. Introduction

This study analyzes the variables affecting the determination of the sale price of non-organic vegetables with the characteristics of fixed selling prices throughout the year in supermarket in Yogyakarta. The initial observation (a 6-month-sale-price observation of vegetables conducted in supermarket) found that some vegetables in the supermarket have a fixed sale price over time, such as water spinach and spinach. A fixed price means a constant sale price of vegetables which is not increasing nor decreasing in a long term. The determination of sale price of a product is an important decision that should be made by a company because the effect on demand and profitability. Company's decision in determining the price of vegetables may affect the development, existence, and its business setbacks [11]. While for consumers, a sale price is something they must sacrifice to get what they want, so it is greatly influence the consumer's decision to buy or not the products offered [1, 5, 12]. Thus, some variables will be considered in determining a sale price of vegetables in supermarket including the perishability, the competitors - traditional markets as the main place providing vegetables, inflation - foodstuffs rose 1.56 percent in 2015 according to Badan Pusat Statistik (BPS) and so forth, in order to maintain the loyalty of consumers shopping vegetables in supermarket as well as making a profit.

Some modern vegetable stall businesses began to appear like Pak Tani with three branches, Ijo Royo-Royo, TOSAGA with two branches, and Warung Ibu Subur. However, these efforts have not been followed by a good method of determining a sale price of vegetables. Unfortunately, those business owners simply determine the price by increasing sum percentage from the purchase price without having any adequate informations how to formulate the sale price. Whereas the main business purpose is to gain benefit. Comparison of the selling price of vegetables in modern vegetable stalls are lower than the supermarkets. This study analyzes the variables influencing the determination of sale price in supermarket in order to provide a clear guide line for the modern vegetable stalls to formulate their sale price of vegetables. For further, if the determination of sale prices in supermarket could be applied at the modern stalls, their business processes would be optimized.

There is a research modelled an EOQ for the inventory management of perishables in small medium-sized grocery stores using a regression approach from the product's holding, lifetime, and markdown policy. It recorded that managers frequently utilize markdown to stabilize demand when the products expiration date nears [7]. While in this sale price analysis study, the markdown policy is not applied in supermarket observed. Yet, the lifetime will be considered as one of the variables influencing the sale price determination. It is along with another research that formulates and analyzes an EOQ model with demand decreasing in the age of the product [6]. The optimal order cycle length incorporates consumer behavior towards freshnes has been derived as the result. Related to consumer behavior, it is acknowledged that consumers will pick the fresher products between products with different ages [3]. It proposes mathematical models that are able to differentiate between different functions of the age dependent demand and/or products with or without a stamped *best-before-date*. The consumer behavior has been analyzed into three perspective: age dependent demand (ADD), customer's eagerness for fresher products (CFP), and product quality risk (PQR). Therefore, based on those three researches of perishability, this study considers the lifetime and consumer behavior affecting the sale price of vegetables in supermarket. Furthermore, there is also a holding cost added in the sale price.

The vegetables price in traditional market is set as a constraint which is not included to the discussion. For instance, a research conducted in Nairobi, Africa provided a detailed analysis of the African Leaf Vegetables (ALV) market such as the promotional strategies of local NGOs and international organizations, increased health awareness, and improved presentation of ALVs in supermarkets and upmarket groceries [9]. There are a few researches employing a quantitative and qualitative analysis related to vegetables. Geneese valley of New York is an exporter of many agricultural products including a number of fruit and vegetable crops. By using questionnaires to collect data on crops, acreage, production volume of leading crops, market channels, and product flow from a total of 227 responses (205 responses are from fruit and vegetable growers, and 22 are from producers of small grains), this study provides a qualitative and quantitative measures of fruit and vegetable production in the Geneese valeey [13]. A qualitative analysis using Williamson's governance structure is also employed to present which type of contractual relationships used by farmers in their business in Romania [2]. While the quantitative analysis using a probit models were used to perceive the determinants of the farmers' market selling preference. The data is collected from 64 farmers and 6 processors in the S-E region of Romania. Another similar study is conducted in Netherlands using questionnaire analyzes fruit and vegetable supply chain between greengroceries and supermarkets. It shows a qualitative analysis to examines Dutch greengroceris position in the market, their competitiveness against supermarkets, as well as to point out changes in consumer habits and define trends in consuming fruit and vegetables. The quantitative method provides an exact statistics data to define ration, set up an order, show layers, examine the partterns and opinion of the average consumers [10]. Furthermore, a study analyzing the impact of supermarkets on traditional markets and retailers in urban centers in Indonesia has been conducted combining a quantitative and qualitative method [4]. The quantitative evaluation uses the different-in-different (DiD) which records two time periods – before and after the treatment of an opening supermarket. Meanwhile, the qualitative evaluations is in the form of in-depth-interviews with key informants.

In fact, the vegetables supplied in supermarket observed are directly from its suppliers with more than 10 suppliers in order to fulfil the vegetables demand in it. This paper employs a cost plus pricing method for the quantitative analysis which is relevant with the informations gained during the interwievs with suppliers and supermarket observed. It is used to find the amount of each variables contributed in formulating the vegetable sale price such as a holding cost and markup. Moreover, in order to obtain the data for qualitative analysis, it does not provide a questionnaire neither to suppliers nor to supermarkets but depth interviews. This study contribution is to investigate a number of variables influencing the determination of vegetable sale price in supermarket using a quantitative analysis of cost plus pricing method and a qualitative analysis of depth interviews with suppliers, supermarkets, and other related parties such as department of agriculture in Yogyakarta. Its objective is to provide a better guide line for modern stalls in determining their vegetable sale price so that their business can be optimized.

2. Methodology

The object of research is non-organic vegetables sold in one national supermarket in Yogyakarta called supermarket X with 8 branches focusing on the fixed sale price vegetables group over time where the study was conducted. There are 24 samples identified as the vegetable group of the fixed sale price during the period of the study. The data collected are the sale price from suppliers to supermarket X and from supermarket X to consumers. However, due to the difficulties in obtaining the suppliers sale price data as it is confidential, the data needed can only be provided by one main supplier of supermarket X, called supplier A, supplying vegetables for the 8 branches of supermarket X in Yogyakarta including several branches of supermarket X in Solo, Magelang, Semarang, and it has established its vegetable supplying business since 2006 until today.

It is referred to Δh as the gap selling price between supplier A and supermarket X. For many kinds of perishable goods, the quality assessed by the consumer gradually decreases as items age. Therefore, one of the qualitative variables analyzed is the vegetables characteristic related to age (vegetable is an aging product). This study also considers another qualitative variable that is the vegetable segmentation defined by consumer behavior in buying vegetables in supermarket. These variables are concluded during the interviews with supplier A. The quantitative analysis of the variables in the determination of vegetables sale prices in supermarket using a method of cost plus pricing in which Δh (%) will be identified into some influenced variables.

Data processing are as follow. (1) Observation and data collection of vegetable sale price. Based on the preliminary interview to relevant parties in Yogyakarta such as government services market, department of agriculture, department of trade economy and cooperatives, and market seller at a main market Giwangan; and literature studies related to the pricing method, it is concluded that the sale prices of vegetables at supermarket approached by cost plus pricing method. There are 24 vegetables samples of fixed sale price in supermarket X. Vegetables price data is collected from one experienced supplier and one national supermarket as a representative of many suppliers and supermarkets in Yogyakarta. (2) Vegetables are classified into two categories based on morphological features: leaf vegetables and non-leafy vegetables [8]. Morphology is part of biology which studies the form and structure of organism and their specific structural features including aspects of the outward appearance such as shape, structure, colour, pattern, and size. The leaf vegetables or green vegetables are plant leaves eaten as a vegetable that spinach, water spinach, kenikir, and green mustard as the examples. While the non-leafy vegetables means the leaves are not consumed as the main course but their root or stalk. A coneshaped objects like corn is part of this category. According to the interviews with supplier A and supermarket X, there are different treatment between spinach and white radish dealing with the freshness, and the purchase order. Therefore, this study concludes that the vegetables sold in supermarket have their own considerations regarding the sale price. Thus, it is important to divide them into two categories based on their characteristics. First, leaf vegetables or vegetable greens with the aging product (a) is 0 as their freshness will decrease fast so they cannot last for more than one day. Moreover, they have no shelf life (t=0), due to the high level of spoilage so supermarket always replaces the leaf vegetables every day. It means that they can only be sold at the shelf on the same day they purchased from supplier A. They must be terminated on the next morning by replacing them with the fressh new vegetables. The second is the non-leafy vegetables with the aging group (a) is a+1 (they can last for more than one day) and their shelf life (t) is t+1. These vegetables will be placed at the shelf for more than one day (t+1) as their freshness can last than a day (a+1) from the day they are purchased. They are sold in packaging such as oyster mushroom and bean sprouts which have an expired date sticked on the packing, and in kilogram without packaging that have a random lifetime such as beet and white radish (there is no expired date). This classification is necessary because there will be different analysis of each category both quantitatively and qualitatively. (3) Consumer behavior in buying vegetables in a supermarket. Each category of vegetable will be analyzed using consumer behavior approach regarding with the aging product and freshness level of the vegetables. Consumer behavior will also have an impact on safety stock of vegetables provided in the shelf and supermarket ordering method to supplier. (4) Determining the sale price using cost plus pricing analysis. It is used for the quantitative analysis where a gap sale price (Δh) between supplier A and supermarket X found. The variable identification of gap amount (Δh) for both vegetable categories is associated with the qualitative analysis of vegetable segmentation and vegetables age. This methodology can be seen clearly in Figure 1 below.

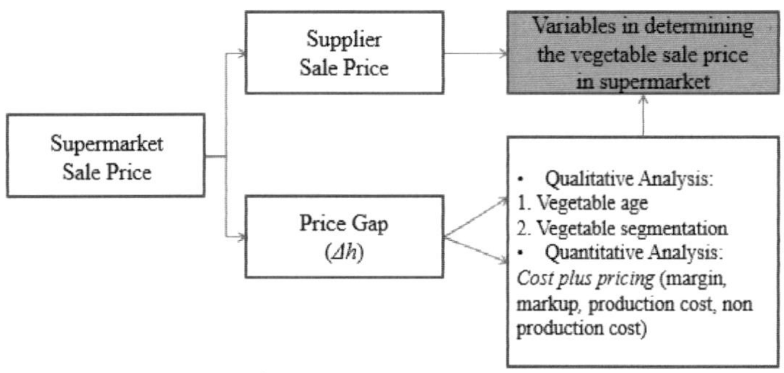

Figure 1. Conceptual model.

3. Vegetable Segmentation in Supermarket

There is a belief among retailers, although it is not documented in research, that consumers do not like a rapid change in price [12]. Sometimes retailers do not alter the sale price consistently along with the change of supply prices from suppliers because a conventional retailer assumes consumers are not comfortable with a frequent price changes condition. This may not be entirely true, but these assumptions affect the decision made by some retailers. When a sale price alters, especially a decline in prices, retailers should not expect consumer purchasing behavior changes quickly. In order to create an effective changing price, consumers must feel the change and respond it. The aim of retailers are not only to maximize sales and profit for each department separately, but also for the entire store. There are many possibilities for substitution of goods between departments, such as frozen vegetables with fresh vegetables, retailers must determine the price and trade policies to benefit the business as a whole system, not just for one particular department.

Based on the observation of the vegetable sale price in 8 branches of supermarket X for 6 months (January – June 2015), it is found that a group of vegetables have the constant price during the observation, some others tend to be constant, and a few of them rapidly changed (including high value economic commodities) such as onions and chilli. This study focuses on the group of vegetables at a constant prices over time that the sale price does not change during the observation conducted. It shows that pakcoy, celery, and spinach are sold in a fixed price of Rp. 2950,-/150gr, and are placed in the same shelf. Genjer leaf, green mustard, water spinach, and kenikir have the same fixed sale price of Rp. 1950,-/150gr which are sold together in another shelf.

The data is clustered to identify the characteristics similarity of each vegetable related to *Δh*. There are four clusters of vegetables. The first cluster consists of beet, kecipir, lencak, parsley, and long bean sprouts have a small *Δh* (<30%) among the clusters. The fourth are white radish, oyster mushroom, and pakcoy have the highest Δh for more than 80%. It can be concluded that these two clusters are kind of rarely consumed vegetable called special segmentation. However, the difference *Δh* of both clusters can be caused by the sales price from supplier A. For example, between beet and kecipir, the average sale price from supplier A is up to Rp. 20.000, - /kg, so supermarket A only increases a little sale price which is less than 15% (9% and 13%). While the second cluster (green mustard, genjer leaf, kenikir, long beans sprouts, okra, string beans, peeled petai, short bean sprouts, chives, water spinach, and peeled pickled corn) have an average *Δh* between 30-40%. Due to the characteristic of high consumed vegetable as a daily main course, and they represent a high daily demand, thus supermarket A has to be adjusted downwards to account for each vegetable. Consider this vegetable segmentation of each cluster, it can be concluded that supermarket A sale price is correlated to the sale price from supplier A with a various level of *Δh*.

4. Vegetable Age Influences Consumer Buying Behavior

According to the observation, there are some vegetables like green mustard, genjer leaf, water spinach, and kenikir are provided at the same shelf with the same sale price of Rp. 1950/150 grams, as shown in Table 1. Spinach, celery, and pakcoy are arranged on same shelf with the sale price of Rp. 2950/bunch of 150 grams. Referring to the variable of safety stock at shelf and spoilage risk, it has been discussed that the fixed sale price vegetable group is divided into two categories: (1) leaf vegetables, and (2) non-leafy vegetables.

In the case of spoilage, the leaf vegetable category has zero inventory because the declining freshness of the perishable product is alligned with the its age as it is kind of a high spoilage vegetable (a=0, t=0). In order to maintain customer satisfaction of freshness, the replenishment occurs every day. The remaining vegetables on the self before supermarket closes will be immediately discarded on the next morning. It is a removal of spoiled product which they will not be sold to consumers. All stock of vegetables ordered by supermarket A each day are assumed to meet consumer demand on day (t =0) [2]. This leaf category does not have a holding cost due to inventory is not provided. While the non-leafy category like beet and string beans, their age can last more than one day (a+1). Average age of vegetable is 4 days [6]. This information is proved by some of the vegetables packed like string beans listed an expired date in supermarket X. The consumer behavior to buy perishable product has a linear correlation with the shelf life of the product [2]. The desire of consumers to purchase a perishable product decreases along with the declining shelf life of the product (willingness to pay decreases). The vegetable purchased meets the demand on day t and still remains some stocks of vegetables on the shelf for the next t + 1 (vegetable age is a + 1). The non-leafy vegetables have an expected lifetime of expired sticked on the packaging (expiration date), and the others have a random lifetime sold in kilogram or without packaging. Therefore, they are assumed to have the holding cost (t+1). Here are the vegetables of each category shown in Table 1 and 2.

Table 1. Leaf vegetable category with a = 0 and t = 0.

Vegetables	Hsupp	Hsm	Δh	Δh%
GREEN MUSTARD 150 GR	1,500	1,950	450	30
GENJER LEAF	1,500	1,950	450	30
KENIKIR 150 GR	1,500	1,950	450	30
WATER SPINACH 150 GR	1,400	1,950	550	39
SPINACH 150 GR	2,000	2,950	950	48

CELERY 150 GR	2,000	2,950	950	48
PAKCOY	1,500	2,950	1,450	97
CHIVES	2,850	3,950	1,100	39
PARSLEY	4,000	4,950	950	24

Table 2. Non-leafy vegetable category with a + 1 and t + 1.

Vegetables	Hsupp	Hsm	Δh	Δh%
STRING BEANS PCK 200 GR	2,000	2,690	690	35
LONG BEAN SPROUTS PCK 200 GR	2,200	2,790	590	27
LENCAK PCK 200 GR	2,500	2,990	490	20
SHORT BEAN SPROUTS PCK 150 GR	2,167	2,990	823	38
LONG BEANS 200 GR	2,300	2,990	690	30
PEELED PETAI 150 GR	2,500	3,390	890	36
OKRA 150 GR	3,000	3,990	990	33
UNPEELED PICKLED CORN 250 GR	2,500	3,990	1,490	60
OYSTER MUSHROOM 250 GR	3,000	5,490	2,490	83
PUMPKIN SIAM WHL	4,000	5,950	1,950	49
PEELED PICKLED CORN 250 GR	5,000	6,990	1.990	40
OYONG WHL	7,000	9,950	2.950	42
WHITE RADISH WHL	6,000	10,950	4,950	83
KENIKIR WHL	15,000	16,950	1,950	13
BEET WHL	22,000	23,950	1,950	9

5. Cost Plus Pricing Analysis

Sale price calculation using Cost Plus Pricing method with full cost pricing approach determines the sale price comprised production costs, non-production costs, and markup. This approach is considered relevant because based on interviews with supermarket X, it is known that the costs such as electricity and water are also taken into consideration in determining the sale price of all products including vegetables which these costs are overhead cost. However, the amount of these costs is not explained by supermarket.

The production costs are those costs incurred for processing raw materials into finished products. In this study, the sale price of vegetables from supplier A to supermarket X is treated as the raw material costs for supermarket X. Each supplier has to follow the regulations related to the vegetables ordered by supermarket X costs such as the packaging, the weight, barcode, and the size. For example, spinach is tied with a weight of 150 grams /bunch, string beans is packed with a weight of 200 grams */pack* and sticked a barcode. The vegetables are ready to serve to the shelves once they arrive at supermarket every morning. Thus, it is concluded that supermarket X does not provide any direct labor costs. All vegetable demand is provided by suppliers so it is assumed direct labor cost at supermarket is 0. The overhead costs are vegetables checker salary, and electricity and water bills. The responsbility of a checker is to check each vegetable supplied from suppliers every morning, the number and the physical condition of vegetables in order to meet the quality standards of supermatket X. The checker also has to check the freshness and neatness of vegetables at their shelves regularly.

For non-production costs such as marketing cost, and general and administration, may be included in the calculation of the sale price. Unfortunately, this information is not obtained due to the confidential issues. Further, it is assumed 0. According to the full cost pricing approach, the sale price will be calculated as follows:

$$\text{The sale price} = \text{Total Production Costs} + \text{Total Non-Production Costs} + \text{Markup} \quad (1)$$

$$Markup = \frac{(Expected\ profit \times (Production\ costs + Non-producton\ costs)) + Non-production\ costs \times 100\%}{Production\ costs} \quad (2)$$

In determining the sale price using cost plus pricing method with full cost pricing approach, there is an expected profit added into it. This expected profit will be treated as a contribution margin rate (CMR) [6].

$$CMR = (p - v)/p \quad (3)$$

Note:
CMR : Contribution margin rate
P : the sale price of vegetables in the supermarket
V : variable costs

5.1 Leaf Vegetable Category with a = 0 and t = 0

The profit expected (CMR) is 30% obtained from the calculation of each vegetable. Because this category is a daily consumed vegetable that consumers have been familiar with them, marketing costs is not needed. Therefore, non-production costs is 0. There is some amount of markup added into the price of vegetables according to full cost pricing approach. Based on the calculation, the markup is as high as the CMR or the expected profit of 30%. Detailed calculation can be seen in Table 3.

Based on the calculation of CMR, pakcoy has the highest value of almost 50%. Within this category, green mustard and pakcoy are in the same family of brassicaceae (biology study) with pakcoy. Yet, demand of green mustard is more than pakcoy related to the taste, so it can substitute pakcoy for a main course. Both of them have the same supplier sale price of Rp. 1500, -/bunch. However, the Δh of the sale price to the end consumer is different. Pakcoy particularly highlighted in yellow in Table 5, it is found a difference amount of sale price (Rp. 1000) between the real sale price *Hsm* and the calculation of sale price using full cost pricing approach. While the other vegetables sale prices calculation using full cost pricing are close to the real sale price of supermarket (*Hsm*). This may be caused by vegetables segmentation whether they are consumed as a daily dish, or they are only for special purpose like parsley which is ordered about 5 bunches/day according to an interview with the supplier A. Most of parsley are served as a garnish instead. Considered the calculation results are close to supermarket price, it can be concluded that the variables of supplier's sale price (*Hsupp*), CMR and Markup are taken into the determination of sale price in supermarket. It should be noted that supermarket applies a unique number to inform the sale price to consumers (Rp. 1,950, Rp. 2,950, Rp. 3,950 and Rp. 4,950).

5.2 Non-leafy Vegetable with a + 1 and t + 1

Because there is stock on the shelf (t + 1), a holding cost must be applied. It is suggested that a holding cost for perishable products is between 15% to 35% [6]. As described in the previous section, vegetables can last for 4 days, unless the freshness will be decrease. Holding cost is set equal to every period t+1 to each vegetable on the shelf. The coloured rows showed in Table 6 differentiate the holding cost adjusted to each vegetable. Meanwhile, the CMR is set at 10% in accordance with the amount suggested for retailers [6]. It is different from the previous category due to holding cost added into the sale price. If CMR is calculated more than 10% or at least as same as the leaf vegetables category of 30%, hence the sale price will be more expensive and it will have an impact on the sales of vegetables.

Generally, vegetables in packaging such as lencak, bean sprouts, and long beans have a 10% *h*. While beet and kecipir who have an expensive raw materials cost (supplier sale price is in average of Rp. 18.000/kg) treated as a low daily consumed. This means that the vegetables are not being consumed by most people but a few so that the supermarket X does not order in large amount (based on interviews with supplier A). Supermarket X committs to provide vary vegetables representing a range of consumers from low daily consumed like beets to high daily consumed. Due to the expensive raw material cost as a low daily consumed, beets and kecipir are adjusted to only 10% holding cost. According to Table 4, white radish and oyster mushrooms are in the same cluster of price. Both of these vegetables have the same Δh of 45%. They are also a category of a low daily consumed vegetable, but the raw material price is quite cheap in average of Rp. 4,500, -, thus, supermarket X can add the highest h of 30%. It is relevant with the recommended holding cost of less than 35% [6]. The other vegetables of 20% h are part of high daily consumed vegetables and the raw material costs is around Rp. 2,500 to Rp. 7,000 (or the value of Δh range between 30% and 40%). The recommendation CMR is 10% for fresh products [6] applied to all vegetables within this category. For the markup estimation, beets and kecipir have no markup due to the expensive costs of raw materials, and they are a low daily consumend vegetable. Therefore, supermarket cannot set a higher markup dealing with an affordable sale price. By using the same consideration in determining the holding cost h, white radish and oyster mushrooms are calculated in the highest markup of 40% among the vegetables. The other vegetables with markup of 20% is a high daily consumed vegetable category.

If we look at the reasons of adjusting h and markup, it shows a pattern of vegetable characteristics consisted of vegetable segmentation (low or high daily consumed vegetable) and the supplier sale price as the raw material costs (from cheap to expensive level). Within this category, there are two groups of vegetable segmentation. It is concluded that for the low consumed vegetable (beet, kecipir, white radish, and oyster mushroom), the highest amount of holding cost (30%) is added into the cheap raw material costs, vice versa. Even though, white radish is sold in kilogram, and oyster mushroom is in packaging of 250 gram. The highest markup (40%) is also added into the cheap raw material costs while the expensive raw material costs have no markup. For the high consumed vegetable, the amount of holding cost is adjusted between 10%-20%. It is incorporated with the 20% markup into the sale price. Finally, the full cost pricing approach has been provided a similar sale price as supermarket did. Table 5 and 6 show the calculation in detail.

Table 3. The determination of sale price of the leaf vegetable with a = 0 and t = 0.

Vegetables	Hsupp	Hsm *P*	CMR *(p-v)/p*	Production Cost Raw Material	Non Production Cost	Markup %	Sale Price (Rp)	Price Gap (Rp)
	(1)	(2)	(2-1)/(2)=(3)	(4)	(5)	(6)	(4)+(6*4)=(8)	(2)-(7)=(9)
GREEN MUSTARD 150 GR	1,500	1,950	0.23	1,500	0	0,3	1,950	-
GENJER LEAF	1,500	1,950	0.23	1,500	0	0,3	1,950	-
KENIKIR 150 GR	1,500	1,950	0.23	1,500	0	0,3	1,950	-
WATER SPINACH 150GR	1,400	1,950	0.28	1,400	0	0,3	1,820	130
SPINACH 150 GR	2,000	2,950	0.32	2,000	0	0,3	2,600	350
CELERY 150 GR	2,000	2,950	0.32	2,000	0	0,3	2,600	350
PAKCOY	1,500	2,950	0.49	1,500	0	0,3	1,950	1.000
CHIVES	2,850	3,950	0.28	2,850	0	0,3	3,705	245
PARSLEY	4,000	4,950	0.19	4,000	0	0,3	5,200	(250)
Mean			0.3					

Table 4. The determination of sale price of the non-leafy vegetable with a + 1 dan t + 1.

Vegetables	Hsupp	Hsm *p*	Δh	%	Production Cost Raw Material	holding cost (%)	*h* (Rp)	CMR 10%
	(1)	(2)	(1)-(2)=(3)		(4)	(5)		(6)
BEET WHL	22,000	23,950	1,950	0.08	22,000	0.10	2,200	2,420
KECIPIR WHL	15,000	16,950	1,950	0,12	15,000	0.10	1,500	1,650
LENCAK 200 GR	2,500	2,990	490	0,16	2,500	0.10	250	300
LONG BEAN SPROUTS 200 GR	2,200	2,790	590	0.21	2,200	0.10	220	264
LONG BEANS 200 GR	2,300	2,990	690	0,23	2,300	0.10	230	276
OKRA 150 GR	3,000	3,990	990	0,25	3,000	0.10	300	360
STRING BEANS PCK 200 GR	2,000	2,690	690	0.26	2,000	0.10	200	240
PEELED PETAI 150 GR	2,500	3,390	890	0,26	2,500	0.10	250	300
SHORT BEAN SPROUTS 150 GR	2,167	2,990	823	0.28	2,167	0.10	216.70	260.04
PEELED PICKLED CORNS 250 GR	5,000	6,990	1,990	0.28	5,000	0.20	1,000	600
OYONG WHL	7,000	9,950	2,950	0.30	7,000	0.20	1,400	840
PUMPKIN SIAM WHL	4,000	5,950	1,950	0.33	4,000	0.20	800	480
UNPEELED PICKLED CORNS 250GR	2,500	3,990	1,490	0.37	2,500	0.20	500	300
WHITE RADISH WHL	6,000	10,950	4,950	0.45	6,000	0.30	1,800	780
OYSTER MUSHROOM 250 GR	3,000	5,490	2,490	0.45	3,000	0.30	900	390

Full Cost (Rp)	Markup (%)	Markup (Rp)	Sale Price (Rp)	Hsm (Rp)	Price Gap (Rp)
(4)+(5)=(7)	(8)	(8)*(7)=(9)	(10)		(1)-(10)=(11)
24,200	0	0	24,200	23,950	250
16,500	0	0	16,500	16,950	(450)
2,750	0.20	550	3,300	2,990	310
2,420	0.20	484	2,904	2,790	114
2,530	0.20	506	3,036	2,990	46
3,300	0.20	660	3,960	3,990	(30)
2,200	0.20	440	2,640	2,690	(50)
2,750	0.20	550	3,300	3,390	(90)
2,383.70	0.20	477	2,860.44	2,990	(129)
6,000	0.20	1,200	7,200	6,990	210
8,400	0.20	1,680	10,080	9,950	130
4,800	0.20	960	5,760	5,950	(190)
3,000	0.30	900	3,900	3,990	(90)
8,400	0.40	3,120	10,920	10,950	(30)
4,200	0.40	1,560	5,460	5,490	(30)

IOP Conf. Series: Materials Science and Engineering **215** (2017) 012042 doi:10.1088/1757-899X/215/1/012042

6. Conclusions

Variables considered qualitatively in determining the sale price of vegetables in supermarket are the vegetable age and vegetable segmentation (leaf vegetable and non-leafy vegetable) which lead to a different estimation of category. The quantitative variables using cost plus pricing method and full cost pricing approach are expected profit called Contribution Margin Rate (CMR), holding cost, markup, production cost, and non-production cost. Since all vegetables are supplied by suppliers, so the supplier sale price to supermarket is treated as the raw material cost in calculating the production cost.

Price is a confidential issue, thus there are some costs such as the salary of checker, electricity, and water bills as the overhead cost assumed 0. There is only one overhead cost calculated, the holding cost h. Non-production cost including the marketing cost is set 0. Finally, it is found a pattern in adjusting the holding cost and markup using 2 variables: the vegetable segmentation whether it is a low or a high daily consumed vegetable and the raw material cost obtained from the supplier sale price. This pattern occurs while the estimation of sale price is being analyzed. There are some informations assumed such as the overhead costs and non-production costs as they are a confidential data. For further research, the data needed could be obtained from more than one supplier and supermarket so it will provide a robust data for analyzing the variables. The sale price of vegetables from modern stalls may also be added in order to gain the gap sale price between supermarkets and the modern stalls. Finally, a sale price calculation can be modelled using the variables influencing the determination of sale price so it will be a better guide line for those moderns stalls in formulating their vgetable sale price.

7. References

[1] Adiyoga W 2011 Faktor-faktor yang mempengaruhi perilaku dan keputusan konsumen untuk membeli kentang, bawang merah, dan cabai merah *J.Hort.* **21**

[2] Alboiu Cornelia 2011 Farmers' choices in the vegetable supply chain: problems and possibilities *Agricultural Economies and Rural Development New Series.* **8** 221-234

[3] Amorim P, Costa A M, and Almada-Lobo B 2013 Influence of consumer purchasing behaviour on the production planning of perishable food *OR Spectrum.*

[4] Daniel S, Adri P, Sri B, Akhmadi, Meuthia R 2007 Impact of supermarkets on traditional markets and retailers in Indonesia's urban centers (Jakarta: The SMERU Research Institute)

[5] Dasipah E, Budiyono H, and Julaeni M 2010 Analisis perilaku konsumen dalam membeli produk sayuran di kota Bekasi *Jurnal Agribisnis dan Pengembangan Wilayah.* **1**

[6] Dobson G, Pinker, Edieal J, and Yildiz O 2016 An EOQ model for perishable goods with age-dependent demand rate *European Journal of Operation Research.* 1-5

[7] Ferguson Mark, Jayaraman Vaidy, and Souza G C 2007 Note: an application of the EOQ model with nonlinear holding cost to inventory management of perishables *European Journal of Operation Research.* **18** 485-490

[8] H D Tindall 1983 Vegetables in the tropics (London: Macmillan Press Ltd.) 533

[9] Irungu Charity 2007 Analysis of markets for African Leafy Vegetables witihin Nairobi and its environs and implications for on-farm conservation of biodiversity (Rome: Global Facilitation Unit for Underutilized Species (GFU))

[10] Katalin J 2012 Competetiveness study of the fruit and vegetable supply chain in the Netherlands (Hungarian: Theses of doctoral (Ph.D.) dissertation)

[11] Kotler Philip and Amstrong Garry 2008 *Prisinp-prinsip Pemasaran* Alih bahasa: Bob sabran MM Edisi 12 jilid 1 dan 2 (Jakarta: Erlangga)

[12] McLaughlin E W 2004 The dynamics of fresh fruit and vegetable pricing in the supermarket channel *Preventive Medicine.* **39** 81-87

[13] Park K S 2012 Quantitative and qualitative measures of fruit and vegetable production in the Geneese valley (New York: Geneese Valley Regional Market Authority and the New York State Department of Agriculture and Markets)

Design of Maternity Pillow by Using Kansei and Taguchi Methods

Fety Ilma Rahmillah[1), Rachmah Nanda kartika[2)

[1,2)] Department of Industrial Engineering, Faculty of Industrial Technology, Islamic University of Indonesia, Jl. Kaliurang Km 14.5 Yogyakarta, Indonesia

fety.rahmillah@uii.ac.id[1)], rachmahnanda@gmail.com[2)]

Abstract

One of the customers' considerations for purchasing a product is it can satisfy their feeling and emotion. It because of such product can enhance sleep quality of pregnant women. However, most of the existing product such as maternity pillows are still designed based on companies' perspective. This study aims to capture the desire of pregnant women toward maternity pillow desired product by using kansei words and analyze the optimal design with Taguchi method. Eight collected kansei words were durable, aesthetic, comfort, portable, simple, multifunction, attractive motive, and easy to maintain. While L16 orthogonal array is used because there are three variables with two levels and four variables with four levels. It can be concluded that the best maternity pillow that can satisfy the customers can be designed by combining D1-E2-F2-G2-C1-B2-A2 means the model is U shape, flowery motive, medium color, Bag model B, cotton pillow cover, filled with silicon, and use double zipper. However, it is also possible to create combination of D1-E2-F2-G2-C1-B1-A1 by using consideration of cost which means that the zipper is switched to single as well as filled with dacron. In addition, the total percentage of contribution by using ANOVA reaches 95%.

Keywords: Maternity Pillow, Taguchi, Kansei, Design

1. Introduction

Process trimester pregnancies from I, II, and III require changes in the anatomy of the body, such as the digestive system, cardiovascular, urinary, and musculoskeletal in pregnant women depends on the pregnancy. Some of the complaints often experienced by women in the third trimester of pregnancy are lower back pain, shortness of breath, varicose veins, haemorrhoid, sleep disorders, diastasis recti, pelvic pain and others [1]. Musculoskeletal discomfort most frequently complained of pregnant women in the third trimester in India is 64.6% suffered a calf muscle cramps, 37.1% reported leg pain, and 33.7% experienced back pain [2]. Whereas in the second trimester, a general musculoskeletal dysfunction experienced by pregnant women are sick calves (47.8%), back pain (42%), and pelvic girdle pain (37%).

Musculoskeletal disorders can be influenced by a variety of physical activity, culture, and environment. Most pregnant women do not seek medical help until the discomfort really began to interfere the activity. The majority of the discomfort associated with the physical changes that occur during pregnancy and the resultant effect on the biomechanical functional movement [2]. Musculoskeletal disorders may be affected by the quality of sleep. Sleep disorders cause depression and stress affects the fetus. Mild stress causes an increase in heart rate of the fetus, but severe stress and old will make the fetus become hyperactive. Further consequence of sleep disorders are depression and babies born have a little time to sleep soundly [3]. Insomnia in pregnant women mostly caused by the anxiety and panic associated with changes in responsibilities as parents. Sleep disorders during pregnancy occurs during the first trimester (13%-80%) and third trimester (66%-97%) [4].

One way to improve the quality of sleep is by using more pillows [4][5] because healthy pregnant women usually require more pillows than non-pregnant cardiac patients, and due to the difficulties in finding a comfortable sleep position in late pregnancy. The primary function of the pillow is to support cervical spine to be aligned with the other parts of the vertebral column in the body [6]. Maternity pillows on the market varies greatly. Sometimes, it confusing to select which one is appropriate. However, there has been no specific studies that examine pregnant pillow products.

In the selection of a product, consumers will influenced by emotions and feelings. The expression of emotions and feelings will cause a nuance, impression, sense of love, and feel of comfort when consumers decide to choose a product [7]. Therefore, it is necessary to know emotions and feelings of maternity pillow that want to be developed. One method that can quantify emotion of the customer in product development is Kansei Engineering [8]. It has been implemented in designing baby bag [9][10], office chair [11], and washing machine [12]. While in designing the products, there are some products' elements need to be evaluated. It is necessary to create several designs as well as combine and test the designs. Taguchi is a method that can be used to find the best combination of elements of maternity pillow that will satisfy the customers by considering the minimum number of experiments. Therefore, this study aims to capture the desire of pregnant women toward maternity pillow desired product by using kansei words and analyze the optimal design with Taguchi method.

2. Research Method

Research was conducted by using literature review, interview and questionnaire. There were some steps to do in this research. First, survey to maternity pillows via world wide web was done to collect kansei words as well as identify basic specification of the product such as the function of each part of the pillow which could be separated as head, neck, back, belly, pelvic, center, and legs support. In addition, direct interview involving eight women were also done to collect kansei words which aimed to explore ideal maternity pillow that fit with their desires. The number of subjects who were interviewed usually around 15±10. This process is the technique of free association that is commonly used in qualitative research to psychoanalysis [13]. Twelve kansei words were collected from this first stage.

Then, first questionnaire consists of various sample of maternity pillows and collected kansei words were shown to fifteen pregnant woman to be selected. The selected Kansei words are chosen by more than 90% of total respondents. Nine kansei words (affordable, durable, aesthetic, comfort, portable, simple, multifunction, attractive motive, and easy to maintain) were picked. Then, second questionnaire was distributed to thirty five respondents to gain data on respondents' opinion of product's importance towards kansei words, identify the initial preferences on model, cloth of pillow case, as well as motive. Each respondent was asked to assess the statements by scaling the importance level which score 1 (very disagree), 2 (disagree), 3 (neutral), 4 (agree), and 5 (very agree). However, kansei word relates to price "affordable" is excluded due to its validity. According to validity test using corrected item-total correlation score, kansei words which has score above 0.4 were valid [14]. While the reliability score is more than 0.8 for all kansei words means that reliable [15]. It is performed by using IBM SPSS version 24. Table 1 shows the validity and reliability score of the kansei words.

Table 1. Product Importance towards Kansei Words

Kansei Words	Description	Validity Score	Reliability Score
Affordable	Reasonably price	0,000*	0,880
Durable	Has strong zipper and durable fabric	0,594	0,852
Aesthetic	Gives aesthetic design	0,675	0,845
Comfort	The pillow case comfy for the skin, fluffy pillow, total body support as well as comfort for breasfeeding	0,672	0,844
Portable	Able to be easily carried or moved	0,779	0,836
Simple	Modest design	0,467	0,877
Multifunction	Support more than one part of the body, can be used for various sleeping positions as well as other activities	0,666	0,845
Attractive motive	Chic appearance	0,694	0,842
Easy to maintain	Easy to clean, insert, and remove the cover	0,779	0,836

The next step is designing the experiment by using Taguchi method. It is to find the best combination of elements of pillow that will satisfy the customers. The elements in designing maternity pillows are generated from expert designers. By grouping items and categories, obtained 7-core elements with 22 categories with predetermined specification (control factors) while the noise is users who are breastfeeding and pregnant women. Generally, the categories chosen are come from material that is easy to find and available as the main component to make maternity pillow. The elements (factors) and levels determined in this study are shown in Table 2. This study did not consider another attribute since the seven-mentioned attribute representing the dominant factors in manufacturing maternity pillow.

Four model of maternity pillows were used for the basic design experiment of Taguchi because they had been represented various products in the market. These are the advantages and disadvantages of the selected pillows. Advantages of U model are convenient to change position at night, total body support (head support, neck support, back support, belly support, leg support), naturally conform to the curves of the body, adjustable reading position, and can be used to breast feeding baby. While the negatives are bulky or consume more spaces, more heavy, and more expensive. Pillow B can support total body (head support, neck support, back support, belly support, leg support), but inconvenient to change direction during the night. It is also ideal for nursing while lying in the bed because easier to move than using U model. However, B pillow also consume more spaces and heavy like U pillow. E pillow can support several parts of the body such as head, bump or back, leg, and central, but if the sleep position is changing, the back is less support. The other advantage is removable central part to easier nursing and portable. Pillow J is more practical or portable, less consume spaces, less heavy, but need more pillow for head and neck support. It is also inconvenient to change position during night sleep.

Table 2. Classification Elements of Maternity Pillow (Control Factors)

	Elements	Level	Category
A	Zipper	1	Single
		2	Double
B	Filling	1	Dacron
		2	Silicon
C	Cloth cover	1	Cotton
		2	Waterproof Cotton
D	Model	1	U
		2	B
		3	E
		4	J
E	Motive	1	Plain
		2	Flowery
		3	Polka-dot
		4	Combination
F	Color	1	Light
		2	Medium
		3	Dark
		4	Gradation
G	Bag	1	A
		2	B
		3	C
		4	D

Respondents asked to evaluate each virtual design with five semantic differential scales. Subjects were thirty-four women who were familiar with maternity pillow such as pregnant women and breastfeeding mothers. The fractional factorial design used is a standard L16 orthogonal array for control factors. It was chosen due to the minimum number of experiment. Each row of the matrix described one design as shown in Table 3. The response is customers' satisfaction which can be assessed through averaging the score of eight kansei words chosen from the previous questionnaire (durable, aesthetic, comfort, portable, simple, multifunction, attractive motive, and easy to maintain) with five semantic differential scale.

Table 3. L16 Orthogonal Array to Design Maternity Pillow

Design	**Control Factors (Elements)**						
	A	**B**	**C**	**D**	**E**	**F**	**G**
	Zipper	**Filling**	**Cloth Cover**	**Model**	**Motive**	**Color**	**Bag Model**
1	Single	Dacron	Cotton	U	Plain	Light	A
2	Single	Silicon	Waterproof	U	Flowery	Medium	B

Design	Control Factors (Elements)						
	A	**B**	**C**	**D**	**E**	**F**	**G**
	Zipper	**Filling**	**Cloth Cover**	**Model**	**Motive**	**Color**	**Bag Model**
			Cotton				
3	Double	Dacron	Waterproof Cotton	U	Polka-dot	Dark	C
4	Double	Silicon	Cotton	U	Combination	Gradation	D
5	Double	Silicon	Cotton	B	Plain	Medium	A
6	Double	Dacron	Waterproof Cotton	B	Flowery	Light	B
7	Single	Silicon	Waterproof Cotton	B	Polka-dot	Gradation	D
8	Single	Dacron	Cotton	B	Combination	Dark	C
9	Single	Silicon	Waterproof Cotton	E	Plain	Dark	D
10	Single	Dacron	Cotton	E	Flowery	Gradation	C
11	Double	Silicon	Cotton	E	Polka-dot	Light	B
12	Double	Dacron	Waterproof Cotton	E	Combination	Medium	A
13	Double	Dacron	Waterproof Cotton	J	Plain	Light	B
14	Double	Silicon	Cotton	J	Flowery	Medium	A
15	Single	Dacron	Cotton	J	Polka-dot	Dark	D
16	Single	Silicon	Waterproof Cotton	J	Combination	Gradation	C

Signal Noise Ratio (SNR)

The response is customer satisfaction which uses Larger the Better (LTB). Equation (1) can be used to calculate SNR.

where:

$$SNR_{STB} = -10\ Log\left[\frac{1}{n}\sum_{i=n}^{n} yi^2\right] \quad (1)$$

n = number of tests in the experiment (trial)

yi = response value for each respondents

3. Results and Discussion

First, normality test is used know whether the data set is well-modeled by a normal distribution. By using Anderson-Darling normality test, the data is considered normal since the significant value is 0.52 (>0.05). While the homogeneity test is used to find out the two or more of data distribution that has the same variation or not. It defines whether the X and Y data is homogen or not. In homogeneity test, if the value of x^2 calculation < x^2 table, then the data is homogeneous. The x^2 calculation is 459.16 and x^2 table is 598.32. Thus, x^2 calculation is less than x^2 table, which means the data is homogeneous.

IOP Conf. Series: Materials Science and Engineering **215** (2017) 012043 doi:10.1088/1757-899X/215/1/012043

Effect of Factors to Responses

The optimal customers' satisfaction can be gained through combination design of maternity pillow D1-E2-F2-G2-C1-B2-A2 which is shown in Table 4. The calculation is using equation (1) for calculating LTB.

Table 4. Response Optimal Combination Table for S/N Ratio (LTB)

	A	B	C	D	E	F	G
1	0,301	0,2989	0,3059	0,4087	0,2697	0,3081	0,3074
2	0,3025	0,3045	0,2975	0,3248	0,3371	0,3258	0,3128
3				0,2774	0,3193	0,2836	0,2919
4				0,1959	0,2807	0,2893	0,2947
Differences	0,0015	0,0056	0,0083	0,2127	0,0674	0,0423	0,0209
Rank	7	6	5	1	2	3	4
Selected	A2	B2	C1	D1	E2	F2	G2

Analysis of Variance (ANOVA)

It is to explore which of the factors significantly affect the customer's satisfaction. Table 5 shows the results of ANOVA's response. Statistically, the F-test can be used to specify which factor has a significant effect towards customers' satisfaction. It can be known by comparing the value of F. When F>4 means that the change of operating factors has a significant effect on the quality characteristics. It can be seen from Table 4 that all the factors significantly contributed toward customer satisfaction: Factor D (Model), motive (Factor E), Bag model (Factor G), Color (Factor F), filling (Factor B), Zipper (Factor A), and cloth cover (Factor C) as much as 78.34%, 9.80%, 3.37%, 2.99%, 0.36, 0.34, and 0.06 respectively. The total contribution is 95% while the other five percent is from unidentified factor.

Table 5. Response of ANOVA Test

Source of Var	**SS**	**DB**	**MS**	**F Stat**	**F Tabel**	**SS'**	**% Contribution**
Factor A	0,19	1	0,19	39,99	3,86	0,18	0,34
Factor B	0,20	1	0,2	41,99	3,86	0,19	0,36
Factor C	0,04	1	0,04	7,47	3,86	0,03	0,06
Factor D	42,27	3	14,1	2989,91	2,62	42,26	78,34
Factor E	5,30	3	1,77	374,94	2,62	5,29	9,80
Factor F	1,63	3	0,54	114,94	2,62	1,61	2,99
Factor G	1,83	3	0,61	129,57	2,62	1,82	3,37
Residual	2,49	528	0				

Total	53,94	543	95

It can be seen from Table 4 and Table 5 that the effect of factor A, B, and C is relatively small, but it doesn't meant that those factors are unimportant. There is no significant differentiation (only 0,0015 for Factor A and 0,0056 for Factor B), whether A1 or A2 and B1 or B2 that will be used will not have significant impact towards customer satisfaction. It is possible to select the products' element by using other consideration such as single zipper (A1) is chosen due to practicality and cheaper cost while dacron (B1) is used because of its cheaper price.

4. Conclusions

Based on the discussion above, it can be concluded that the best maternity pillow that can satisfy the customers can be designed by combining D1-E2-F2-G2-C1-B2-A2 means the model is U shape, flowery motive, medium color, Bag model B, cotton pillow cover, filled with silicon, and use double zipper. However, it is possible to create combination of D1-E2-F2-G2-C1-B1-A1 by using consideration of cost which means that the zipper is switched to single as well as filled with dacron. In addition, the total percentage of contribution reach 95%.

Recommendation

It is necessary to do further research such as in increasing the quality of the maternity pillow.

5. References

[1] Wahyuni, & Ni'mah, L. (2013). Manfaat Senam Hamil untuk Meningkatkan Durasi Tidur Ibu Hamil. *Jurnal Kesehatan Masyarakat*, 145-152.

[2] Ramachandra, P., Maiya, A. G., Kumar, P., & Kamath, A. (2015). Prevalence of Musculoskeletal Dysfunctions among Indian Pregnant Women. *Journal of Pregnancy*.

[3] Sharma, S. & Franco, R. (2004). Sleep and Its Disorders in Pregnancy. *Wisconsin Medical Journal,* 103(5), pp. 48-52

[4] Jeon, M. Y., Jeong, H., Lee, S., Choi, W., Park, H., Tak, S. J., et al. (2014). Improving the quality of sleep with an optimal pillow: A randomized, Comparative study. *Tohoku J. Exp. Med.*, 183-188.

[5] Farine, D. & Seaward, G. (2007). When It Comes to Pregnant Woman Sleeping, Is Left Right?. *JOGC Octobre*

[6] Verhaert, V. (2011). Ergonomic Analysis of Integrated Bed Measurements: Towards Smart Sleep Systems. *Dissertation*. Arenberg Doctoral School of Science, Engineering & Technology Faculty of Engineering Department of Mechanical Engineering

[7] Norman, D. (2004). *Emotional Design: Why We Love (or Hate) Everyday Things*. Prentice Hall, Singapore.

[8] Nagamachi, M. (1995). Kansei Engineering: A New Ergonomic Consumer-Oriented Technology For Product Development. *International Journal of Industrial Ergonomics*, 15(1), 3-11.

[9] Janari, D., Rakhmawati, A. (2016). Developing Baby Bag Design by Using Kansei Engineering Method. *Materials Science and Engineering* 105 doi:10.1088/1757-899X/105/1/012031

[10] A. H. Soewardi and B. M. I. Nasution. (2016). Design of Kansei Baby Bags by Using Fuzzy Linguistic Principles. *IEEE International Conference on Knowledge Engineering and Applications (ICKEA),* Singapore, pp. 165-171, doi: 10.1109/ICKEA.2016.7803012

[11] Park J, Han SH. (2004). "A fuzzy rule-based approach to modeling affective user satisfaction towards office chair design". *International Journal of Industrial Ergonomics* 34. Pp. 31-47.

[12] Ishihara S, Ishihara K, Nakagawa R, Nagamachi M, Sako H, Fujiwara Y, Naito M. (2010). Development and improvement of a washer-dryer with kansei ergonomics. *Proceedings of The International MultiConference of Engineers and Computer Scientists.*

[13] Steinar., (2007) *Doing interviews,* London: SAGE Publications.

[14]Fayers PM, Machin D. (2000). Quality of Life: Assessment, Analysis and Interpretation. Chichester: John Wiley & Sons. pp. 72–79.

[15] Bland J, Altman D. (1997). Statistics notes: Cronbach's alpha. BMJ. 314:275.